ADVANCED COMPOSITE
MOLD MAKING

ADVANCED COMPOSITE MOLD MAKING

John J. Morena
Consultant

Sponsored by the Society of Plastics Engineers, Inc.

VNR VAN NOSTRAND REINHOLD COMPANY
New York

Copyright © 1988 by Van Nostrand Reinhold Company Inc.
Library of Congress Catalog Card Number 87-18867
ISBN 0-442-26414-3

All rights reserved. No part of this work covered by the copyright hereon may be reproduced or used in any form or by any means—graphic, electronic, or mechanical, including photocopying, recording, taping, or information storage and retrieval systems—without written permission of the publisher.

Printed in the United States of America

Van Nostrand Reinhold Company Inc.
115 Fifth Avenue
New York, New York 10003

Van Nostrand Reinhold Company Limited
Molly Millars Lane
Wokingham, Berkshire RG11 2PY, England

Van Nostrand Reinhold
480 La Trobe Street
Melbourne, Victoria 3000, Australia

Macmillan of Canada
Division of Canada Publishing Corporation
184 Commander Boulevard
Agincourt, Ontario M1S 3C7, Canada

16 15 14 13 12 11 10 9 8 7 6 5 4 3 2 1

Library of Congress Cataloging-in-Publication Data

Morena, John J., 1937–
  Advanced composite mold making.

  "Sponsored by the Society of Plastics Engineers, Inc."
  Bibliography: p.
  Includes index.
  1. Plastics—Molds. 2. Composite materials.
I. Society of Plastics Engineers. II. Title.
TP1150.M57   1988   668.4'12   87-18867
ISBN 0-442-26414-3

TO GEORGE LUBIN
WITH SPECIAL GRATITUDE

Foreword

Advanced Composite Mold Making is the first volume to treat, in-depth, the fabrication of molds using nonmetallic materials. As the only contemporary book covering this burgeoning field, it outlines a technology for making prototype and production tooling which currently has widespread applications in the aircraft, aerospace, and automotive industries. Future applications cannot help but touch a broad spectrum of other industries.

This volume has been written to serve a broad readership, as both a definitive text and a reference source. Among its potential readers and users are advanced tool designers, engineers involved with nonmetallic structuring, industry management concerned with the subject and its industrial ramifications, and students.

This work is unusually well-ordered, with a detailed introduction to all aspects of advanced composite mold making setting the stage. The chapters sequentially follow the entire materials and fabrication process of mold and tool building. The effects of statics and strengths of materials are examined in-depth, along with the basic elements involved in the design of a mold and the engineering aspects of the molds and mold materials. There is a particular emphasis on the large number of thermal elements influencing the mold or tool and its performance.

The most extensive treatment in the volume is that of fabricating master models and patterns, metallic and nonmetallic molds, mold accessories, mold assembly, and the actual tools involved. The vital, but often overlooked, aspects of mold handling, storage, safety, and quality control are also discussed. It is because of this exhaustive coverage of the subject that the Society of Plastics Engineers (SPE) is particularly pleased to recommend this book and add it to its list of sponsored volumes.

ROBERT D. FORGER
Executive Director
Society of Plastics Engineers

Preface

Advanced Composite Mold Making is a definitive volume, compiled to collect, collate, and present the engineering, design, materials, and processes required to fabricate a composite mold or tool.

The use of composites today in all sectors of industry is vast. Given this fact, the scope of the book includes information for producing the molds and tools to form both the nonmetallic parts and bonded structures which are used in a multitude of industries, including military, aerospace, marine, transportation, leisure time, and commercial industrial applications.

Advanced Composite Mold Making also establishes the ground rules for the design of educational programs at all levels, as well as providing guidance for the engineer, designer, technician, or craftsman, through state-of-the-art methods of mold making.

The understanding of composites is widespread, and so the concepts covered in the book are intended for use in both domestic and international industries.

In addition, the detailed information is useful and practical. Thus, it is presented in a "hands-on" manner — a form that will appeal to both experts and beginners in the composites field.

It must be stressed at the outset that composite mold and tool making, because of the continuing advances in structural composite technology, has also become an "art of quality." For this reason, certain innovative, advanced materials and processes will be introduced in a thorough, easy to understand, form.

Locating concise and reliable data regarding the formation of composites is difficult. Locating similar information for the fabrication of molds and tools is sometimes an impossible task. This book endeavors to provide that information, answering questions as they arise, (and sometimes before they arise), in order to prevent errors in design or fabrication. It also utilizes extensive data to inform the reader so that mold and tool problems, in the area of both design and materials, can be solved expeditiously.

The early chapters acquaint the reader with the design and engineering tools required to produce a mold that will provide quality parts and remain trouble-free over a lengthy economic and production life.

The materials chapter examines the quality and state of the art of the materials of composites, mold, and tool making. The fabrication chapters illustrate the innovative processes involved, and the means by which state-of-the-art materials can be put to creative use for advanced composite mold making are immediately recognizable.

Finally, an extensive tabular data bank, along with a listing of additional tool innovations in related metallic part, assembly, and forming areas, is provided.

Advanced Composite Mold Making represents the culmination of many years of trial and error experimentation, resulting in an extensive and detailed collection of proven information. This information is a collection and expression of the varied experiences of many associates and friends to whom the writer is indebted. Among them are: Lee Davis, prominent writer of varied interests and talents, whose guidance, assistance, and professional expertise helped to make this book a quality presentation of technical data; Bill Forster, mold and tool designer, researcher and technical assistant, for his continued support throughout the entire writing process; Paul Petervari, Art Nelson, Robert J. Sanderson, Bill Pagels, Bob Semprini, and Richard Chance, of Grumman Aircraft Systems, for providing assistance and data for certain sections of the manuscript; Lou Reidell, program manager and manager of composite mold and tool design, Grumman Aircraft Systems, for technical information, Dimitrios Maltezos, dean and professor, State University of New York, for reviewing sections of the manuscript, and especially Susan Munger, Gail Nalven, Alberta Gordon, Ray Kanarr, and the staff of Van Nostrand Reinhold, for their guidance in the preparation of the manuscript. Finally, to John and Terry Morena, who supported every aspect of the text.

Contents

Foreword, by *Robert D. Forger* / vii
Preface / ix

1. Introduction / 1
2. Strength of Materials and Mold Design / 5
3. Mold Engineering / 28
4. Mold Materials / 73
5. Preparation of Masters / 129
6. Fabricating Master Molds and Nonmetallic Molds and Tools / 153
7. Fabricating Metallic Molds / 231
8. Conventional or Permanent Reusable Vacuum Bags / 253
9. Mold Fittings, Accessories, and Support Tools / 330
10. Facility Requirements / 357
11. Inspection and Quality Control / 381
12. The Future of Composite Mold Making / 399

Bibliography / 411
Index 413

1
Introduction

When fabricating a mold or tool to form a simple hand lay-up, or an advanced composite part, assembly, or thermoplastic composite material, the best materials and processes to use are the ones that result in the production of the best quality finished shape or contour at the lowest cost.

Molds, tools, and generally the models or masters, are required prerequisites for any molded part or assembly. With the continued increase in the use of composite part materials comes the necessity for more advanced materials and processes to form molds and tools. Composite forming molds and tools of the best quality can be produced from a mixed array of both metallic and nonmetallic materials. Innovative, as well as standard, fabrication methods are currently being employed to process the mold materials. The most advantageous mold or tool should also provide dimensional stability during use, be easy to fabricate, maintain, and repair, and be convenient to handle and store.

In the same sense that the attempt to obtain the lowest cost to produce a mold, tool, or part is coupled with the effort to ensure the lowest cost of operation, the factors of thermal mass and thermal uniformity in the heat-up rate play an important role in mold material choice and design.

Aluminum has always been a basic material for use in molding laminated fiberglass reinforcements and assemblies. When graphite-reinforced materials appeared in the aircraft and aerospace industries, new requirements developed. The old standby, steel, with its mass and high cost to process, fit hardly any application other than simple contours and loose tolerance applications.

Fiberglass-reinforced, low-temperature laminating materials serve well for fabricating checking and inspection tools, trim fixtures, prototype and intermediate molds, and low-temperature-curing molds with limited production rates.

The use of reinforced, high-temperature composite materials, which was just gaining acceptance in mold and tool fabrication, took a downward turn when the U.S. government declared certain constituents harmful to the health and safety of users. It was at this point that industry started to look not only at nonmetallic mold and tool materials, but to reexamine the materials used to form metallic molds and tools as well.

New machining methods were developed to form contoured electroformed or plated nickel tools because this material demonstrated a low coefficient of thermal expansion (CTE). The need for low CTE materials directed mold and tool researchers to examine other materials such as monolithic bulk graphite, castable reinforced cements and ceramics, specially formulated mass-cast resin compounds, graphite-reinforced low-temperature-precure epoxy prepreg materials, and high-temperature structural foams.

The introduction of the use of reinforced thermoplastic composites broadened the requirements necessary for low CTE mold and tool materials by increasing the molding temperatures of the part materials from the average of 350°F to 650°–750°F. This narrowed the mold material choices, leaving steel and some of the nonmetallic materials as acceptable options in this temperature range.

It is for the engineers of the future to decide which high-temperature mold and tool materials will be available for use in producing parts and assemblies made from state of the art and newly-developed reinforced thermoset composites or reinforced and self-reinforcing thermoplastic composites.

The management of industry, no matter what sector is involved, must realize that the cost of producing a mold, tool, or fixture for the formation of a composite material part must be disregarded if the part material and configuration dictate what that mold or tool material must be. Management should, at that point — especially if nonmetallic material is required — use every available resource to assure that the highest quality nonmetallic mold or tool is produced. Manual versus automated design, inexpensive or expensive masters, fair or good thermal and physical properties of mold materials, short cuts or careful procedures in mold fabrication, and close or detached management during mold use and storage are just some of the considerations that must be examined. The nature of mold making is such that there is little room for sacrifices to be made with respect to quality.

The art of mold making in the area of advanced composite molds can be divided into a number of sequential units. The expert mold or tool designer can look at a conceptual part design or model and visualize how the pattern or master for the mold, tool, and finished item should be fabricated.

It is the intent of this book to assist in formulating these concepts by presenting them in a logical and understandable form:

Mold Materials: An extensive overview of all of the important materials used in the formation and production of the masters, molds, tools, and fixtures utilized to produce state-of-the-art and advanced composite parts and assemblies is given. These materials include solid and resinous primary and secondary forms, both of which are examined here.

Strengths of Materials and Mold Design: A review of physical and thermal considerations that affect the design of molds and tools used in the composite molding process, and an examination from the point of view from the structural designer of the strengths of materials with the static aspects of these materials related to mold and tool design.

Mold Engineering: The entire engineering process is examined, presenting an overall approach to mold and tool cost, the projected life of the mold or tool during production use, and the various approaches in automated mold and tool design. Comparisons between low- and high-production molds and tools are made followed by statements regarding mold and tool refurbishment and replacement.

Preparation of Masters: Up-to-date procedures for preparing required master patterns, models, and masters to be used in the formation of advanced composite molds are detailed. Master and mold treatments, releases, and parting agents are thoroughly described.

Fabricating Molds: An encyclopedic amount of information regarding the production of master and expendable molds, and suggesting materials and tolerances needed for quality results, is given. Mass cast rigid, flexible, and laminated nonmetallic molds to be used to contour mold and bond structural composites, unique and innovative production methods to produce laminate molds and substructures, and metal-faced laminated molds and tools are all examined. The evolution of metallic molds and tools is investigated, beginning with conventional aluminum and steel, and finally resulting in the inexpensive approaches of electroformed or plated nickel mold or tool forms. The developing areas of nonsilicone and reversion-resistant silicone caul, pressure and intensifier elastomers, and the methods by which they are replacing conventional bagging materials as reusable production aids are described. An introduction to semiautomatic vacuum bagging systems, elastomeric expansion rubber molding for out-of-autoclave and out-of-press forming of composites, and a complete listing of mold fittings and accessories, support tools, and efficient production aids are all given. Finally, the necessary procedures to maintain and repair production molds and tools are detailed.

Facility Requirements: An introduction to the important area of the safe handling of the materials, equipment, molds, and tools of the composite craftsman is given. The overall industrial engineering aspects of storage and transportation, as well as the facilities requirements of waste disposal and safety are fully examined.

Inspection, Quality Control, and the Future of Composite Mold Making: Methods are suggested for materials acceptance, and for establishing and controlling manufacturing inspection procedures relating to the production and acceptance of a high quality mold or tool. The projected direction that advanced composite mold making may take in the future is investigated in the closing section of the book.

As the reader proceeds, it will become immediately apparent that the presentation format is directed toward educating the beginning as well as the advanced mold and tool maker. This book is designed to be used as an individual or group educational aid for engineers, designers, and craftsmen. It is hoped that, utilizing the information gained from this book, the reader will be better able to choose the most appropriate materials and processes to be used when producing an advanced composite mold or tool.

2
Strength of Materials and Mold Design

Design sections of technical manuals or books usually present complicated and cumbersome design parameters that are difficult for the reader to understand and/or apply.

In an effort to ameliorate this situation, this book is designed with a "hands-on" approach to the materials and processes of advanced composite mold making. For this reason, the present chapter is structured in such a way that interpretation of the data can be made by readers possessing different levels of design background.

Since statics deals with forces and reactions on rigid bodies, and the strengths of materials expands the science of statics to include internal reactions, these two subjects are fundamental tools needed for proper mold design.

To complement the understanding of strengths of materials, stress-strain relationships and diagrams will be presented. These stress-strain relationships provide the equations necessary to arrive at a safe stress strength. The stress-strain diagrams will characterize a material's strength properties. Thus, by combining statics, strengths of materials, the stress-strain diagrams, and stress-strain relationships, proper loading conditions and safe stress can be calculated.

Lap joint strength and the theory of elasticity determine the maximum stresses a material can withstand without failure. These maximum stresses are adjusted using the stress-strain relationships. Once this is accomplished, a final formula is used to determine the size of load-carrying members.

Since molds or tools are rarely used at the temperature at which they were created, and differences exist between ambient and cure temperatures, thermal effects and heat transfer rates are presented in this chapter. Proper understanding of heat transfer modes is imperative in calculating tool heat-up rates in the autoclave, press, and in oven heating. These will also be covered in detail in the following pages.

2.1. STATICS AND THE STRENGTHS OF MATERIALS

The degree of tolerance, reliability of performance, and strength characteristics are the global design parameters in tool design.

In terms of tolerance, a designer should know what loads will be placed on the mold or tool in order to design this mold or tool with the proper strength characteristics. These strength characteristics will have safety factors associated with them to insure reliability, in spite of any irregularities in materials or unanticipated mold or tool loading.

A designer should also possess an understanding of statics and strength of materials, for these are the primary analytical methods used in tool design. Statics deals with externally applied loads and their reactions on rigid bodies. Strengths of materials is a more in-depth analysis than statics, and examines the internal reactions within deformable bodies when external loads are applied.

2.1.1. Statics

Statics studies forces acting upon bodies at rest. When a body is pushed or pulled, a force is exerted upon that body to either deform it or move it. A force consists of two components: magnitude and direction. For example, if a cable lifts a tool and the force exerted by the cable is considered positive, then the downward force exerted by the tool on the cable is negative. This exemplifies Newton's third law, which states that for every action there must be an equal and opposite reaction.

Although the choice of direction or sense is arbitrary, a good rule of thumb is: If a force acts upon a mold, it is positive, and the reaction of the mold on that force is negative.

Since forces have both a magnitude and direction, a body at rest is not viewed as having zero forces acting upon it, but rather, the body has many forces acting upon it which cancel each other because of the sense of the forces.

A force which deforms a body by decreasing the body size is a compressive force, (for instance, the squeezing of a sponge). The complementary force, a tensile force, tends to increase the body size (for example, stretching a rubber band).

A diving board, fabricated from a reinforced plastic laminate, experiences both tensile and compressive forces (Fig. 2.1).

If the diving board is of uniform rectangular cross section, then the board is in tension above the center line, and the board is in compression below the center line.

This diving board example can be used to demonstrate important properties of materials:

If the board were fabricated from a cementitious material and subjected to an end load, the board would break. To transfer this to our frame of reference: Although cementitious compounds have excellent compressive properties, their tensile properties are poor, and thus failure can occur under heavy end loads.

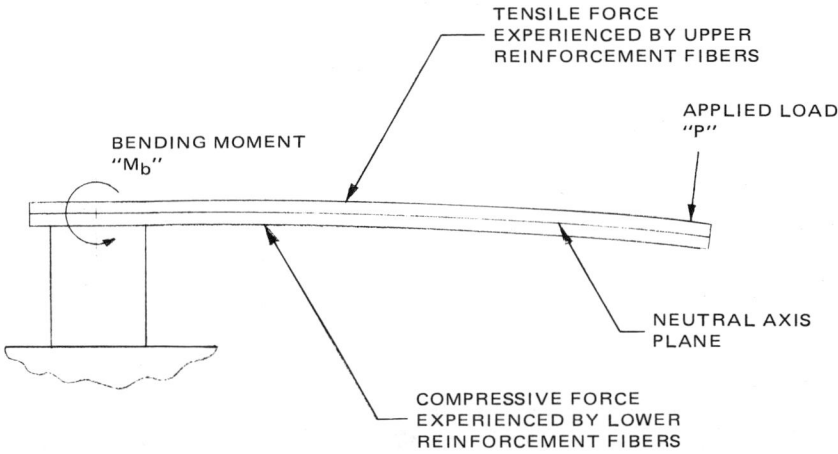

Fig. 2.1. Diving Board.

If the board were fabricated from rubber and subjected to an end load, the board would deform greatly. The analogy is that rubber contains excellent tensile properties, but poor compressive ones.

The force that the diving board exerts to counter the applied end load and still remain stationary is called a moment. A moment is defined as the force which causes rotation of a body at rest. The magnitude of the moment is equal to a force multiplied by a distance. Thus, a proper understanding of the loading conditions of a structure should be achieved before a design structure and the materials for that structure are selected.

Equilibrium occurs when a body subjected to forces and moments is at rest. Furthermore, if equilibrium is to occur, the sum of all forces must equal zero and the sum of all moments must equal zero. Hence, utilization of the equilibrium conditions and their associated equations will allow the reaction forces to be solved.

2.1.2. Strength of Materials

In order to successfully design a mold or tool which will make finished parts reliably throughout its intended service life, a designer must be familiar with both engineering mechanics and strength of materials. Therefore, this section is designed to present these fundamental principles in a clear and concise manner.

Mechanics is the study of forces which act on rigid bodies. But, since a rigid body neglects internal reactions, the strength of that body cannot be determined

8 ADVANCED COMPOSITE MOLD MAKING

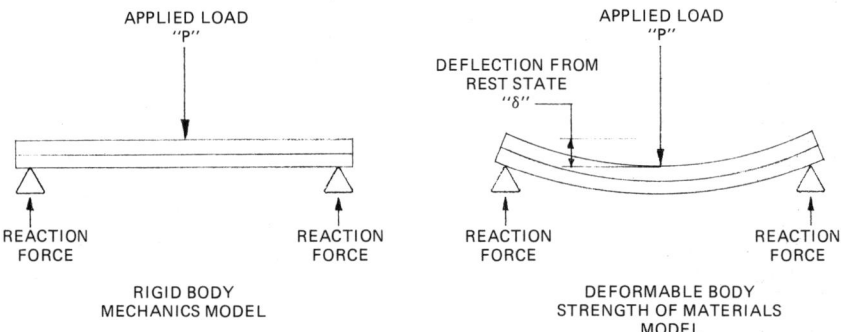

Fig. 2.2. Simply Supported Beam.

through mechanics. Thus, the strengths of materials, which is the study of externally applied forces on a body and the corresponding internal reactions of that body on the applied forces, must be utilized.

To exemplify the difference between statics and strengths of materials, a loaded, simply supported beam is considered in Fig. 2.2.

In statics, the beam is viewed as being a rigid body that does not bend or deflect. Hence, through the use of statics, the reaction forces can be found by summing moments about either supporting point and then summing forces.

Since the strengths of materials is a more in-depth study of forces and internal reactions, it provides useful extended solutions. Among these is the maximum allowable load prior to failure and the deflection distance compared to the rest state. For a further explanation of this, refer to Table 2.1.

Stress and Strain. To properly design a structural mold or tool member, the characteristics of that member must be known. Among these characteristics are strength, strain, and stiffness. In engineering applications, strength is defined in terms of stresses. Stresses are defined as force per unit cross-sectional area:

$$\sigma = P/A \qquad (2.1)$$

where:

σ = force per unit area (psi) or (N/cm^2)
P = applied load (lb) or (N)
A = cross-sectional area (in.^{2}) or (cm^2).

The usual unknown in tool design is the cross-sectional area. In most cases, the applied load (P) is known. Once the material is chosen, the maximum stress σ_{MAX} (otherwise known as the proportional limit) can be read from a chart of material properties.

STRENGTH OF MATERIALS AND MOLD DESIGN 9

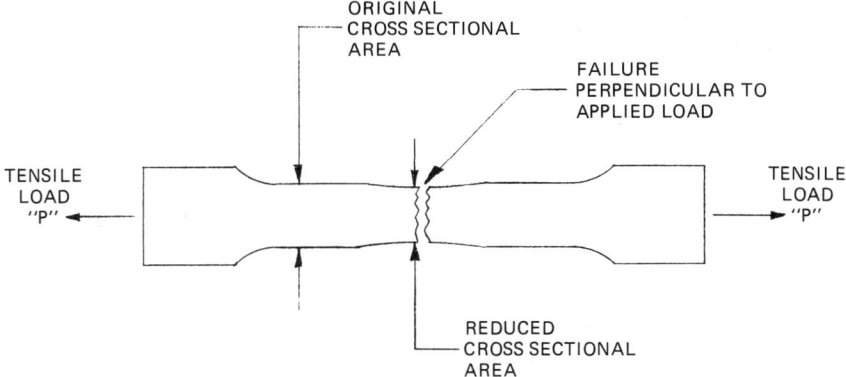

Fig. 2.3. Failure of a Tensile Test Specimen.

It is common practice to use the proportional limit as the maximum allowable design stress. If, by chance, the cross-sectional area was underrated and material failure occurred, the plane of failure would be normal to the applied stress. This is shown in Fig. 2.3.

To avoid the failure described above, a safety factor is usually introduced. The above equation can be rearranged and altered (Eq. 2.2) to incorporate a safety factor and the proportional limit.

$$A = \frac{N}{\sigma_{AL}} \times P \qquad (2.2)$$

where:

A = cross-sectional area (in^2) or (cm^2)
N = safety factor
σ_{AL} = allowable design stress (proportional limit)
P = applied load (lb) or (N).

The safety factor (N) can be assigned various values for different circumstances. For well-defined stress, a value of 2 is standard, but for stress of uncertainty or stress concentrations, a value of 5 should be used.

A decision must be made when brittle material, impact loading, or cyclic loading is involved, for these conditions cannot be calculated and an additional factor of safety may be warranted.

Strain (ϵ) is defined as the change in length divided by the original length of a given specimen.

$$\epsilon = \delta/L \qquad (2.3)$$

where:

δ = change in length (in.) or (cm)
L = total original length (in.) or (cm).

It should be noted that this strain is only valid in the elastic region of the stress-strain curve of Fig. 2.4.

Within this region, a load can be applied which deforms the material, and once the load is relieved, the material assumes its original shape. Therefore, the following conditions should hold true for the above equation to be valid:

- The applied stress should not proceed past the proportional limit.
- Uniform stress is produced by axial loads.
- The specimen is of constant cross-sectional area.
- The material is homogeneous.

It is also important to remember that, if a material is strained past its elastic limit, (plastic) permanent deformation occurs. This is seen in Fig. 2.4. in the area of failure. The area in question experiences a reduction of area, which is called necking. Hence, the actual strain experienced by the material is much higher, if the reduction of area is accounted for. For work in tool design, the reduction of area can be neglected in order to ease calculations.

Although there are significant property differences between metallic molds and composite molds, both will be presented in a general form. Since the stress

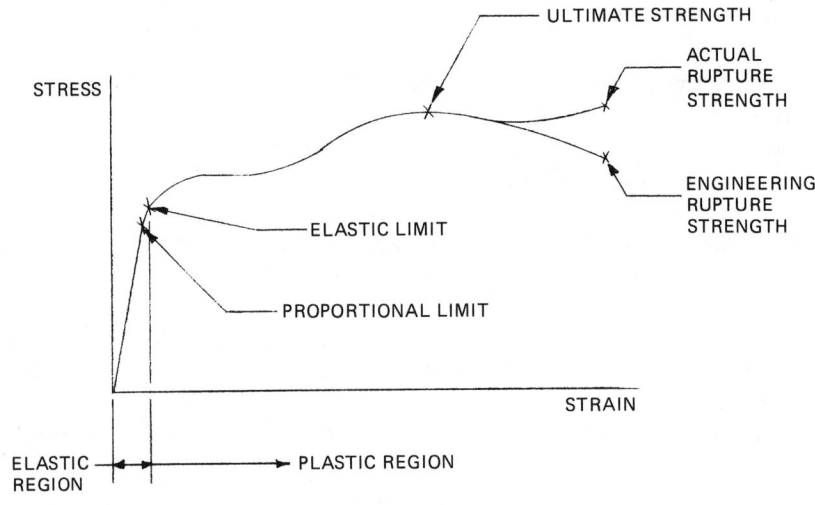

Fig. 2.4. Stress-Strain Curve.

STRENGTH OF MATERIALS AND MOLD DESIGN 11

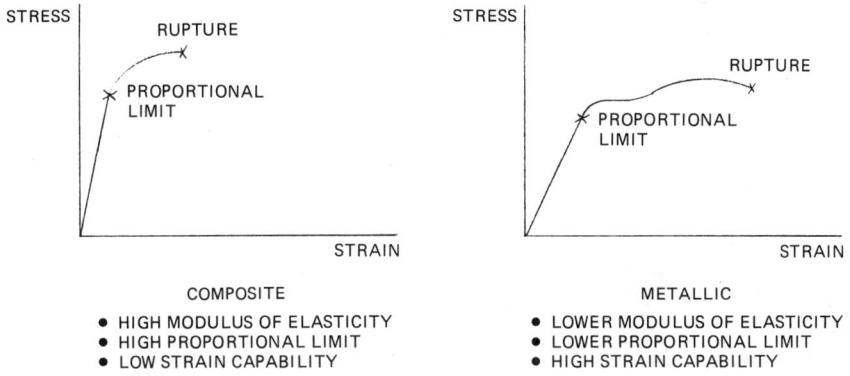

Fig. 2.5. Metallic vs. Composite Stress-Strain Curve.

strain-strain curve characterizes the relationship between stress and strain, it is important that the following discussion and Fig. 2.5 be understood.

Figure 2.5 shows a stress-strain curve for both a metal and a composite member. The region between the origin and the proportional limit shows that stress is linearly proportional to strain, according to Hooke's Law.

$$\sigma = E\epsilon \qquad (2.4)$$

where:

σ = stress
E = Young's Modulus
ϵ = strain.

Equation (2.4) may be rearranged to express the slope in terms of stress and strain:

$$E = \sigma/\epsilon \qquad (2.5)$$

The slope, or Young's Modulus (modulus of elasticity) is actually a measure of stiffness or resistance to deflection. For an isentropic material such as metal, which has the same properties in all directions, the modulus of elasticity is constant throughout the material.

On the contrary, in an anisentropic material such as a composite, the modulus of elasticity will vary and is a function of the fiber orientation.

A composite member, when loaded in the longitudinal direction, has the highest modulus of elasticity when all of the internal fibers run longitudinally. The traverse modulus is at a minimum under these conditions.

Trade-offs can be made between the longitudinal and traverse moduli by varying the ply orientation within a laminate mold or tool. In tool design, it is customary to have equal numbers of 0° and 90° layers of fibers, compared to the 45° and 135° layers of fibers. This provides a relatively uniform property distribution, even though the laminate is still anisentropic.

Equations (2.1) and (2.3) can be substituted into Eq. (2.4) to yield the following:

$$\frac{P}{A} = E\frac{\delta}{L}. \tag{2.6}$$

By rearrangement, a useful expression for the change in length as a function of stress is obtained.

$$\delta = \frac{PL}{AE} = \frac{\sigma L}{E} \tag{2.7}$$

It should be noted that Eq. (2.7) is an expression for axially loaded members only. Since the stress-strain graph is linearly elastic up to the proportional limit, all tool designs should use the proportional limit as the maximum allowable stress as previously mentioned.

The remaining terms of Fig. 2.5 and the contrasts between metals and composites are given below:

Elastic limit: The stress limit beyond which permanent deformation occurs within a material. When the deforming load is removed, residual stress remains within the material. This permanent deformation or set is known as *inelastic action* and is typical of metallic members. Composites have an elastic limit, but beyond the elastic limit, permanent deformation occurs and catastrophic failure is imminent.

Yield point: The point at which a material displays an increase in strain without an increase in stress. Metals exhibit this property and composites do not, due to the fact that composites cannot experience high strain without catastrophic failure.

Ultimate stress: The highest value of stress observed on the stress-strain curve. In general, metallic members can be strained past the ultimate stress point, but composites cannot. For composites, the yield point is virtually the same as the ultimate stress point, beyond which catastrophic failure occurs.

Rupture strength: The point at which failure occurs. An illustration of this is given in Fig. 2.6.

STRENGTH OF MATERIALS AND MOLD DESIGN   13

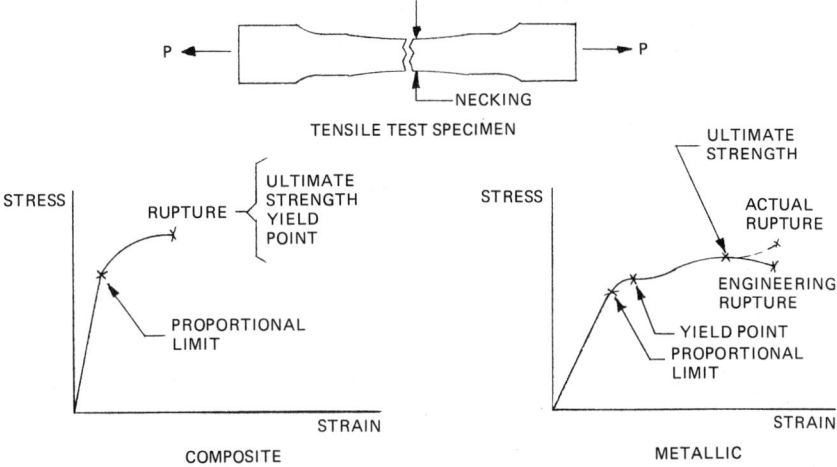

Fig. 2.6. Actual vs. Engineering Rupture Strength.

Figure 2.6 shows an actual rupture strength which is higher than that of the engineering rupture strength. This occurs through necking of metals, which was previously discussed and shown in Fig. 2.5. Here again, rupture strength for composites is synonymous with ultimate stress and the yield point.

Safety Factors. To avoid failure in tool design, the maximum value of shear stress or normal stress should never be used. Therefore, it is advantageous to have an acceptable scheme for safety calculations.

For pure shear or pure tension, it is easily deduced that the maximum stress value can be divided by a desired safety factor to arrive at a usable design stress. But what happens when a bolt experiences both shear and tension?

Figure 2.7 illustrates the safety calculation procedure for the unique case of a bolt experiencing both shear and tension. Figure 2.7 shows a straight line drawn from the maximum shear to the maximum tension. This represents a failure line, since any value above it exceeds the maximum strength limits. By dividing both maximum values by a factor of safety (N), the line is shifted downward and a safe stress line is created. Any value which lies on or below the safe stress line is a usable value of stress.

Two equations can be written for the safe stress line and any value which is equal to or less than the usable shear stress (τ_u) or the usable tensile stress (σ_u) can be employed in design calculations.

$$\tau_u = \tau_A - \frac{\tau_{MAX}}{\sigma_{MAX}} \sigma_u \qquad (2.8)$$

$$\sigma_u = \frac{\sigma_{MAX}}{\tau_{MAX}} [\tau_A - \tau_u] \qquad (2.9)$$

14 ADVANCED COMPOSITE MOLD MAKING

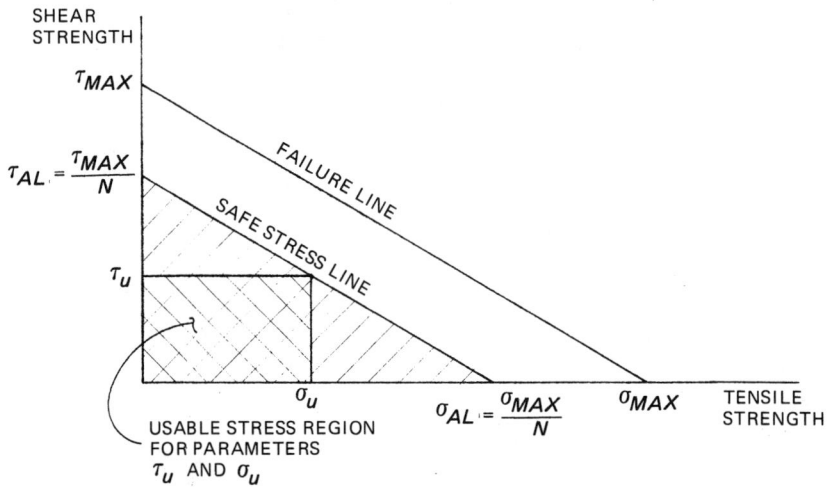

Fig. 2.7. Allowable Bolt Shear-Tension Diagram.

where:

τ_{MAX} = maximum shear stress
σ_{MAX} = maximum tensile stress
τ_{AL} = allowable shear stress
σ_{AL} = allowable tensile stress

It should be noted that, if the usable shear or tensile stress is exceeded, it would be necessary to add more bolts.

Lap Joint Strength. Since bolted and riveted connections are present in almost every tool design, it is important to understand their various failure modes.

In metallic molds, both the metal bolt and the metal mold material possess the ability to plastically deform without failure. Composite mold material, on the other hand, usually experiences catastrophic failure, such as shear tear-out, when extremely large stresses are applied. It should be noted, however, that local deformations can occur within the fastening area without causing catastrophic failure of the composite.

The stress region between initial plastic deformation of the connecting area and catastrophic failure is very small; hence, a good tool design should not contain stress figures that fall into this stress region.

Whether the bolted material is a metal or composite, there are three basic failure modes that must be considered: shearing of the bolts; tensile failure of the plate; and bearing failure of the plate or bolt.

The strengths of the materials are based upon the maximum allowable shear, and upon tensile or bearing failure for the given conditions.

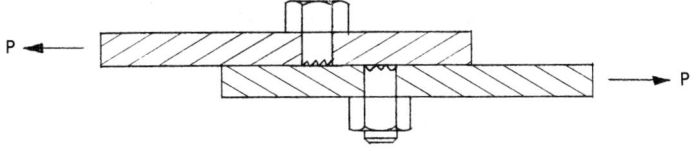

Fig. 2.8. Bolt Shear Failure.

A separate examination of each of these failures follows below.

Shear of the bolt occurs when the shear strength (τ_{MAX}) of the bolt is exceeded. The load (P) which initiates failure is defined as:

$$P = AS = \frac{\pi d^2}{4} \tau_{MAX} \qquad (2.10)$$

where:

A = cross-sectional area of bolt
d = diameter of bolt.
S = shear stress.

An illustration of bolt shear failure is given in Fig. 2.8.

Tensile failure of the plate, sometimes known as plate tear-through, occurs at the minimum width of the plate ($p - d$), where p = the plate width and d = the diameter of the bolt. Tear-through is therefore a function of the effective plate width ($p - d$) and the plate thickness (t). The tensile failure load (P) is thereby defined as:

$$P = AS = (p - d)t\,\sigma \qquad (2.11)$$

where:

A = cross sectional area of the plate at the minimum width.

Plate tear-through is illustrated in Fig. 2.9.

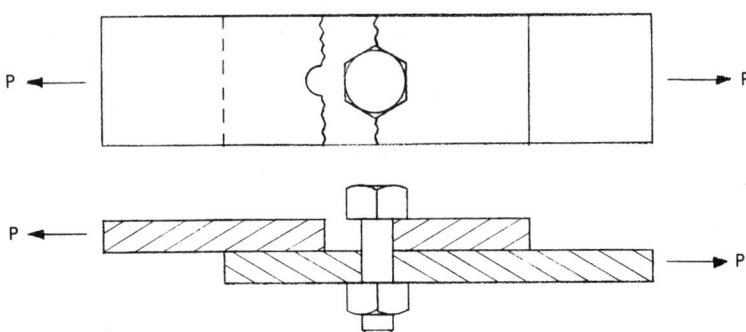

Fig. 2.9. Plate Tear-Through.

16 ADVANCED COMPOSITE MOLD MAKING

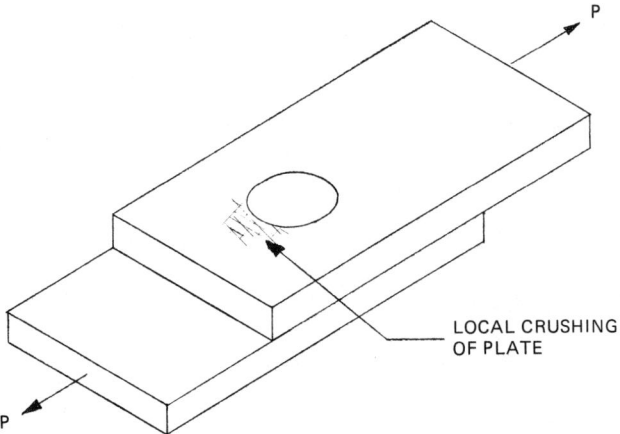

Fig. 2.10. Plate Bearing Failure.

Bearing failure of the plate or bolt occurs when the local compression strength is exceeded. Hence, permanent plastic deformation is experienced by either the bolt or the plate. Although the actual bearing stress (σ_b) intensity is not uniform over the entire bolt, in tool design this stress intensity is considered to be uniform over the projected area (dt) where d = bolt diameter and t = plate thickness.

The bearing failure load (P) is defined below:

$$P = AS = dt\sigma_b \tag{2.12}$$

where:

A = projected area of the bolt.

Bearing failure of the plate and bolt is illustrated in Fig. 2.10.

Other joint failure modes are possible, such as tear-out and behind-the-bolt shear (Fig. 2.11A and B), but these can be alleviated with proper tool design. A correct tool design would be one that contains a distance from the bolt center to the edge of the plate equal to twice the bolt diameter.

Finally, it is important to note that holes in structural members reduce the member's effective strength. Hence, a stress concentration factor (K) can be utilized to reduce the allowable design stress.

$$K = \frac{P - d}{P} \tag{2.13}$$

where:

STRENGTH OF MATERIALS AND MOLD DESIGN   17

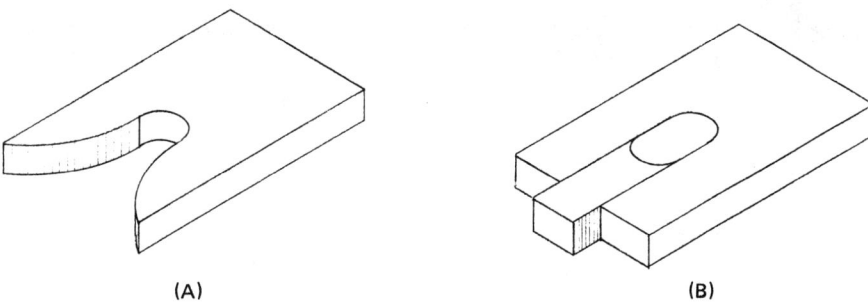

Fig. 2.11. A. Tear-out; B. Behind-the-Bolt Shear.

d = diameter of hole, and
P = width of plate.

This stress concentration factor is based upon the smallest effective cross-sectional area through the joint, and can be used in conjunction with other concentration factors.

Beams. The analysis of static loading on tools may be accomplished through the use of theory of elasticity. It should be noted at the outset that, although stress-strain data can be used in tool design, creep-rupture data is more accurate for composites and plastics. Because of viscoelasticity, the loading strength of plastics diminishes over time. Hence, the design strength of the mold is dependent upon the magnitude of the applied load and the duration of its application. Furthermore, increased temperatures decrease the strength of composite or plastic molds and tools.

All of these factors are important and should be considered in order to avoid designing a mold or tool that will experience critical loading. Since large safety factors are always incorporated into a mold or tool, critical loading should never present itself.

The following discussion and Table 2.1 are provided so that the designer can analyze almost any loading condition.

The equations in the table are governed by the rules of superposition, which implies that the equations are additive. For instance, if a problem occurs with more than one of the loading conditions in the table, all that needs to be done is to break the problem into its smallest parts and add the results. Thus, no matter what the loading conditions, the table and rules of superposition can be used to achieve a solution.

To model a problem, a designer must understand the loading conditions, as previously described. Consider the following problem of a composite tape-laying machine carrying out its designed task of dispensing a composite fiber

18 ADVANCED COMPOSITE MOLD MAKING

Table 2.1. Bending Moment, Vertical Shear, and Deflection of Beams of Uniform Cross Section under Various Conditions of Loading.*

P = concentrated loads, lb.
R_1, R_2 = reactions, lb.
w = uniform load per unit of length, lb per in.
W = total uniform load on beam, lb.
l = length of beam, in.
x = distance from support to any section, in.
E = modulus of elasticity, lb per sq in.

I = moment of inertia, in.^{4}
V_x = vertical shear at any section, lb.
V = maximum vertical shear, lb.
M_x = bending moment at any section, lb-in.
M = maximum bending moment, lb-in.
y = maximum deflection, in.

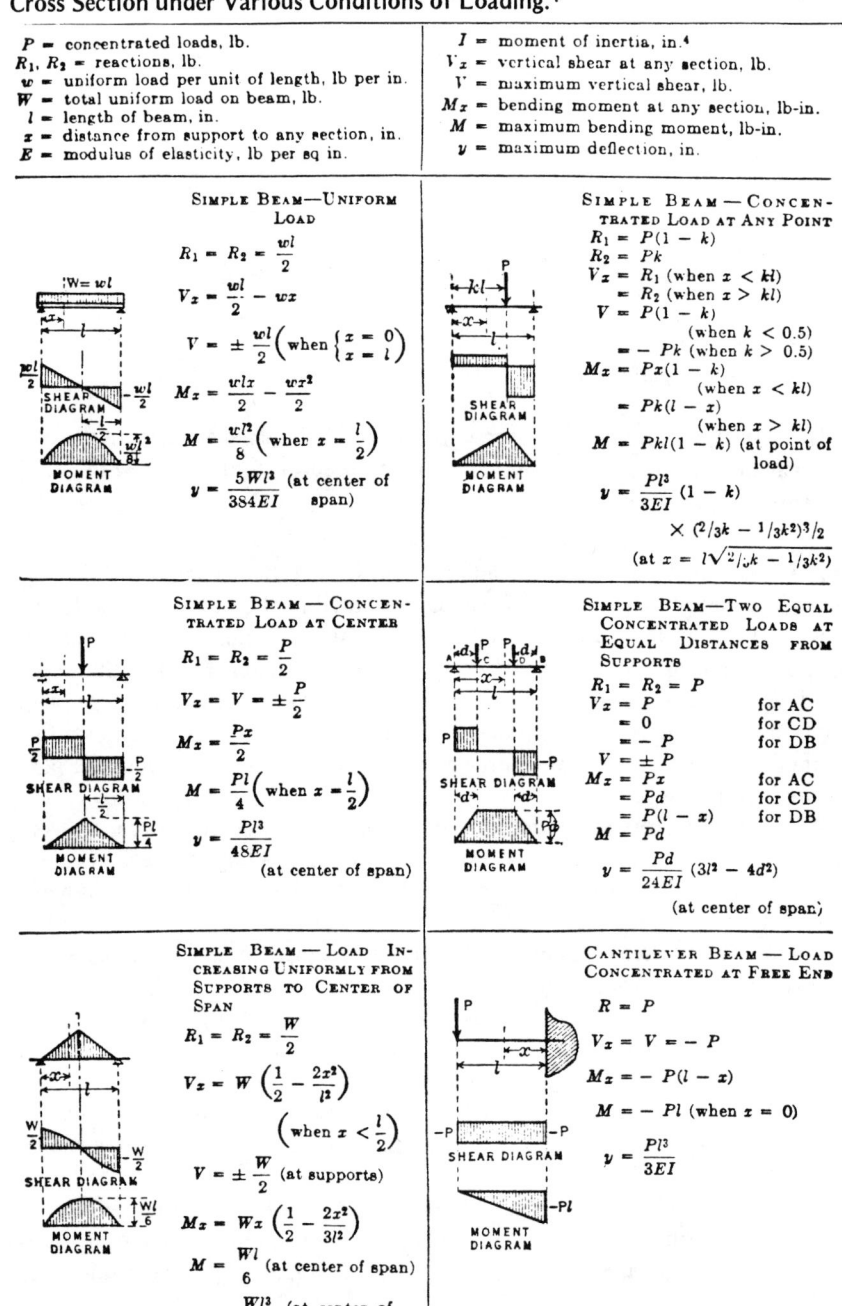

SIMPLE BEAM—UNIFORM LOAD

$R_1 = R_2 = \dfrac{wl}{2}$

$V_x = \dfrac{wl}{2} - wx$

$V = \pm \dfrac{wl}{2}$ (when $\begin{cases} x = 0 \\ x = l \end{cases}$)

$M_x = \dfrac{wlx}{2} - \dfrac{wx^2}{2}$

$M = \dfrac{wl^2}{8}$ (when $x = \dfrac{l}{2}$)

$y = \dfrac{5Wl^3}{384EI}$ (at center of span)

SIMPLE BEAM—CONCENTRATED LOAD AT ANY POINT

$R_1 = P(1 - k)$
$R_2 = Pk$
$V_x = R_1$ (when $x < kl$)
   $= R_2$ (when $x > kl$)
$V = P(1 - k)$
   (when $k < 0.5$)
   $= -Pk$ (when $k > 0.5$)
$M_x = Px(1 - k)$
   (when $x < kl$)
   $= Pk(l - x)$
   (when $x > kl$)
$M = Pkl(1 - k)$ (at point of load)

$y = \dfrac{Pl^3}{3EI}(1 - k)$
   $\times (^2/_3 k - 1/3k^2)^{3/2}$
   (at $x = l\sqrt{^2/_3 k - 1/3k^2}$)

SIMPLE BEAM—CONCENTRATED LOAD AT CENTER

$R_1 = R_2 = \dfrac{P}{2}$

$V_x = V = \pm \dfrac{P}{2}$

$M_x = \dfrac{Px}{2}$

$M = \dfrac{Pl}{4}$ (when $x = \dfrac{l}{2}$)

$y = \dfrac{Pl^3}{48EI}$
   (at center of span)

SIMPLE BEAM—TWO EQUAL CONCENTRATED LOADS AT EQUAL DISTANCES FROM SUPPORTS

$R_1 = R_2 = P$
$V_x = P$ for AC
   $= 0$ for CD
   $= -P$ for DB
$V = \pm P$
$M_x = Px$ for AC
   $= Pd$ for CD
   $= P(l - x)$ for DB
$M = Pd$

$y = \dfrac{Pd}{24EI}(3l^2 - 4d^2)$
   (at center of span)

SIMPLE BEAM—LOAD INCREASING UNIFORMLY FROM SUPPORTS TO CENTER OF SPAN

$R_1 = R_2 = \dfrac{W}{2}$

$V_x = W\left(\dfrac{1}{2} - \dfrac{2x^2}{l^2}\right)$
   (when $x < \dfrac{l}{2}$)

$V = \pm \dfrac{W}{2}$ (at supports)

$M_x = Wx\left(\dfrac{1}{2} - \dfrac{2x^2}{3l^2}\right)$

$M = \dfrac{Wl}{6}$ (at center of span)

$y = \dfrac{Wl^3}{60EI}$ (at center of span)

CANTILEVER BEAM—LOAD CONCENTRATED AT FREE END

$R = P$

$V_x = V = -P$

$M_x = -P(l - x)$

$M = -Pl$ (when $x = 0$)

$y = \dfrac{Pl^3}{3EI}$

STRENGTH OF MATERIALS AND MOLD DESIGN   19

Table 2.1. *(Continued)*

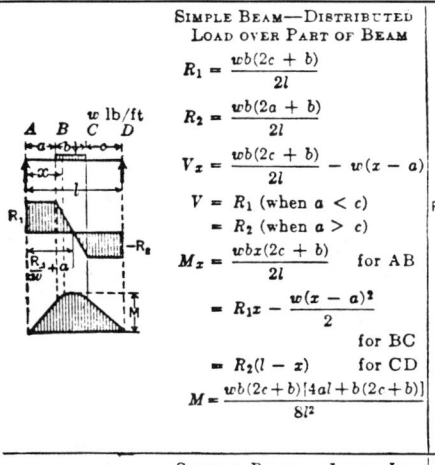 SIMPLE BEAM—DISTRIBUTED LOAD OVER PART OF BEAM $$R_1 = \frac{wb(2c + b)}{2l}$$ $$R_2 = \frac{wb(2a + b)}{2l}$$ $$V_x = \frac{wb(2c + b)}{2l} - w(x - a)$$ $V = R_1$ (when $a < c$) $ = R_2$ (when $a > c$) $$M_x = \frac{wbx(2c + b)}{2l} \text{ for AB}$$ $$= R_1 x - \frac{w(x - a)^2}{2}$$ for BC $$= R_2(l - x) \text{ for CD}$$ $$M = \frac{wb(2c + b)[4al + b(2c + b)]}{8l^2}$$	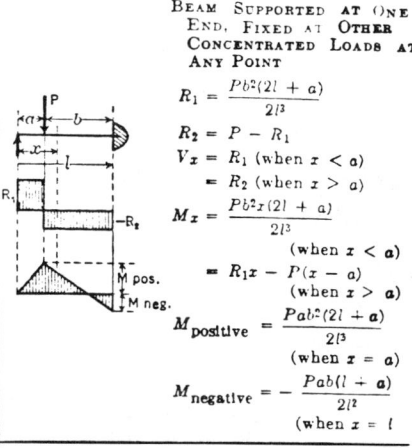 BEAM SUPPORTED AT ONE END, FIXED AT OTHER CONCENTRATED LOADS AT ANY POINT $$R_1 = \frac{Pb^2(2l + a)}{2l^3}$$ $$R_2 = P - R_1$$ $V_x = R_1$ (when $x < a$) $ = R_2$ (when $x > a$) $$M_x = \frac{Pb^2 x(2l + a)}{2l^3}$$ (when $x < a$) $$= R_1 x - P(x - a)$$ (when $x > a$) $$M_{\text{positive}} = \frac{Pab^2(2l + a)}{2l^3}$$ (when $x = a$) $$M_{\text{negative}} = -\frac{Pab(l + a)}{2l^2}$$ (when $x = l$)
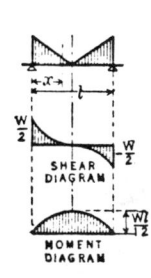 SIMPLE BEAM—LOAD INCREASING UNIFORMLY FROM CENTER TO SUPPORTS $$R_1 = R_2 = \frac{W}{2}$$ $$V_x = -W\left(\frac{2x}{l} - \frac{2x^2}{l^2} - \frac{1}{2}\right)$$ (when $x < \frac{l}{2}$) $$V = \pm \frac{W}{2}$$ $$M_x = Wx\left(\frac{1}{2} - \frac{x}{l} + \frac{2}{3}\frac{x^2}{l^2}\right)$$ (when $x < \frac{l}{2}$) $$M = \frac{Wl}{12} \text{ (at center of span)}$$ $$y = \frac{3}{320}\frac{Wl^3}{EI} \text{ (at center of span)}$$	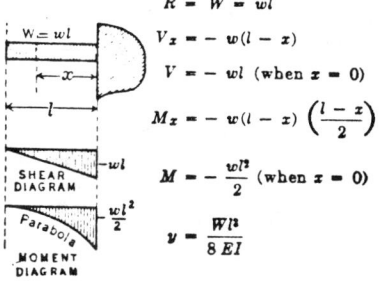 CANTILEVER BEAM—UNIFORM LOAD $$R = W = wl$$ $$V_x = -w(l - x)$$ $$V = -wl \text{ (when } x = 0\text{)}$$ $$M_x = -w(l - x)\left(\frac{l - x}{2}\right)$$ $$M = -\frac{wl^2}{2} \text{ (when } x = 0\text{)}$$ $$y = \frac{Wl^3}{8EI}$$
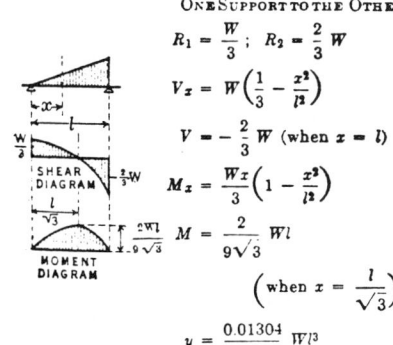 SIMPLE BEAM—LOAD INCREASING UNIFORMLY FROM ONE SUPPORT TO THE OTHER $$R_1 = \frac{W}{3}\ ;\ R_2 = \frac{2}{3}W$$ $$V_x = W\left(\frac{1}{3} - \frac{x^2}{l^2}\right)$$ $$V = -\frac{2}{3}W \text{ (when } x = l\text{)}$$ $$M_x = \frac{Wx}{3}\left(1 - \frac{x^2}{l^2}\right)$$ $$M = \frac{2}{9\sqrt{3}}Wl$$ $$\left(\text{when } x = \frac{l}{\sqrt{3}}\right)$$ $$y = \frac{0.01304}{EI}Wl^3$$	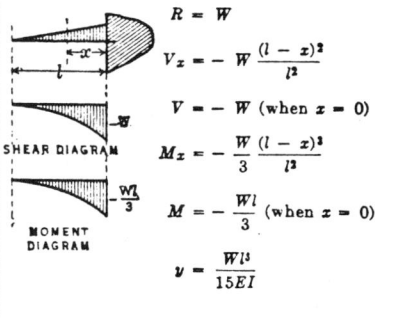 CANTILEVER BEAM—LOAD INCREASING UNIFORMLY FROM FREE END TO SUPPORT $$R = W$$ $$V_x = -W\frac{(l - x)^2}{l^2}$$ $$V = -W \text{ (when } x = 0\text{)}$$ $$M_x = -\frac{W}{3}\frac{(l - x)^3}{l^2}$$ $$M = -\frac{Wl}{3} \text{ (when } x = 0\text{)}$$ $$y = \frac{Wl^3}{15EI}$$

Table 2.1. *(Continued)*

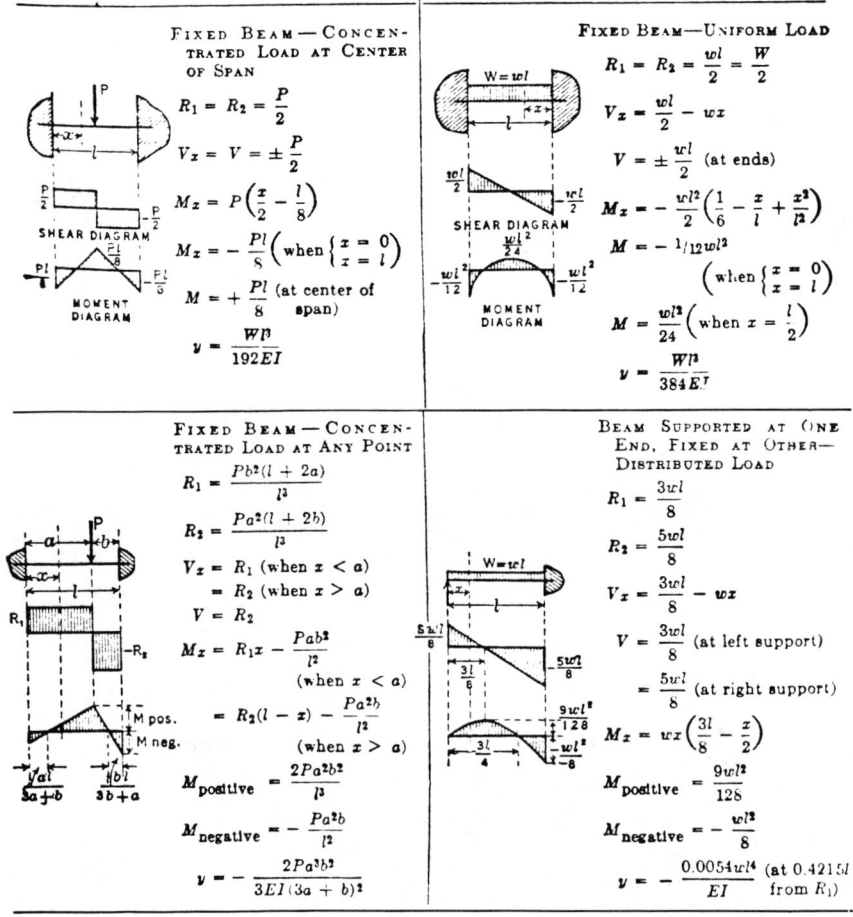

From *Handbook of Engineering Fundamentals*, O. W. Eshbach and M. Souders, Wiley Engineering Handbook Series. Copyright © 1975 John Wiley & Sons, Inc. Reprinted by permission of John Wiley & Sons, Inc.

tape. The problem can be modeled as in Fig. 2.12, if the structure has an OML (outside mold line) plate and an eggcrate back-up structure, which would be the reaction points within the model.

The maximum tensile and compression stresses are:

$$\sigma = \frac{MD}{2I} \qquad (2.14)$$

where:

STRENGTH OF MATERIALS AND MOLD DESIGN 21

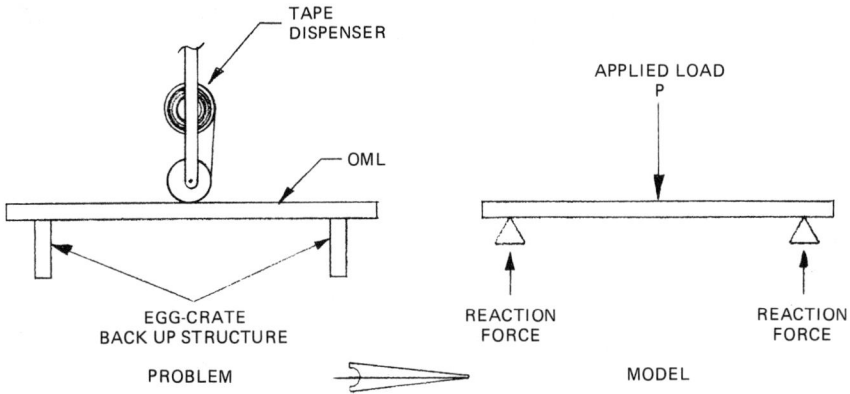

Fig. 2.12. Beam Subject to a Concentrated Load.

M = maximum bending moment (lbs)
D = beam thickness (in.)
I = moment of inertia (in.^{4})
σ = stress (lb/in.^{2})

Although there can be numerous loading conditions, Eq. (2.14) is valid as long as the maximum bending moment is determinable. The maximum deflection (δ_{max}) would occur at the center of the beam in Fig. 2.12 and would be equal to:

$$\delta_{max} = \frac{PL}{48EI} \qquad (2.15)$$

where:

P = load (lb)
L = length of beam (in.)
E = Young's Modulus (lb/in.^{2})
I = moment of inertia (in.^{4}).

Designing structural members within a tool is basically a bookkeeping operation, whereby "plugging in" the different variables yields a solution, to which safety factors may be applied.

2.2. THERMAL EFFECTS AND DESIGN CONSIDERATIONS

Since most composites are cured at elevated temperatures, as high as 350°F, a tool designer must know the effects associated with these temperatures. The three modes of heat transfer — conduction, convection, and radiation — should be understood. Heat-up rates and autoclave heating should also be understood in order to determine the duration of the cure cycle.

It becomes apparent, then, that a designer must consider statics, the strength of materials, and thermal effects in order to create a tool which is reliable throughout its service life.

2.2.1. Heat Transfer Modes

When temperature gradients exist within a body, energy is transferred from the high-energy region (hot portion) to the low-energy region (cold portion) via molecular action. Heat transfer via molecular action within solids is called conduction, and conductive heat transfer is governed by the following equation:

$$Q = kA \frac{(T_1 - T_2)}{L} \qquad (2.16)$$

where:

Q = quantity of heat transmitted per unit time (Btu/hr)
k = thermal conductivity (Btu/hr ft °F)
A = cross sectional area (ft^2)
$(T_1 - T_2)$ = temperature gradient (°F)
L = length of the specimen (ft).

In general, materials that are good thermal conductors are also good electrical conductors. The material property which is responsible for conduction is the thermal conductivity (k).

Various values of thermal conductivity are listed in Table 2.2.

Convection is also a process of heat transfer. An example used here is a fluid, where fluid particles of different temperatures are present and mixed together. If moved rapidly, the moving fluid will cool an object faster than a stagnant fluid. The following equation, known as Newton's *Law of Cooling*, governs conductive heat transfer:

$$q = hA (T_S - T_F) \qquad (2.17)$$

where:

q = quantity of heat transmitted per unit time (Btu/hr)
h = convective heat transfer coefficient (Btu/hr – ft^2)
T_S = temperature of the solid (°F)
T_F = temperature of the fluid (°F).

Table 2.2. Thermal Conductivity.

PROPERTY	WATER	FIBERGLASS/ EPOXY	GRAPHITE	SOFT STEEL	ALUMINUM	NICKEL	ALUMINA 96% α–Al_2O_3
Thermal conductivity, k Btu/in.·ft²·hr·°F	(68–420°F) 4.4	1.2–2.9	1220	328	(122°F) 842–1510	(32–212°F) 464	244
G-cal/cm·cm²·sec·°C	(20–200°C) .0014	.0004–.0010	.42	.113	(50°C) .29–.52	(0–100°C) .14	.084

Since there is no fluid motion at a solid-fluid boundary, conduction takes place within this film layer, after which convective fluid motion cools the fluid layer.

Radiation is the transfer of energy due to electromagnetic waves, which can carry energy through a perfect vacuum. A good example of radiation is the energy transfer from the sun to the earth, resulting in heat, light, and ultraviolet radiation reaching the earth. All bodies emit and absorb radiation. The energy (E_b) radiated from a black body is given by:

$$E_b = \sigma A T^4 \qquad (2.18)$$

where:

σ = Stefan-Boltzmann constant (σ = .1713 × 10^{-8}) Btu/ft^2-hr R^4)
T = temperature of body (°R).
A = area of surface

The energy radiated by a gray body is given by:

$$E = e A \sigma T^4 \qquad (2.19)$$

where:

e = emissivity (always lower than 1)

The rate at which heat is emitted from a body is given by:

$$q = A\, e\, \sigma\, (T_H^4 - T_C^4) \qquad (2.20)$$

where:

A = surface area of the body
T_C = temperature of cold body
T_H = temperature of hot body.

These are the three modes of heat transfer, with conduction and convection being the most important in tool design.

2.2.2. Tool Heat-up Rate

A uniform specific temperature rise is usually desirable for the tool heat-up rate. The three parameters which govern the heat-up rate are temperature, specific heat of the material, and the mass of the material.

Equation (2.21) governs the heat-up rate, where q is the amount of heat per unit time.

$$q = m c_p \Delta T \qquad (2.21)$$

where:

q = amount of heat required per unit time (Btu/hr)
m = mass of the tool (lb)
c_p = specific heat of the material (Btu/lb °F)
T = intended temperature rise ($T_{FINAL} - T_{INITIAL}$) per unit time.

The temperature difference (ΔT) is really the difference between the final and ambient temperatures. Hence, for a graphite epoxy composite, whereby $T_{FINAL} = 350°F$ and $T_{INITIAL} = 68°$, then $\Delta T = 282°F$.

Therefore, the only controllable factors are the specific heat and the mass of the material. A list of specific heat is shown in Table 2.3.

It should be noted that, of the possible choices for mold materials, aluminum has the highest required heat per pound.

2.2.3. Autoclave Heating

The properties which determine the heating time in an autoclave are the tool mass, the tool material, and the heat transfer properties of the tool material. It is always desirable to utilize the smallest specific heat possible and the least possible mass.

For the most part, the size and the mass of the tool are determined by the size and configuration of the finished part. In addition, process requirements will usually determine the tool material.

Thus, the autoclave heat-up rate is determined by convective heat transfer from the autoclave to the tool.

$$q = h A (T_S - T_F) \qquad (2.22)$$

Here again, the temperature gradient is governed by heat-up rates; the surface area of the tool is governed by the part, and the convective heat transfer coefficient is dependent upon the air flow within the autoclave.

The only controllable factor is the placement of the tool in the autoclave to provide the largest contact area without causing uneven heating. One way of achieving this is by the strategic placing of a heat blanket.

26 ADVANCED COMPOSITE MOLD MAKING

Table 2.3. Specific Heat.

PROPERTY	WATER	FIBERGLASS/EPOXY	GRAPHITE	SOFT STEEL	ALUMINUM	NICKEL	ALUMINA 96% α–Al$_2$O$_3$
Specific heat, C_p							
Btu/lb	(60°F) 1.00	(RT) .19	(79–169°F) .165	.114	(32–212°F) .226	.13	.185
G-cal/g	(16°C) 1.00	(RT) .19	(26–76°C) .165	.114	(0–100°C) .226	.13	.185

Table 2.4. Thermal Expansion.

PROPERTY	WATER	FIBERGLASS/EPOXY	GRAPHITE	SOFT STEEL	ALUMINUM	NICKEL	ALUMINA 96% α–Al$_2$O$_3$
Thermal expansion, α							
in./in.·°F	—	$6.1-19.4 \times 10^{-6}$	(122°F) 4.4×10^{-6}	(60°F) 6.6×10^{-6}	(68–575°F) 14.1×10^{-6}	(75–212°F) 7.6×10^{-6}	(75–575°F) 3×10^{-6}
cm/cm·°C	—	$11-35 \times 10^{-6}$	(50°C) 7.9×10^{-6}	(16°C) 11.9×10^{-6}	(20–300°C) 25.5×10^{-6}	(25–100°C) 13.7×10^{-6}	(25–300°C) 6.4×10^{-6}

2.2.4 Thermal Expansion of Tools

Since composite parts have a much smaller coefficient of thermal expansion than metallic tools, tool design must accommodate these differences.

Thermal adjustments can be made to a tool to provide for differences in the coefficient of thermal expansion (CTE). A tool design which has CTE provisions incorporated into the tool is considered to be thermally shrunk.

Since composites cure at elevated temperatures, a tool should be designed to make a finished part at that temperature. This is extremely important when trim and scribe lines are placed on the tool.

In order to compensate for the CTEs of various mold and tool materials, Eq. 2.23 should be worked out for both the part material and the tooling material:

$$\Delta L = \alpha \Delta T \qquad (2.23)$$

where:

ΔL = per unit change in length (in./in.)
α = coefficient of thermal expansion (in./in.°F)
ΔT = temperature difference between the ambient and gelation temperatures (°F)

Once the per unit change in length is solved for both materials, the difference should be calculated to arrive at a thermal shrink factor. This factor should then be multiplied to all dimensions from the thermal shrink point, which is the mass centroid of the tool.

Coefficients of thermal expansions are listed in Table 2.4.

3
Mold Engineering

Before the Second World War, the use of plastics as a material for mold making, rather than as a molding material, was relatively unknown. Since then, there has been a steady increase in the use in the construction of dies, fixtures, molds, and tools, and in their use as materials in the fabrication of mold making articles from glass, plastics, wood, and metal.

The widespread establishment and acceptance of tooling with plastics is based upon the following clear advantages over tools of metal and/or wood:

1. relatively light in weight and easy to handle;
2. resistance to corrosive atmosphere, lubricants, and weather;
3. the ability to be cast or laid up to the desired final shape in one operation, thus saving on labor costs;
4. more frequent changes in design are feasible;
5. relatively inexpensive equipment can be utilized for fabrication;
6. the delivery of finished tools is faster, thus reducing the required lead time between design and production of the end-use article;
7. where several tools are required, duplication is easier;
8. revisions and repairs are simpler.

In a word, rapidity and economy make the utilization of plastics practical in mold making today, provided there is a careful evaluation of the end use when a plastics material is selected for each specific application.

Further development and growth of tooling with plastics must be based upon sound engineering standards and mold design.

It must be stressed that economy in mold making may become extravagance in the production of molded parts. The reasons are firm ones: A well-organized, durable mold is essential for any successful molding operation. The length of the run, the rate of production, and the accuracy of detail and dimensions of the part are some of the major factors to be considered in the selection of a suitable mold.

Because of the specialized nature of mold making, a large number of mold and tool makers have been established throughout the industry for producing the molds and tools for parts and assembly fabrications. However, in the larger companies and plastic manufacturing facilities, one of the more important locations is a well-equipped and manned mold and tool shop.

Materials. Understanding the limitations as well as the characteristics of materials is essential, whether the mold or tool designer is designing a product or the mold and the tooling to produce it. The selection of materials for a design is a complex procedure, involving constant evaluation and reevaluation. Neither polymers, metals, ceramics, wood, or other materials are panaceas for all problems. A good design is a compromise, using the best possible materials for a specific purpose.

Considerations. A number of considerations should be taken into account before proceeding with the engineering of a particular mold or tool. First, some knowledge of the part design, as well as the manufacturing methods to be used, should be obtained.

Second, the number of parts or assemblies that will be made from the mold, and the intricacy of each part, should be studied.

When a part is to be produced with fine details and close tolerances, designers almost universally go to a nonmetallic mold or tool that is made from the same material as the structural part. In fact, in some instances, the prototype mold might well be the production mold, due to mold engineering constraints.

Further help is provided by new techniques for recording designs. Once produced manually, new designs are now automatically transferred to computer memory. Existing designs can be duplicated (digitized) for the record, while new designs can be created with computer assistance (CAD) and presented in three-dimensional form (CAM) and computers can assist with the transformation to machining (N/C) and inspection. These terms will be discussed later in the chapter.

Third, cost and fabrication feasibility must be considered. Some of the more important details to be examined are:

The choice of a metal or nonmetal material for the mold or tool;
The temperature at which the mold is to be used;
Part and mold tolerances;
The size of the mold or tool;
Undercuts, hardware, and inserts (if any);
The mold wall thickness;
Tapering and draft angles;
Inside and outside radii of the tool or mold;
Special reinforcements, such as ribs, fillers, etc.;
Warpage due to substructure support;

Table 3.1. A Typical Cost Breakdown.

MATERIAL	COST/FT^2 OF SURFACE AREA
Monolithic graphite block N/C machined (2–3 in. thick)	$200–$300
Electroformed nickel (.350 in. thick nickel)	$300
Integrally stiffened graphite epoxy prepreg (¼ in. thick laminate integrally stiffened)	$200–$300
Cast ceramic (2–3 in. thick)	$50

Parting and trim lines;
Holes, rails, and attachments;
The surface finish;
Post-mold making; and
Handling and storage.

The least expensive and most accurate materials for making a light-weight mold or tool with complex contours has proven to be impregnated monolithic graphite block material (N/C machined), electroformed nickel, or graphite fabric reinforced epoxy-type materials and mass-cast ceramic. A comparative cost study of these materials is given in Table 3.1.

A final consideration in a good mold or tool design should be the rate at which the mold material will heat up. Table 3.2 shows the general rates at which various materials will conduct heat.

There are certain fundamental truths in mold engineering, and now is a good time to introduce them:

- There is no such thing as low-cost tooling — unless it produces a quality part at low cost.
- The materials used in the molds and tools account for the smallest part of the cost.

Table 3.2. General Mold Material Heat-up Rates.

HEAT-UP RATE	MOLD OR TOOL MATERIAL
Fast	Impregnated monolithic graphite block (N/C machined), Graphite fabric reinforced epoxy Mass cast/conductive fillers Aluminum Nickel
Slow	Steel

- It costs less to correct a tool or mold than to correct the molded part or bonded assembly.
- Good molds and tooling encourage good workmanship. Poor tools encourage inferior workmanship.
- The damaging effects of a poor quality mold will linger long after the benefits of its initial low cost are gone.
- Quality cannot be inspected into a part. It can only be built into the mold or tool.

Another truth, which deserves a place of its own, preferably framed and hung in the workplace, is: The quality of a part is only as good as the quality of the mold that produced it.

3.1. ESTIMATING MOLD AND TOOL MAKING COSTS

Many variables are present in the evolution of a mold or tool design. Because of this, the cost of producing a complicated mold may vary significantly from that of producing a simple contour of the same size.

Thus, it is important for designers to be familiar with a number of practices and methods for estimating the cost of molds and tools, in order to effectively determine whether or not to manufacture a certain mold or tool in-house, establish a sale price, and/or record the cost information for future quoting purposes.

Generally speaking, the definition of a cost estimate is: *A procedure for compiling a cost of molds and tools made or to be made, where experience alone cannot supply complete figures.*

The basis for all mold and tool designs is similar, and general estimates are presented to make sales proposals, bids on new and repeat work, and for other special contractual needs.

Actual cost figures (past or present), facts concerning available facilities and equipment, current labor and burden rates, the present and future market prices of materials, and knowledge of the mold-making process to be performed may all be used in compiling an estimated cost.

The estimator's function in estimating costs is to consider the best methods to estimate the most accurate mold or tool sales price and establish the most accurate standard for both accounting purposes and to determine if the company should fabricate the mold, tool, and/or parts and assemblies.

Generally, it is good practice to staff the estimating group, team, or department with individuals who are specialists in various tool design and fabrication functions, or in general and cost accounting, as well as engineers, designer draftsmen, and a staff for clerical, stenographic, and filing work.

In all cases, the mold or tooling estimator should consult with the designers and master mechanics on matters relating to the making of new molds, tools, jigs, and fixtures, as well as the repair and reconditioning of them.

In general, the estimator cannot also be expected to be the tool designer and maker. The estimator usually doesn't know more than the manufacturing department regarding plant layout, facilities, available machinery, and equipment. Therefore, consulting with the production tool departments in order to determine what type of equipment, class of labor, and tool shop capacity exists for all new molds and tools is strongly recommended. In a similar vein, the rate-setting or time-study group in the company may be extremely useful in providing data needed to establish labor costs.

3.1.1. Preparation

Before proceeding with an estimate, the following factors should be considered:

- Existing previous estimates
- Existing actual cost records
- Available production time
- Quantities of parts to be produced
- Availability of materials
- Availability of space, equipment, machinery, and tools
- Anticipated future rates of labor, burden, and material.

Although changes in labor, material, equipment, and associated factors can realistically be expected to occur, an examination of previous estimates and existing actual costs should be a priority consideration to any cost estimation process.

The reasons for this are multiple and logical: First, time can be saved by reviewing existing formats, and comparing and checking previous costs. The records may indicate an estimate for a similar or exact-scaled version of the mold or tool to be priced.

Next, cost and time-consuming problems can be avoided if previous estimates are studied for records of production data, problems experienced in mold or tool fabrication, rejected parts, material spillage, tool repairs, and poor or problem mold and tool designs.

Third, help in avoiding unforeseen problems can be received if the records indicate the use of other tooling to assist in part fabrication or short cuts to completed part manufacture, or the requirement of special tools as a result of first part production.

Also, personnel problems can be predicted and avoided, since most records document the learning curves of inexperienced operators and technicians.

Finally, past mistakes that can cause additional handling or other increases in cost can be avoided if information concerning products produced with molds and tools not especially designed or laid out for larger or continuous runs of the subject product is contained in records. Molds and tools that are especially planned for the production of the article and its quantities generally result in greater economy in the long run.

3.1.2. The Process of Estimation

In the estimation of actually costing molds and tools, the costs must be divided into three major divisions: labor, burden, and material.

Labor. Future labor costs should always be considered when choosing personnel to fabricate the tooling required. The complexity of a new or existing design should be a factor in choosing who will fabricate the mold or tool components.

When calculating the mold or tool designers' time, do not use weekly or monthly salary rates. Hourly rates for the various types of designers should be estimated with the designers' rates fixed and grouped by the standard time rate.

Besides having a knowledge of the operations which will be performed on that mold or tool, the tools or assembly machinery that will be required to complete the part, and the departments that will be required in the fabrication of the tooling, the estimator must also know the cost of labor in that department and who will be assigned to each task in the mold fabrication, so that the class of the operator can be determined.

For experimental work, the class of labor is an important consideration. In some cases, the services of highly trained and skilled personnel is required. Materials and process engineers, chemists, and even management personnel must be considered in the costing.

Estimators should also consider the use of more than one mold or tool on more than one machine operated by the same operator. In fact, in summarizing costs, multiple tooling may be the key to obtaining an order, since the time required to complete a part fabrication or assembly operation will dictate the number of molds or tools required to produce a fixed number of parts. One operator may be able to operate two pieces of equipment with two molds inside more cheaply than one operator operating one piece of equipment with one mold inside. Further consideration must be given to doubling shifts, overtime shifts, and bonus shifts.

In summation, labor is probably the most complicated and variable area in the cost estimating procedures. It involves more computations than estimating the cost of material and burden, since it requires a knowledge on the part of the estimator of every operation involved in the task to be performed.

To estimate the labor needed with any degree of accuracy, each individual operation should be written down in detail on specially prepared sheets. The sheet design should be based upon the complexity and logistics of the estimating team. If the item or assembly to be manufactured is composed of various minor

34 ADVANCED COMPOSITE MOLD MAKING

parts or assemblies, each one should be broken down into its component parts to reveal the tooling and labor required, and the operations should be listed in detail.

When designing a labor cost estimating sheet, the following facts and items should be addressed:

- The estimate I.D. number
- Date of the estimate
- Customer name
- Part or assembly number
- Associated assembly numbers
- Drawing numbers
- Description of the part or assembly (size, function, materials, etc.)
- Department numbers where work will be performed
- Names of operations, in proper sequence
- Number of operations, in proper sequence
- Equipment to be used, in proper sequence
- Setup costs for operations and machinery in rates per hour for each operation
- A subtotal of each department's costs, so that overhead and burden can be calculated easily for setups performed in various departments. Standard setup costs should be available in the estimating files, and in extraordinary cases the estimator should consult with the production department.
- The burden percentage and burden rate in hours should be considered separately from the money for setup and actual mold fabrication
- Setup cost based on each mold or tool setup operation, not on the number of molds or tools to be made.

Burden. The proper area for tool or mold fabrication should be chosen by considering the past and future quantities of parts to be produced from the tooling. Increased quantities may warrant the fabrication of additional tooling, as indicated by the estimate records. Partial fabrication by outside vendors at fixed costs, consideration of energy expenses such as power, heat, the addition or reduction of lighting, and the addition or reduction of indirect labor or equipment may warrant production in separate or isolated parts of the plant or tool shops. The absence of these factors may indicate that production is occurring in areas where similar molds are produced or overhead expenses are less.

Burden should be applied to the estimated cost of a mold or tool as it would be applied to the cost of a product or assembly by the costing department or costing system of any department. This is recommended because, upon completion of the mold or tool, the actual cost can be compared by that department to the estimate prepared by the estimator.

Material. The use of mold masters, tools, jigs, and additional equipment must be considered in the same manner as any other machinery. The availability of existing molds and tools, machinery, and the time necessary to produce the mold or tool will effect cost. The quantity of parts to be made, as well as the time available to make these parts, will determine the design configuration of the mold. Therefore, the estimator should have a thorough understanding of mold and tool design.

The estimator is charged with the responsibility of including special jigs, fixtures, dies, tools, patterns, masters, etc. that are required for the production of the part into the probable cost of the part or product.

The quantity to be made or sold must be divided into the total mold or tooling costs to determine the amount that may be recovered in the price of the item sold.

The estimator must also add in the cost of maintaining the tooling during use, since the maintenance of molds and tools during use sometimes becomes a costly factor.

All ordinary molds, tools, jigs, and fixtures that the estimator has previously calculated can be safely estimated without any assistance. However, for more complicated designs, the estimator should consult with an experienced tool designer, mold maker, pattern maker, or others who, through their individual expertise, are uniquely equipped to provide accurate estimating data.

The cost of experimental work, new concepts, products, or molds for items never produced before must be based upon the judgment of the cost estimator, designer, or engineer. Allowances must be made for necessary work to be done to assure the fulfillment of the design concept. Thus, time, material, equipment, and the labor needed to perform the experimental investigation must be allowed for.

In conducting experiments to determine the fabrication feasibility of a product mold, it is necessary to use materials or products which, after use, will become worthless. Thus, this waste material, and any salvage therefrom, should be calculated into the estimate.

Lastly, allowances must be made and consideration of additional equipment such as ovens, autoclaves, and transportation of models and molds to other locations for post-processing must be calculated.

It must be emphasized that material estimation requires good personal judgments. Material in inventory bears a price different from material to be purchased. If the mold or tool is to be made in the future, then material needs and costs must be predicted. Allowance for material waste, spoilage in cutting (fabric patterns are casualties here, for instance), or mold assembly must be considered.

Some material pricing considerations when examining material costing are:

- The type of mold materials (metallic, nonmetallic)
- The total amount of materials in the finished mold order

- The price of materials: in inventory; at market price; and at a projected future market price
- Estimates on spoilage and waste
- All other mold or tool identification information, which is carried over from the labor estimate form list on page 34.

A summary of costs is sometimes documented on a cost summary sheet. If it is necessary to calculate the costs of a number of molds, tools, or components (and parts thereof), a summary sheet of labor, material, and burden costs should be completed. If new tools are required, in addition to existing molds or tools, the individual costs should be calculated on the labor, material, and burden forms and summarized on a summary sheet. In this way, the total labor, material, and burden costs for previous and new tooling can be documented and presented, set burden and mold operating burden can be shown in detail, and the total cost of the molds or tools needed itemized and presented.

The mold or tool cost is usually shown separately in the total overall estimate. In this way, the costs of tooling can be adjusted by varying the total number of parts to be made in each mold, thus reducing the costs for future orders of finished parts and assemblies, or reducing tool costs in order to be more competitive in finished part cost.

In addition to aiding in estimating the costs of production, final pricing of the article, part, or assembly will be easier if a summary sheet for the estimate is used. Furthermore, the summary sheet will assist in predicting costs for future orders of larger or smaller quantities.

Repeat orders should never be considered to be part of the basis for projecting the cost of the original tool. Repeat orders can be counted and relied upon only for anticipation of space and machinery requirements for future orders.

It may be well to note at this point that good judgment, although elusive as a costing factor, should not be overlooked. One may compare this to an old time horse trader or a talented gem buyer. Each has the ability to look at and determine the value of what they are seeing immediately. An estimator, through his experience, should be able to look at an existing design, mold or tool, new drawing, design or sketch, and generally establish a cost without the use of special formulas or data.

Final Estimates. The final estimated mold or tool cost does not usually depend entirely upon figures prepared by the cost estimator. Modifications may, out of necessity, be made in order to conform with market conditions, the need to be competitive, the financial condition of the company, and the prospect of future business relating to all of these factors.

In an ideal situation, estimators should not be influenced by the sales manager or sales departments, the financial group, production managers or departments, or quality control functions and departments. Reality, of course, often dictates otherwise, and sometimes the estimator can be found to be under the supervision of a group that can negatively influence mold or tool costs.

Section Summary. The cost estimator must have specific knowledge, background information, abilities, and awareness, and these qualities are best applied in the following areas:

Cost accounting — Read cost analysis sheets and cost statements and make proper conclusions.

General accounting — Determine the relative burden charges which prevail at the time of the estimation. Know why burden rates go up and down, what increases and decreases in labor and material apply, and what figures to use in the estimate.

Engineering — Estimate costs for molds and tools to produce products that have never been produced before. *Note:* These are simple engineering abilities, which should allow the estimator to determine if further mold or tool engineering is required before costing the tooling.

Plant layout, production methods, machinery and tools — If a new mold or tool design is to be estimated, the estimator should know where the tooling will be made, what mold-making process will be used, what labor costs will be involved (by department), the types of material that will be required, and the equipment that is available.

3.2. Mold Life

One of the most controversial topics discussed over the past years in aircraft and aerospace circles has been the expected life of nonmetallic high-temperature oven and autoclave molds.

The original materials used to make molds were formulated for use at much lower temperatures than those required to cure the materials used to mold parts from their surfaces. As a result, designers of composite molds and tools have pursued various other metallic mold and tool design approaches to produce production molds.

As advances have occurred in the formulation of composite part materials, it is only natural that advanced composite mold materials should follow. Therefore, nonmetallic molds and tools could now safely be compared to metallic ones, and through unique approaches in nonmetallic mold maintenance and repair, have resulted in a less expensive production mold material for the future.

With new advances have come new considerations. Since the CTE can be matched almost exactly to that of the nonmetallic part or assembly material, accuracy must be guaranteed in the production of a mold. This is defined more precisely in Table 3.3.

Table 3.3. A Fabricated Mold, Tool, and Bonding Fixture Life

TYPE OF MOLD	AUTOCLAVE CYCLES AT 350°F BEFORE REPAIR IS NEEDED	
	MOLDS/ TOOLS	BONDING FIXTURES
Sheet, Cast, or Machined Steel	700	1000
Sheet, Cast, or Machined Aluminum	300	500
Epoxy, Fiberglass, or Graphite Fabric Reinforced		
Room Temp. Wet Lay-up (Vacuum Bagged)	30	50
Room Temp./High Temp. Wet Lay-up (Vacuum Bagged)	50	80
High Temp. Wet Lay-up (Vacuum Bagged)	60	90
*High Temp. Prepreg, Wet Surface Coat (Vacuum Bagged)	100	150
*High Temp. Prepreg, Film Surface Coat	200	300

*Fabricated in an autoclave at 85-100 psi

The number of cycles for each mold in Table 3.3 are shown to the point at which repairs to the surface or substructure will have to be made. The number of cycles shown does not indicate the total life of the mold or tool, but rather indicates the number of uses before replacement or major repair might occur. The numbers shown are culled from reports of various industry manufacturers of aircraft and aerospace parts, and are for design use only, as they are average/ approximate.

When considering the mold life of a nonmetallic mold or tool, one should be careful to choose a tooling system that will be processed at the use temperatures. In other words, a mold or tooling laminating system should be chosen by selecting the surface coat or film and laminating resin or prepreg that cures at the temperature at which the mold or tool will be used. Co-curing surfaces, mold and tool laminates, and — when applicable — substructures or eggcrates assure that the mold or tool will expand and contract uniformly throughout the entire structure. Of course, the materials chosen to fabricate the mold or tool must be capable of withstanding continuous cycling at these temperatures.

3.2.1. Extending the Life of Nonmetallic Molds and Tools

Good storage and handling practices are considered vital to the maintenance of the molding surfaces. Storage of production tools with reusable vacuum bags in place over the molding surfaces protects the finish and stores the mating accessories in place over the tool. Designs which include laminate mold or tool edges that are recessed or flush with the egg crate or backup structure edge (Fig. 3.1) eliminate structural failure during handling along laminated edges by eliminating delamination of plies and leaks in the mold laminate.

The placement of a reject part or bonded assembly into the mold cavity, or a 2- to 3-ply room-temperature-cure laminate lay-up which could serve as a mold

MOLD ENGINEERING 39

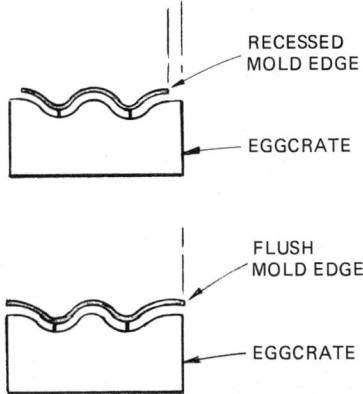

Fig. 3.1. Recessed and Flush Mold and Tool Edges.

contour test sample (greenie), could travel with the mold during fabrication and remain in the storage area compartment while the mold is in use. This can also double as a shield for the tool surface during storage.

Rolled mold or tool edges, although time-consuming and expensive to produce, also assist as "bumpers" on tools in use or storage (Fig. 3.2). The edges, furthermore, serve as a protective surface along the periphery of the tool and add to the stiffening of the laminate.

Operators who mold the parts sometimes create shortcuts to increase daily productivity. This often results in the careless practice of using the mold or tool surface as a table on which to trim part patterns.

To eliminate the scoring or cutting of mold or tool surfaces, "slip sheets" are sometimes used between the fabric/prepreg and tool surface during the finish cutting of the ply edges. Some tool makers have also been known to use fillers

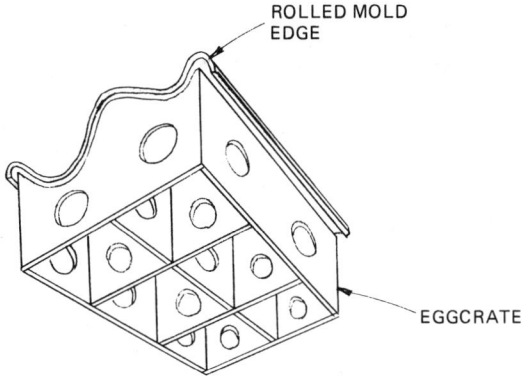

Fig. 3.2. Rolled Mold or Tool Edge.

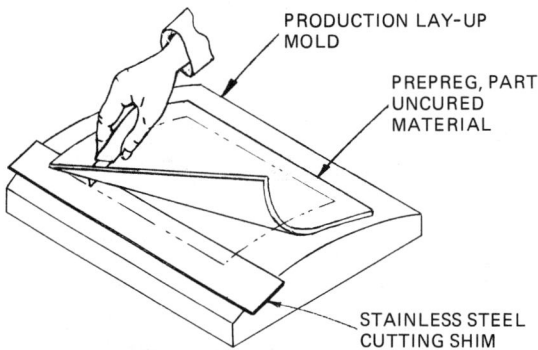

Fig. 3.3. Pattern Cutting Aid to Protect Tool Surface.

in the surface coat along the tool edges which are highly abrasion resistant, in order to reduce scoring of the surface by the cutting knives of operators (Fig. 3.3).

Most mold making materials are not formulated for outdoor use. Therefore, it is recommended that molds and tools not be stored in direct exposure to sunlight and weather. If this is absolutely necessary, tools should be stored with the molding surfaces inverted, upright and away from direct exposure to sunlight. Covers will also help to reduce the damage from exposure to outdoor weather conditions.

Although the effects of aging due to improper storage can effect the surface of a tool, refurbishment and selective replacement of molding surfaces can return the molding surface to like-new condition. Mold and tool refurbishment will be explained further in Section 3.6.

3.3. AUTOMATED DESIGN AND MANUFACTURING

Traditionally, skilled mold and tool designers have been required to perform the daily designs, and craftsmen to manually-execute the production of simple or complex mold and tool cavities and contours. Until recently, however, some demands for precision in both design and production have been impossible to achieve manually. This has sometimes resulted in conceptual shapes rather than exact contours, since only the general positions of part and assembly components could be made. In addition, conventional hand-constructed plaster models and masters, although state-of-the-art, have resulted in problems of dimensional control and accuracy, the imposition of extreme lead and turnaround times, and limitations in the number of design changes.

The introduction of automated design does not erase the basic approaches and systems required to complete a design. In fact, many of these basic approaches, whether done manually or automatically, overlap and must be applied in all cases.

Thus, the following preparatory steps must be taken, whether the designer is proceeding manually or with automated assistance: (1) The mold or tool designer can create and generate the required geometric shapes with extreme accuracy, but the basic parameters must still be reviewed. The governing factors for part designs lead the designer to the ultimate completed mold or tool production materials and processes; and (2) In some instances, the input of and interface with the structural part designer occurs prior to the start of the mold or tool design. In other instances, the mold or tool designer must create around an existing design. Thus, being aware of the part or assembly materials and the form in which they will be supplied becomes of the utmost importance.

Certain questions must be asked by the designer:

- Are the materials metallic, nonmetallic, or combinations of both?
- If they are nonmetallic, are the reinforcements fabric or tape?
- Are the resins thermoset or thermoplastic?
- What is the cure or process temperature?
- What are the sizes, shapes, and complexity of parts and contours?
- What are the tolerances for each?
- If various details exist, how will they be located and to what required degree of accuracy?

Material, labor costs, and the time required to move the tooling through the production cycle also become a part of the designer's responsibility, and are equally important whether the mold or tool design is done manually or automatically. Availability of materials needed to produce the tools, the required labor, the shop workload (shoploading), and facilities needed to transport and store molds and tools are included in this process.

Once these decisions have been reached, the mold or tool materials must be chosen. Unlike the structural designer, the mold and tool designer is faced not only with the considerations of part/assembly and mold/tool design, but the final responsibility of choosing the materials.

In this process, the most significant parameter is the coefficient of thermal expansion (CTE) or coefficient of expansion (CE). The CTE can be referenced when part materials and part tolerances are unique and close. The form of the material used to produce the mold master or production tool must be chosen with careful consideration to the available sizes, temperatures at which it can be used, and the length of its useful life. In some instances, the form of the material must be studied to determine if it will, in fact, fit on the required production equipment in the "as available" size, shape, and mass.

Once all of this preparation has been completed — the part designs reviewed and studied, the available mold/tool materials have been investigated, the production constraints in processing explored — the design can proceed to its automated phase.

42 ADVANCED COMPOSITE MOLD MAKING

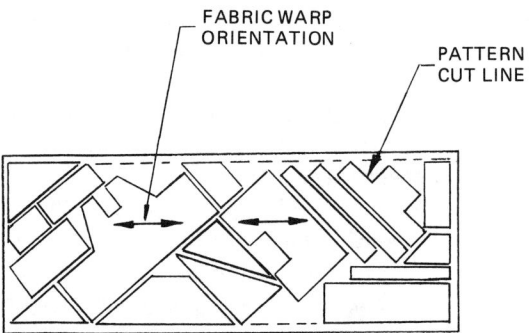

Fig. 3.4. Efficient Layout of Typical Nest Pattern.

3.3.1. Digitizing

Today's fabricator of flat template parts, mold, or tool loftings, and headers can realize a dramatic savings in manufacturing costs through the use of digitizers. The benefits are many, and include increased accuracy, reduced setup times, manual tracing, reduced materials handling, and improved management control.

In general, the most important economic advantage derived from digitizing is the potential savings associated with the preplanning, manual nesting, and recording of existing designs into automated equipment. As a matter of fact, nesting, whether automated or manual (i.e., by automatic plotter or digitizer) may offer the single greatest production potential for automation payback available to the composites manufacturer today (Fig. 3.4).

Digitizing is the semiautomatic drafting and design interface to a number of automated programmable design, drafting, and machine systems. The input to automated systems is digital and the available work areas are comparable to conventional manual drafting stations (42 in. X 60 in.).

Accuracy in the range of ±.005 in. with 0.001 in. resolution (1000 lines/in.), translucent work tables, English or metric outputs and complete computer control of all functions are just a few of the features the mold or tool designer gains as options.

The more accurate digitizers translate point and line locations to a microprocessor by comparing the phases of the signals received by the stylus or cursor from activation of the grid lines on the work table (Fig. 3.5).

In addition, some digitizers can calculate the area of a closed figure, define geometric shapes through the location of two or three points, and locate specially marked points on a work area.

Digitizers, as graphics input tools, are available for cursor control or to complement a computer-aided design (CAD) configuration.

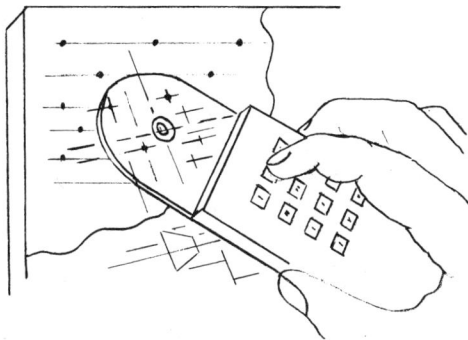

Fig. 3.5. Digitizer Shown Recording Drawing Coordinates.

3.3.2. CADAM* (CAD-CAM) Computer-Aided Design and Manufacturing

Just as advanced composite materials and processes have evolved for mold, tool, and structural part making over the past ten years, so have the materials, methods, and equipment with which the molds, tools, and parts have been engineered and designed. Possibly the most significant advance in the design and manufacturing area has been the development of computer-aided design (CAD) and computer-aided manufacturing (CAM).

Although these systems are only a decade old, they have already proven themselves. In location after location, they are replacing the conventional manual design and manufacturing approach with a series of sophisticated, automated systems.

The future is as equally exciting as the recent past has been: Continued major advances in productivity over the next ten years seem inevitable, as innovative alterations to the basic CAD/CAM systems continue. These productivity improvements, sorely needed by U.S. industry in order for it to keep up with the advanced designs and manufacturing requirements in the various international industrial sectors, simply could not have been accomplished using manual design processes.

Let us examine each of these automated systems individually and in depth, and then proceed to describe their various utilizations in the following subsections.

CAD (Computer-Aided Design). This is a multipurpose, high level function design and drafting package, based upon classical descriptive geometry. The graphics display terminal produces mathematical models in a multiplicity of drawing types by retrieving stored geometry, which is recorded in a modular set

---

*CADAM is a registered trademark of Lockheed California Co.

of interactive computer programs that can operate on a wide variety of available computers.

A CAD work station normally contains a computer and monitor, a keyboard, a hand-operated light pen, display and scanning devices, plotters, microfilm devices, and N/C tapes.

If the system is "turnkey," (i.e., a number of terminals feeding into one central computer) the main processing center will be equipped with a computer and associated system software for drawing and program storage. In this configuration, a series of similar terminals for group design functions is frequently found. This allows for the simultaneous design of all phases of the same product, as well as the creation of a number of different designs of the same product at the same time.

CAM (Computer-Aided Manufacturing). This is an interactive numerical control computer system, designed to direct a machine tool to automatically produce a mold or tool contour swiftly and accurately, with a minimum amount of program errors and inaccuracies. This combination of design engineering and manufacturing allows for a unified, combined process to take place from the actual mold or tool design to the completed machined contour.

CAM interacts and interfaces directly with numerical control (N/C) machining centers that contain such distinct features as the display of cutter paths, styles and types of cutters, speed, feeds, and in some cases, the entire machining program prior to machining itself. All of this allows the draftsman or designer to verify the accuracy of the program and remove any potential errors before they occur.

Recently, in an attempt to reduce material costs of model material that is machined to verify a tape or program, manufacturers have cast, or purchased precast, slabs of machinable polyurethane structural foam and machinable wax as test model materials. Running their N/C milling or turning equipment at full speed (sometimes 600 in./min), machined checking models were produced and then dimensionally measured or used as prototype models to check the engineered program.

CAM systems are capable of handling from two to five axes requirements and can be used for testing, analysis, fabrication, and production of molds and mold masters. This versatility provides for an increased quality of tooling product as well as reduced cost and increased productivity.

There are other advantages to this system as well:

Because the computer performs all of the necessary calculations through prestored data, even an unpracticed operator can complete a complicated mold or tool design. This frees up the expert designer for involvement with more advanced tool design, mold creation, and the supervision of a semiskilled staff.

The combination of skilled operator and automated design opens up hitherto unexplored vistas, and allows for conceptual designs showing the mold or tool in a solid or wire frame shape on the viewing screen. This shape can then be rotated in any direction at any angle for viewing purposes. Two- and three-dimensional colored pictures of the mold form can then be created from points or lines as wires to define the outer surfaces through combinations of cones, spheres, toroids, hexagons, triangles, rectangles, strings, splines, and other geometric forms.

Then, these two- and three-dimensional forms can be reduced, magnified, cut into cross-sections, attached to other shapes that are stored for recall and then translated to a printed copy at any time during the design sequence.

In addition, the mold weight, surface area, volume, and any other geometric function can be traced, maintained, and recorded during all design processes.

Finally, all of the thermal, physical, and chemical standards affecting a functional design can be stored and called in to be matched to a final design at any time.

Numerical Control Machining (N/C). Numerical control machining will be covered in much greater detail in a later section, but must now be referenced as part of the CAM portion of this chapter.

The use of N/C machining evolved in the aerospace and aircraft industry in the 1950s. Although the automated machining processes has involved tens of thousands of machines cutting parts, it was not until recent years that the process was used to produce mold and tool masters, models, and contours for aircraft structures.

Now, N/C milling machines shape the models for thermoset lay-up masters, electroformed nickel plating mandrels, and prototype tools. They are also used to produce prepreg patterns on textile-type cutting machines and to make molds or parts on automated prepreg tape-laying machines.

Combining CAD and CAM into One System. In a combined CAD/CAM system, a central equipment area, consisting of a computer and its stored software, drawings, and other programs forms the central focus point.

The mold designer's work area consists of a terminal, a digitizer (either alone or combined with a monitor), graphic tablets and keyboards for the input of alphanumerics and functions, and a printer to produce various forms of hard copies.

The keyboards are used to input and call back data — such as standard symbols, previous related drawings, and reference calculations — needed to complete a particular mold design.

Once the designer calls back this data, it can then be specified where on a drawing the input symbol should be placed — adding depth if a three-dimensional form is required, showing all views when needed, producing a mirror-image of the form, increasing or reducing the image scale, storing the input data for later reference, etc. The accuracy achieved through the use of these systems, along with the quality of the finished drawing, give the CAD/CAM approach abilities which cannot be matched manually.

Selection of a System. When CAD/CAM systems were first introduced, the few available systems were structured so that a choice of system could be made simply. But these computer networks, with their associated hardware and software, have proliferated dramatically, necessitating more thorough research and matching to the immediate and projected needs of individual purchasers before a choice is made.

Thus, research is the first step in the selection of a CAD/CAM system. Individual mold and tool design requirements, coupled with considerations for savings in time and money will determine the final choice.

Once this is done, individuals from the user organization should visit companies who have purchased systems similar or identical to those contemplated for acquisition by the organization. The opinion of an active user will usually produce information that will add to the chosen system's versatility and expand the knowledge of those responsible for choosing, installing, expanding, and operating and maintaining the system.

These same individuals should then visit the suppliers of the proposed system. Special attention should be paid to the suppliers' background and experience in providing a system that will cover the specific design requirements of the user organization.

Economics is a vital factor, and various alternatives should be thoroughly explored. The value of the system per dollar can be measured by spreading the cost over the predicted lifetime of the equipment. The decision to buy, lease, or rent might be a function of the user's ability to acquire all of the hardware and software that is required to complete the system.

One-time costs such as conversion of manual data to automatic, transferring stored data from another system, training of operator personnel, maintenance of the system, and all other unidentified expenses should be totalled, divided by the life of the equipment, and added to the monthly cost of the lease, rental, or purchase agreement.

Before the final choice of a system is made, the user should require a series of successful on-line tests and demonstrations by the supplying company's personnel.

Two Dramatic Developments. Already evolved from the original forms of CAD/CAM systems are the CADAM and CATIA systems. These are the creations of two major aircraft companies.

The CATIA system, because of its specialized characteristics, will be covered in detail in Section 3.3.

The CADAM system, developed as an advanced tool system by Lockheed California Co., is specifically designed for the aerospace and aircraft sector of design and manufacturing industries. However, the other sectors (such as automotive, marine, etc.) will without a doubt be added to the list of industries able to be serviced by this system, as their use of new tool innovations increases.

CADAM begins with package segments expected and necessary in most packaged systems. It functions on the IBM system 360 and 370 CPUs, and contains display terminals producing graphics that are based upon classic descriptive geometry.

Its basic series of programs generate the database necessary for numerical controlled (N/C) tapes, input directly to N/C files, generate compound and complex curved surfaces, allow for various geometric drawing functions, the production of designs, and the supplying of many areas of database management.

As with conventional CAD-CAM systems, the CADAM system is supported by input equipment which includes the refresh-type display and scanning devices, digitizers, etc. On the output side, the system produces hard copy derived from plotters, various types of printers, N/C tapes, and computer-type microfilm devices.

CADAM systems also employ conventional functional equipment options through combinations of various attachments such as alphanumeric and function keyboards, digitizer inputs, cursors operated on cathode ray tubes, light pens, etc. Input data is retrievable, as with most CAD-CAM systems, and is stored in the central computer as mathematical models.

From this base, the system adds features such as common database and data management systems for all types of user disciplines. Going beyond the alphanumeric database and file management systems, a geometric data system using both two-dimensions and two-and-one-half-dimensions is provided, which can protect existing files and aid in the release and dissemination of authorized data, while at the same time preventing accidental deletion of drawings and the entering of unauthorized changes.

CADAM output models are automatically assembled into drawing frames for microfilm filing, or into sheets for flat-bed plotting. Multiple models can be assembled which result in a single machine operation when N/C machine programming is required. The designed models can be stored, after immediate retrieval, on the drawing perimeter. The model views can be separated, stacked, grouped, juxtaposed, etc. by interfacing them with other drawing details, components, assemblies, or subassemblies.

The error detecting logic of the CADAM system can eliminate errors caused by various configurations of human design, allowing at least six-digit precision during a rapid changing of dimensions and model sizes. Exploded views produced from separate details are also possible.

Another major advantage of the CADAM system is the ability to interface with databases and programs worldwide. A faster CAD-CAM response than with basic systems is a key factor in the CADAM system design, thus increasing programming efficiency, system functionality, and output accuracy.

Future Progress. As more adroit CAD/CAM systems are developed, and the need for manufacturers and design companies utilizing these systems increases, the demand will be filled by lower cost, faster response equipment with increasingly sophisticated software graphics. As time goes on, end-users will be able to store more data, produce more complicated and resourceful programs, perform more nongraphic functions, process more data, and perform more complicated design functions more quickly at a reduced cost.

As an area distinctively devoted to software emerges, the design of molds and tools will be complemented by specific designs in software packages. These packages will be custom-designed for use in specific areas of mold and tool design and provided by either the equipment or software manufacturer.

As the technology continues to develop, semiskilled operators will be able to call up stored data or produce a segment of a mold or tool design by voice command or using a plug-in cartridge or modular preprogrammed input unit.

Utilizing Numerical Control Machining (N/C) in CADAM. Computer assistance in the programming of the complex contours of recent aerospace and aircraft structures and substructures is now a necessity. The automatically programmed tool (APT) programming language, developed at the Massachusetts Institute of Technology, allows mold and tool designers to program five axis contour milling or mold surfaces in any configuration, from simple to complex. The language, a conventional, easily learned, remembered, and written method of creating custom programs, provides a well-defined, universal interface between the operator and the machine.

Variations of the APT language, such as interactive graphics, have evolved as the need for more complex designs separated the aerospace and aircraft industries from the simpler needs of other manufacturing industry sectors. In the aerospace industry particularly, the programmer, even though dealing with a simple mold or tool design, had to possess the ability to see geometric forms in the mathematical equations they developed and also see the form of the mold or tool rotate in space. Due to this, thinking in three dimensions has come to be considered natural.

During early programming advances at a number of aerospace and aircraft companies, interactive graphics were developed. Using this programming, the mold or tool contour programmer used a light pen or cursor to drive images on a CRT in order to debug programs or edit cutter locations and paths prior to machining. This departure from the manual engineering approach brought the computer-assisted design directly to the level of a mathematical model which became the link between CAD and CAM.

A design engineer using CADAM could now create cutter paths, cutter locations, see the cutter tool relative to the mold or tool, and set and adjust machine/cutter spindle speeds and feed rates.

The design engineer could now prepare the CAM phase of programming in its initial segments, preparing rough profiles and shapes to finished profiles and contours by calling up on-screen the associated listings of drawings stored in the computer-memory. The drawings could then be modified, and represented for the manufacture of the associated molds or tools, leaving the basic geometric forms to be altered to prepare for introduction of cutter paths and locations. Bond lines, fixtures for positioning molds, inserts, and clamps could be added to the geometric representation to reveal machining interference. The recording of these representations could then become a manufacturing or tool design file entry separate and apart from the engineering file data.

Now, ready for the N/C phase of the mold or tool design CAM function, the design engineer could use CADAM technology to produce cutter paths, after selecting the associated subroutine. The CADAM programs would automatically feed information and programming data to the design engineer instructing the user regarding machining start points, cutter radius and diameter, RPM, and all other necessary data. This information could be expressed in any international measuring language system.

The CADAM system is highly flexible, and thus, the design engineer can produce contours, cutter paths, and program the manufacturing equipment, using various logical and reasonable approaches. Therefore, this program can be defined as a general framework, with the designer actually able to steer the cutter form in any direction necessary to avoid obstacles, such as mold or tool accessories.

In addition, N/C punch tapes are generated after the design engineer verifies the machining program. Most programming errors are avoided, and those that might occur are graphically seen in the replay.

N/C programs available for digitally driven machines might include prepreg fabric cutters, robotic trimming equipment, plotting, welding, flame cutting, PC boardcutting and drilling, wire wrap connecting, and three- and five-axis machines and lathes. Unlike conventional N/C mold and tool programming, which must be verified by the use of actual machined models or a printed hard copy, engineering changes, additions, deletions, and substitutions can be verified immediately.

As mentioned earlier, the design engineer need not learn most of the involved formats required in conventional programs and can produce, with less training, more tool and mold design functions accurately.

It is obvious, therefore, that the interactive graphic technique of N/C programming will eventually replace conventional methods of programming once the various industrial sectors realize that this method leads to better quality CAM programs at a lower cost.

The choice of the CADAM approach to mold and tool design will also broaden the sector in which automated design and manufacturing can be used, since previously chosen conventional methods of programming have proven to be more costly. CADAM, an example of the variations and extensions of the CAD-CAM process, is one of the evolving, available program systems of choice today.

3.3.3. CATIA (Computer Graphics Aided Three-Dimensional Interactive Application)

CATIA*, an innovation in the CAD-CAM system, has been developed by Dassault Systems, Paris, France. This system is a highly interactive, high function three-dimensional geometry system for computer-aided design and manufacturing. The result of development in the areas of complex shape definition and N/C machining, this automated method of mold and tool design uses the IBM 3250 Graphics Display as its high function interactive work station that employs available superior human factors as standard equipment.

This interactive applications system is used for direct construction of three-dimensional objects, with provisions for viewing, analyzing, manipulating, and modifying these entities. The system automatically produces the necessary machining data and instructions to drive N/C equipment of various kinds.

The resulting data can then be fed through special interfaces to the CADAM** system described in Section 3.3.2. Through this interface integration the entire mold and tool design process, from initial concept to final drawings, is facilitated, allowing the mold or tool designer to use various design disciplines at every phase of contour and shape development.

In addition, the CATIA system will accept input geometry from other systems or programs prepared by the user.

A unique quality of the system is its ability to analyze and define mechanisms with one degree of freedom, through its kinematics provision. The formed mechanism must be planer (two-dimensional kinematics), taking into account the capability of defining a set of solids linked together. The participating solids may have been created by any of the following other applicable methodologies: wire frame geometry, polyhedral solid geometry, or surface geometry analysis and N/C surface machining.

These objects may be three-dimensional and presented as solids, surfaces, or wire-frame configurations. Using English or metric dimensions, the analysis function computes volumes, lengths, and areas and inputs changes to any mold

---

*CATIA is a Reg. TM of Dassault Systems, Paris, France.
**CADAM is a Reg. TM of Lockheed California Co.

MOLD ENGINEERING 51

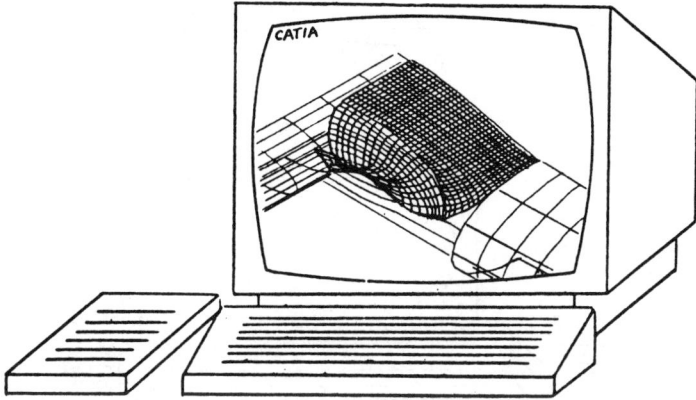

Fig. 3.6. Kinematics.

or tool view, while simultaneously and automatically altering all the views in the three-dimensional object. The resulting image can then be automatically viewed from any angle. Precise geometric definition, accompanied by the corresponding graphic representation, then becomes a part of the database.

The transfer of data into the CADAM system is performed easily. This data and any other design geometry information contained in the database allows for N/C programming of three-axis and multiaxis equipment, and the display of cutter paths through graphically displayed 3D animation.

This and other output data of the CATIA system is an APT source part program with hard copy drawings produced by pen or electrostatic plotters.

Kinematics. When the kinematics function is employed, motion studies and interference checking can be performed, along with dynamic removal of hidden lines, which helps the designer to conceptualize the graphic presentation of a solid object. Input data from other systems and/or analytical programs can be input into CATIA, and this data along with other stored information can then be extracted through user-written programs (Fig. 3.6).

Wire Frame Geometry. Wire frame geometry is reported to be the basic CATIA product, and as such, acts as the prerequisite for the kinematics, polyhedral solid geometry, and surface geometry products. While providing manipulating and analyzing capability, this function also provides various general system management capabilities also. Complex curves are defined, using Bezier polynomials, whose degree capability varies between 1 and 15. Other entities covered include planes, text strings, lines, points, and simple curves. This is illustrated in Figs. 3.7, 3.8, and 3.9.

52 ADVANCED COMPOSITE MOLD MAKING

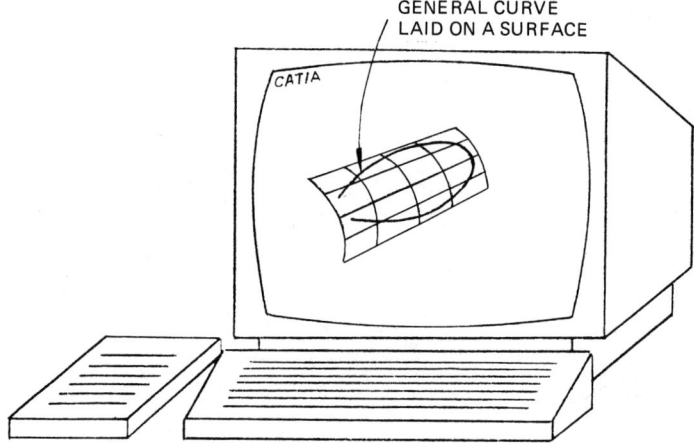

Fig. 3.7. Two-Dimensional Wireframe Geometry.

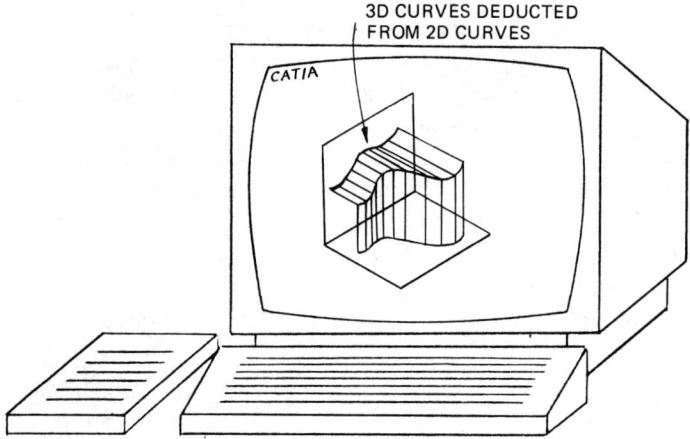

Fig. 3.8. Wireframe Geometry, with Three-Dimensionality Deducted from Two-Dimensional Curves.

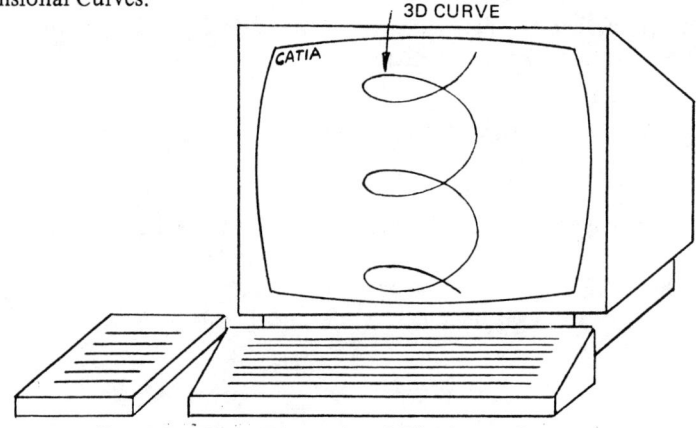

Fig. 3.9. Three-Dimensional Wireframe Geometry.

MOLD ENGINEERING 53

Polyhedral Solid Geometry. Polyhedral solid geometry allows for the manipulation and analysis of solid figures, which are represented by polyhedral approximations. This definition of simple solids or volumes, defined by planer or pseudoplaner facets, covers such geometric shapes as toruses, cylinders, cones, spheres, prisms, pyramids, cubes, and parallelepipeds (Figs. 3.10 and 3.11).

Surface Geometry Analysis and N/C Surface Machining. Surface geometry analysis and N/C surface machining products adds to the wire frame geometry base data the capability of defining complex three-dimensional surfaces and volumes.

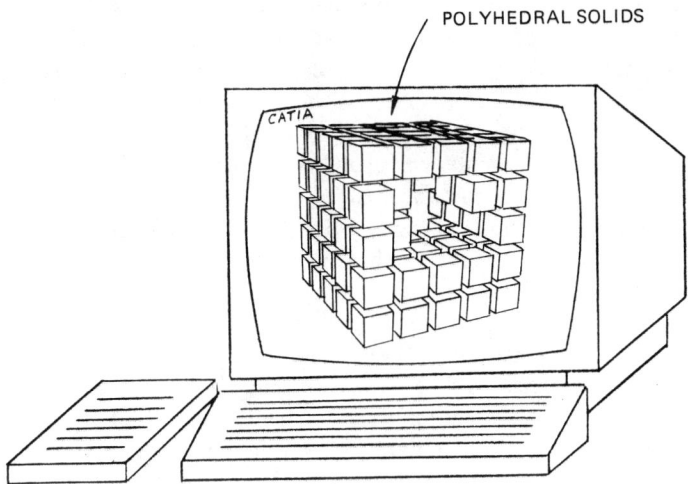

Fig. 3.10. Polyhedral Hidden Lines.

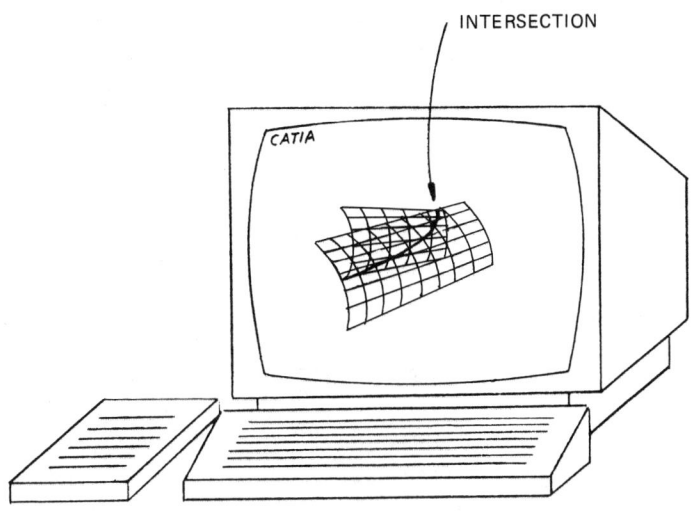

Fig. 3.11. Polyhedral Elements.

Analysis and generation of N/C machining programs are included in the function, and by viewing the defined program movements, proof of program accuracy can be determined without actually machining material to model contours.

The surfaces are presented in the form of patches (Bezier-type) which can pass through or near arcs or points. Deformation of the surface patch around the patch boundaries and center is also possible.

The rectangular array of patches is then formed into simple surfaces, ruled surfaces, ducts, pipes, and surfaces of revolution, complete with fillets between surfaces. Planer faces can be defined with limitations of surfaces by curves into one or more closed domains.

A final function of this mode is the analysis of surface curvatures (geometric) as well as boundaries of a face and the base surface (logical analysis). The closed regions of space, limited by a set of faces (volume) can also be presented (Fig. 3.12).

N/C surface machining capabilities of design fall into three main interactive phases: pass creation, replay, and editing or modification.

Within this, there is a design capability and assistance in defining various ways of driving a cutter, such as showing its tip motion or center motion (if different from the tip motion), side motion, motion along lines or curves, point to point movement, movement within a region with zig-zag or one-way cutter tracking, and three- or multiaxis machining movement.

A definition of filleted end or ball-end cutter shapes is also available with multiple-named N/C passes allowed on a single model.

Finally, the regions of surface areas to be cut can be checked and limited by a feature that allows a sphere, rolling along the intersection of any two surfaces, to stop at a programmed point. This mode is complemented by the replay and editing feature, which can be used to program the real cutter locations according to the tolerances established.

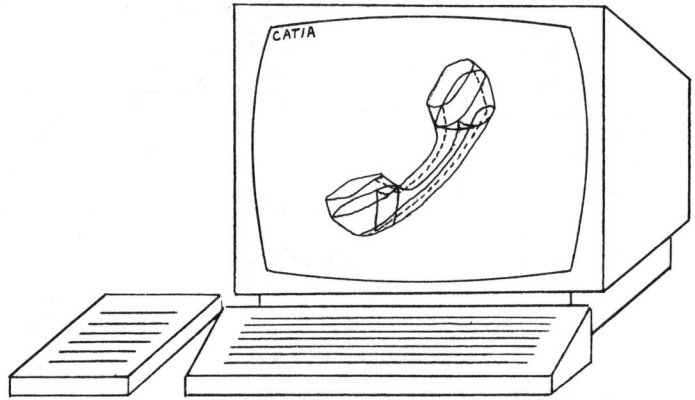

Fig. 3.12. Surface and N/C Example of Three-Dimensional Contour.

MOLD ENGINEERING 55

Section Summary. The preceding computer-aided design and manufacturing systems (CADAM and CATIA) were chosen to demonstrate the flexibility and complexity of design and manufacturing possible through automation. Because of the rapid advancement of technology, it would be impossible to mention all of the available advanced programs that have been developed or will be developed in the near future.

When the moment comes for the utilization of one, some, many, or all of these systems, it would be well to remember that careful consideration and review should be made of as many programs as possible before mixing them with proposed or existing in-house equipment.

3.3.4 Numerical Control (N/C)

Numerical control is a system that automatically operates equipment and machines by introducing discrete numerical information that has been stored as input data on various forms of tape. Its effectiveness depends upon a dimensionally sound database, since it cannot effectively operate with partial or incorrect information regarding shape, contour, and dimensions.

N/C, which originated in the late 1940s, has classically been thought of and referred to as a production tool. Actually, it is a way of directing equipment or a machine tool process through precoded instruction sequences.

Besides the automated manufacturing of molds and tools, N/C is used in electronic placement, insertion and wiring of components, cutting, welding and attaching part and subassemblies, finishing of various surfaces, filament winding, and the milling, turning, boring, drilling, and trimming of all manufactured parts and products.

N/C systems are usually more accurate than the equipment they control. Conventional equipment must rely on the tracing of a cam or a model, or depend upon the dimensions derived from mechanical stops. With the evolution of CAD/CAM computer graphics and the reduced complexity of programming, the accurate manufacturing of molds, tools, and one-of-a-kind prototype parts is now possible. N/C, because of its quick and simple setup time, is suited for small production lots. Thus, it makes sense to utilize it in producing a prototype part, mold, or tool (Fig. 3.13).

The major parts of an N/C setup are few and simple to describe:

Numerical Control System. This is the mechanism that automatically reads, analyzes, and feeds the coded information to the equipment or machine. It is the major component in the N/C system, and the most expensive, but its untiring ability to operate one or more machines, 24 hours a day if need be, quickly justifies its cost, and its accuracy far outstrips that derived from mechanical systems.

56 ADVANCED COMPOSITE MOLD MAKING

Fig. 3.13. N/C Machined Plastic Master Model. *(Courtesy of Grumman Aircraft Systems)*

Console. The console used by the operator to set up work pieces and manually drive the equipment is usually located near the N/C system. In some machines, such as a drill press, it may be part of the equipment head.

Cabinet for Magnetics. This controls the hydraulic pumps or electric motors that operate the cutting heads, spindles, and tables.

Devices to Monitor Movement. This equipment sends signals back to the N/C system, which in turn analyzes the return signal, compares it to the sending signal, and adjusts the messages until the final dimensional message is balanced.

There are two major forms of N/C: point to point numerical control, and continuous path N/C machining.

Point to Point Numerical Control. Point to point numerical control is very much like the movement of a simple drill press, whose head moves from point to point (Fig. 3.14). It is used to contour cut headers or loftings of a mold, tool, or fixture (these are the near-shape plates or boards that are attached to and support the rear of a laminate mold or tool).

MOLD ENGINEERING 57

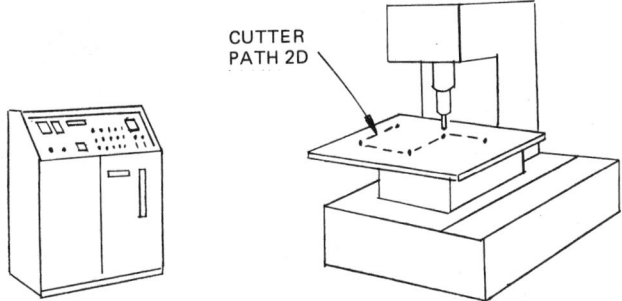

Fig. 3.14. N/C Point-to-Point Machining Movement.

Although point to point numerical control can be utilized for the drilling or cutting of flat panels, its basic control movement information is in no way limited to these simple uses. On the contrary, it can be sent to all types of equipment to draft, draw, cut prepreg patterns, sew and stitch, lay prepreg tape, and even control a punch press, lathe, or tube bender.

At present, the control of milling machines is the most widespread use of N/C. Other popular uses are: the machining of mass-cast models, precast model stock, laminated model stock, honeycomb stock, and various other aircraft, marine, automotive, and aerospace mold and part components (Fig. 3.15).

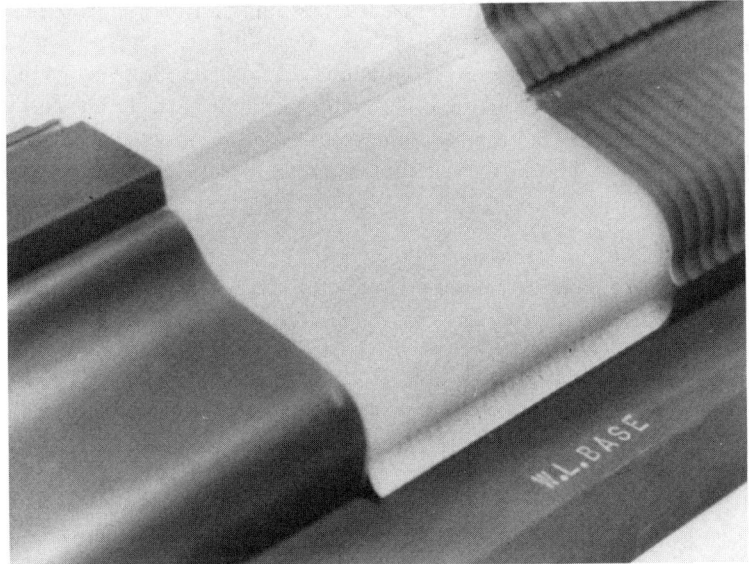

Fig. 3.15. Rough and Fine N/C Machine Cuts Into Precast Model Stock. *(Courtesy of Ren Plastics Div. of Ciba Geigy Corp.)*

58 ADVANCED COMPOSITE MOLD MAKING

As referenced earlier in this chapter, the drafting machines of CAD systems are perfect examples of the numerical control system output. N/C data (on a point to point basis) fed from the same system that controls cutting equipment and machines, can also control the drafting machine, and, if desired, print out the cutter path and produce a drawing.

Continuous Path N/C Machining. Continuous path machining, unlike point to point machining, can operate to produce shapes in any plane. As can be seen in Figs. 3.13 and 3.15, continuous contour machining is the method used to produce the molds, tools, and fixtures of advanced composite parts. This technology can also be utilized for N/C machining of the contours of thermoform molds made from the mass-cast or precast plastic materials presently available to industry. These thermoform molds can be used to form conventional thermoplastic materials or can be used to produce contours and shaped parts from advanced composite thermoplastic materials and advanced self-reinforcing polymers (Fig. 3.16).

Since continuous path machining can produce shapes in any plane, multiaxis continuous path contouring can produce complex shapes with a high degree of accuracy.

The material used determines the velocity at which a machine master or part is machined. The mold master materials available today are formulated to be cut to shape and contour at high speeds (RPM and path movement of cutters), sometimes up to 200 in./min (Fig. 3.17).

Because of all of these factors, continuous path or contouring equipment controls are more expensive than point to point machining. Again, the cost should not be the controlling factor in equipment choice. Instead, quality and accuracy of the finished product should rule the decision.

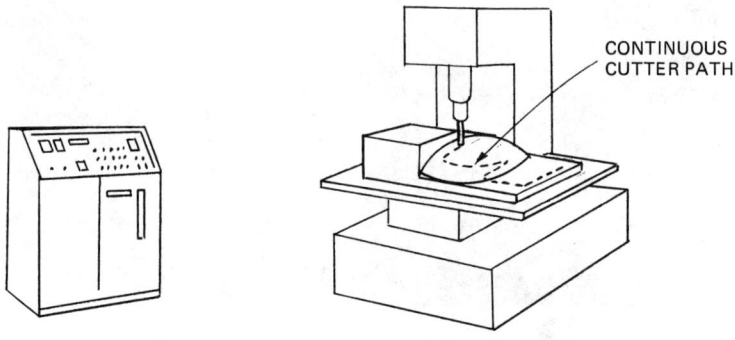

Fig. 3.16. N/C Continuous Path Machining Contour.

Fig. 3.17. Intricate Master Model N/C Machined from Precast Plastic Model Stock. *(Courtesy of Ren Plastics Div. of Ciba Geigy Corp.)*

New Developments. The numerical control processor is an application program that describes the N/C operations to be performed by analyzing the input of user-oriented language statements. Then, the cutter location data, usually of a language common to other types of automated equipment is fed into the post-processor, which is another computer applications program. The post-processor produces the code for the particular N/C equipment or machine to be controlled on punched tape. The easily stored punched tape is then called up for use and read by the N/C system of the equipment or machine.

In order to keep abreast of the technology of advanced mold, tool, and fixture contouring, IBM and other companies have developed equipment to analyze, convert, store, and feed information to machines in the form of a punched tape in the language required by these machines.

This equipment, in the form of N/C processors (specifically referring to the IBM System 360 and 370, Automatically Programmed Tool (APT)), is a member of the family of N/C processors. Advanced contouring (APT-AC) allows full three dimensional machining operations to be performed. The continuous path or contouring machining uses two or more axes simultaneously (multiaxis) to produce the finished contour or shape.

Other IBM systems not only provide output from the APT, but also cover the spectrum by controlling operations from the autospot and AD-APT processors. These processors only allow operations and contouring in two dimensions.

N/C users worldwide are supporting the committees of ANSI and ISO in their efforts to create a common language standardization.

Significant changes that affect the N/C environment have occurred and will continue to occur. As these changes multiply, a reassessment of N/C processors should follow.

Implications. N/C machined models, although an innovation over the years, will not totally supplant the use of splined plaster models. N/C will obviously only work when the master's dimensional data is available. The cost effectiveness and quality of the finished item must be taken into account before choosing the manufacturing method.

In addition, N/C machined models have distinct advantages and disadvantages when compared to splined plaster models, and both should be considered in choosing a model-making method. Some of the advantages of N/C machined models are:

- Lower material and labor cost
- Accuracy of finished model
- Less lead time necessary for fabrication
- Reusable contoured or shaped surfaces vs. one-time expendable plaster materials
- Better vacuum integrity of surfaces when fabricating or molding interim and high-temperature molds
- Lighter weight than plaster models
- Stronger and more durable than plaster models
- Easily transportable
- More desirable and controllable surface finishes obtainable
- Availability of the process to N/C inspect the finished contour and shape. This can be performed using an inspection probe. It is recommended that measurements be taken at the lowest point on the cusp, scallop, or ridge of the machined contour.

Some of the disadvantages of N/C machined models are:

- Varying thermal coefficients of model materials
- Thermal and physical limitations of materials

- Expense of N/C machining facilities
- Higher material costs of model stock
- Slightly longer fabrication flow through the shop than for plaster models
- N/C equipment machine loading and schedule limitations
- Model size limitations because of limitations in the size of the N/C machine envelope (This can be overcome through the use of outside mold and tool fabricators).

There are also other considerations to be made before deciding to install and/or use N/C machining.

Speed of Machining. Although speeds up to 200 in./min have been successfully achieved, 60 in./min is usual. For low tolerance requirements, a mass-cast, precast, or laminated model stock material could be machined in one operation to the finished contour. Better quality and accuracy is obtained through two or more rough cut and finished cut operations. The rough cut or mass-cast near-shape should be around .050 to .080 in. oversize. The cutting patterns or pass width (resulting cusp size) need not be the same as that used for the finished cut.

A basic mold engineering consideration is the filling of voids discovered during the N/C machining phase, so that hand blending will not be necessary after the model has been completed. Model material suppliers can recommend the necessary quick-curing materials of choice, which can be applied to areas not being machined, thereby allowing the patched area to cure as machining is taking place.

Cutters. The cutters used are usually carbide ball cutters, approximately 9 in. in diameter, with two, four, or eight flutes. Model material suppliers should be consulted to recommend the proper cutter and machining speeds when processing their particular type of material.

Choice of Model Materials. Care should be taken when choosing N/C mass-cast, precast, or prelaminated materials. The recommended machining method of cutting without a lubricant might result in the generation of dust during machining operations. Thus, it is best to choose an N/C model material that has been engineered to produce machined curls, instead of dust, as waste (Fig. 3.18).

If this is not possible, a strong vacuum, with the intake located next to the cutter, is recommended.

It is also advisable to protect N/C equipment against the abrasive nature of some model materials. Some of the fillers used in model materials, such as glass, ceramic, and phenolic microballoons or spheres, have a tendency to abrade machine tables after repetitive machining operations. The application of removable plastic coatings or films to machine tables allows for the collection and

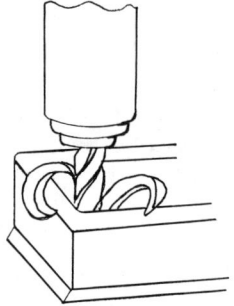

Fig. 3.18. Pattern Stock Machining Waste Should Form Curls, Not Dust.

subsequent disposal of machining chips and dust not collected by the vacuum intake.

Handling of N/C Machined Models. Past experience has proven that, when handling N/C mass-cast, precast or laminated models, transportation by air-ride trailers is a must.

A rigidized frame over which high-density elastomeric pads have been placed is necessary for mounting during transportation. Snug but not tight fasteners are suggested to secure the model during movement from one plant to another. A covering of some sort over the entire model will protect the machined surface.

To eliminate the possibility of environmental shock, the mold master should be sent directly to the processing point. The chance of damage occurring because of layovers in the shipping terminal, etc. is thereby minimized. If this is not possible, and if the model size allows, a shipping container should be built around the model, or the model transported in a covered and — if possible — insulated truck.

The model master's base should, of course, be totally supported during transport.

3.4. PROTOTYPE AND LOW-PRODUCTION MOLDS

Although time- and money-consuming, the making of prototype and low-production molds are required by some companies. In general, this process is justified in the following circumstances:

- When it is necessary to check different plastic materials for the mold
- When it is necessary to choose between metallic or nonmetallic mold materials
- When it is necessary to test the physical or technical feasibility of a design

- When new mold-making concepts and/or untested approaches to tool design are being tried
- When there are market research and cost justification aspects to be taken into account
- When it is necessary to determine the moldability of a particular part design in a specific material
- When the development phase of a product cannot wait 20–30 weeks for production samples
- When quantities do not justify production tooling costs
- When existing production capacity is fully occupied
- When production tooling cannot produce acceptable parts
- When early entry into a new market becomes important.

It should also be noted, in support of prototype mold fabrication, that mold variables, shrinkage, and dimensional stability cannot always be determined during the final mold design stage. Prototype fabrication can provide all the variables.

Also, the prototype mold approach is almost always used to fabricate short runs of parts (one to ten). In some cases, these parts are used for structural tests, environmental evaluations, and for demonstrating or marketing a new product concept and appearance.

In these instances, the mold or tool produces only the outside physical appearance of the part or assembly and may lack internal structural members. The part might be fabricated from a low-temperature thermosetting material either cast or laminated on the surface contour of the prototype master.

Prototype and low-production molds and tools should *not* be made:

- When a part is very simple
- When similar parts have been made
- When moldability is not in question
- When the cost of a very large part is a consideration, and fabrication of an inexpensive low-production mold will aid in that consideration (*Note:* The cost of the low-production mold should be approximately 1/3 to 1/2 of the cost of the production tool. Estimation should be precise in this respect: low-production molds have often been known to cost the same as, or more than, the production mold or tool.)

Certain questions must also be answered before proceeding to the making of prototype or low-production molds:

1. Is the article a skin or bonded assembly?
2. Are there severe draft angles, undercuts, imbedded hardware, or other features that would require a unique mold design or multiple mold segments?

3. What types of material must be used for the part in order for the mold and/or tool to withstand cure temperatures and pressures?
4. What are inspection and quality requirements?
5. Must the surface be smooth to a paintable degree?
6. Can the dimensions in some areas deviate from the basic design?
7. How many parts will the mold or tool produce?

3.4.1. Producing Prototype or Low Production Molds

In some instances, a model or master — whether N/C machined or handmade — can produce an acceptable prototype part or assembly.

In others, the part must be generated by constructing a temporary substructure consisting of headers or contour boards covered with wire screen or mesh. In this case, a room-temperature/high-temperature (partially cures at room temperature, fully cures after post-cure) laminating resin system is used to fabricate the prototype lay-up over a release film or high temperature sheet wax that has been applied to the screen overlay on top of the headers or contour boards. This release film might be paper, Tedlar*, or even peel ply material.

The film releases with the part and the screen and film are removed from the part at that time (Fig. 3.19).

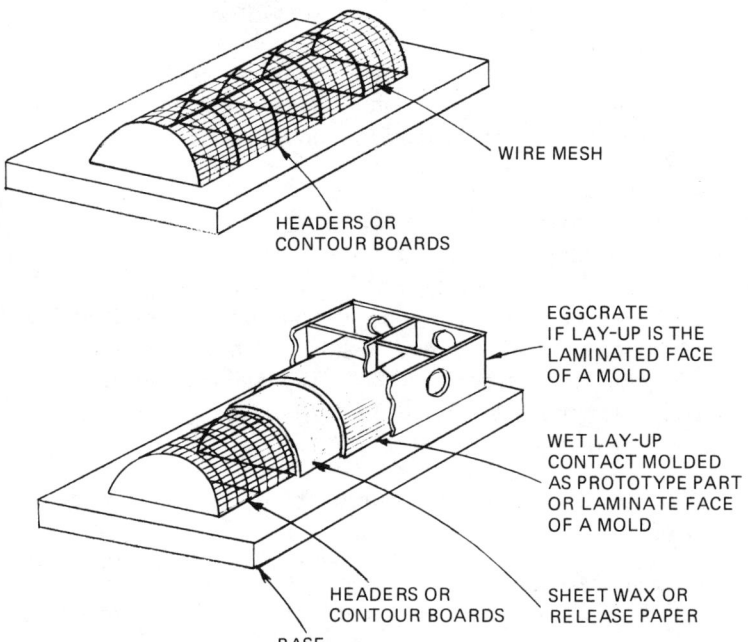

Fig. 3.19. A Part or Mold Fabricated from an Inexpensive Master.

---

*Tedlar is a Reg. TM of Dupont Co.

3.4.2. Further Considerations

Sometimes, questions arise in the making of prototype or low-production molds that are not easily answered.

One of these questions deals with engineering changes, a highly unpredictable situation. We would be able to make correct material choices for prototype tools if we knew the answer to that question. Some prototype molds are easier to change than others. If, for instance, the "mass-casting" technique was chosen to fabricate a prototype or low-production master, one could simply recast the new contour on top of the existing mold surface. Cosmetically, a metal prototype mold would show more indications of alteration than a nonmetallic one.

Another point that is difficult to answer deals with the question of multiple use: Must the mold produce preproduction parts, or just the prototype? An example of this would be the case of a "mass-cast" mold being fabricated to make hand lay-up parts until the design is finalized, and then the contour is used to fabricate an electroformed nickel tool or autoclave mold (Fig. 3.20).

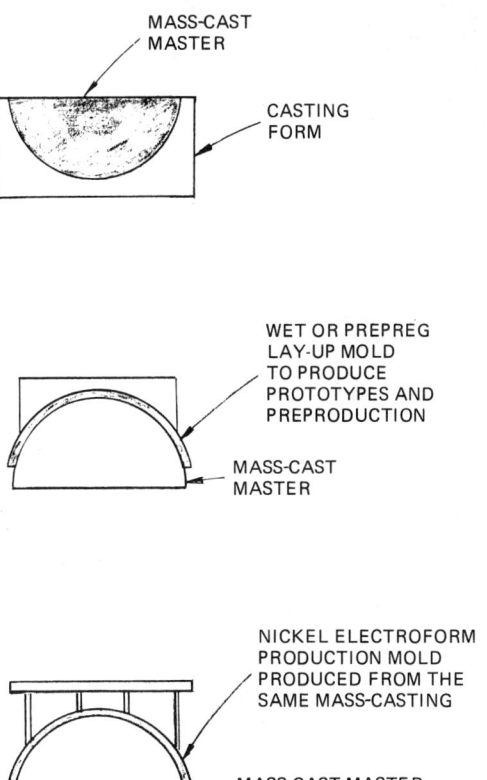

Fig. 3.20. Preproduction and Production Molds from the Same Mass-Cast Master.

66 ADVANCED COMPOSITE MOLD MAKING

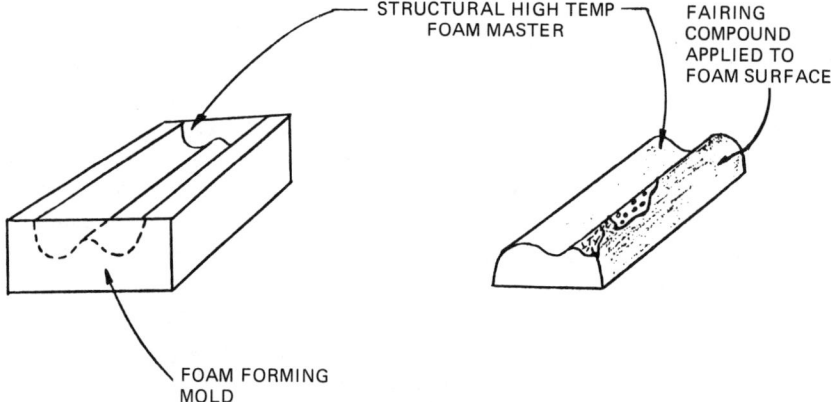

Fig. 3.21. Inexpensive Mold or Part Forming Master.

As mentioned earlier, prototype or low-production molds, because they are usually easier and faster to make, should be less expensive to fabricate than production molds. Sometimes, however, they are not.

It is justifiable to disregard the cost factor if an engineering design is being proven and the design cannot be proven without the fabrication of the tooling.

Sometimes, the cost of producing a three-dimensional pattern or master cannot be considered as a part of the prototype tool cost, since the master might be used in forming the production tooling as well.

Another cost-saving method is to form an inexpensive mold or prototype part from an inexpensive prototype master by making the master from cast or machined high-temperature structural foam (Fig. 3.21).

The foam can be randomly reinforced to reduce expansion during mold or part fabrication and can be coated with an epoxy or polyester fairing compound to provide a smooth finish. Parting films or releases can be applied to the master's surface for easy removal of the cured mold or part. Should the foam master adhere in certain areas, the foam can easily be removed by grit blasting with plastic or glass shot. This blasting media will not damage the internal surfaces of the part since only low-pressure air is required to propel the low-abrasive media.

3.4.3. Mold Making Materials Suitable for Prototype and Low-Production Molds

In general, the available prototype and low-production mold-making materials are (with the exception of N/C machining) all shaped to contour using a master or pattern, which, depending upon the tool material, may be one of the following materials:

- Ceramic
- Plain or impregnated wood
- Machined "mass-cast" or precast slab stock (syntactic foam)
- Wax
- Spray metals
- Other forms of machined metal
- Thermoset laminating and prepreg (200°F precure) materials, both graphite and fiberglass reinforced
- Plaster
- Various sheet metal alloys tack-welded to headers or contour boards to produce a part surface contour
- Paper
- High temperature moldable and machinable foams (chemical foams)
- Flexible or elastomeric nonmetallics
- Reinforced or nonreinforced room temperature vulcanizing (RTV) rubber
- Polysulfide casting material

In summary, the prototype mold or tool must be judged to be worthwhile on its own merits. If the prototype tool approach produces successful, acceptable quality parts in a short period of time, relatively inexpensively, then this approach is cost-efficient.

3.5. HIGH-PRODUCTION MOLDS (ELECTROPLATED NICKEL, STEEL, AND ALUMINUM)

Mold and tool designers have, in many instances, resorted to designing molds made from various metals, due to the deterioration of poor-quality materials that were available to form nonmetallic molds and tools.

In terms of life expectancy, metal molds and tools far surpass nonmetallic molds, in all cases.

The trade-off is the expense in operating and maintaining the tools, because of extensive heat-up rates in ovens and autoclaves, expansion due to greater thermal differences than in nonmetallic molds, the added cost in handling and storage, long fabrication lead times, and higher costs to produce the molds — which, in some instances, cannot be made in-house and can thus only be secured from outside suppliers.

Therefore, it is important, when engineering high-production metal molds and tools, to always consider reducing the metal mass and weight by designing headers and supports with sufficient lightening holes and the selective placement of supports.

The use of high-production mold materials provides a mold life expectancy far beyond that of the nonmetallic mold materials referred to in Section 3.2. In

Table 3.4. Thermal Expansion

MOLD MATERIAL	CTE $\times 10^{-6}$ (in./in./$°$F)
Ceramic Casting	.4
Graphite Epoxy Prepreg	3–4
Graphite Epoxy, Wet Lay-up	5–6
Electroformed Nickel	7.4
Aluminum	12.5
Fiberglass Prepreg	11
Fiberglass Epoxy, Wet Lay-up	12–14
Steel	6.7

some instances, such as steel tooling, the predicted life is infinite, with unlimited cycles as the expected use frequency factor.

Generally speaking, electroformed nickel, machined and sheet metal steel, and aluminum are considered high-production materials. Nickel metal molds and tools are preferred for highly contoured parts and assemblies.

Aluminum, because of its low cost and lower density than either nickel or steel, is the preferred engineering material. However, care should be taken when choosing aluminum as a mold or tool material, since its thermal coefficient of expansion (CTE) is high and this is sometimes an undesirable trait when forming advanced composite parts (see Table 3.4). Although the CTE of aluminum is twice that of steel, the factors of its lower density, ease of machining and welding, and excellent heat-up rate make it a desirable mold material, even though extensive dimensional correction factors might be necessary.

It is important to note that, unlike nonmetallic molds, electroformed nickel and metal molds must be engineered to more exacting specifications, due to distortions which may develop from the plating process, the welding of the substructure, N/C machining of the required contours, and accelerated heating in ovens and autoclaves.

Vast differences can occur because of varying mold thicknesses in metallic-type molds and thus the heat-up rates must be altered.

Ovens and autoclaves heat by convection to mold surfaces. Therefore, heat passage through the mold is critical and the path of circulation to the surface molding the contour or bonding the structure should be well planned. As a continuous face on the front of the mold is important, it is necessary in all cases to avoid having any hole or hardware which penetrates all of the way through the mold or tool face. Vacuum integrity is of prime concern, and this will almost assure a vacuum-tight mold design.

Also of importance is the consistency of design from one metal production mold to the next. Because molds and tools are grouped, stacked, and processed together in ovens and autoclaves, it is important not to design or engineer molds that have uncommon heat rates, since this will increase the cure times of parts which rely on the recorded temperature of a lagging thermocouple attached to a mass of a mold and not the temperature in the cavity of an oven or autoclave.

3.5.1. Comparison of Metallic Mold Materials

Aluminum. The CTE of aluminum makes the material desirable for forming fiberglass-reinforced parts, but undesirable for graphite, where the CTE is less than one-third that of aluminum. Therefore, only small, simple-shaped graphite-reinforced epoxy parts should be considered for fabrication on aluminum surfaces, since improperly shaped angles, poor inside and outside radii, and undesirable part growth can occur.

In addition, cast aluminum has proven to cause poor vacuum integrity, because of the porosity present in various areas of a casting.

Steel. Steel has one of the lowest CTE of all metallic production tool materials, and is therefore used for large, complex shapes, panels, and spars, assuming that the mold mass, weight, and heat-up rate is calculated to be acceptable (Fig. 3.22).

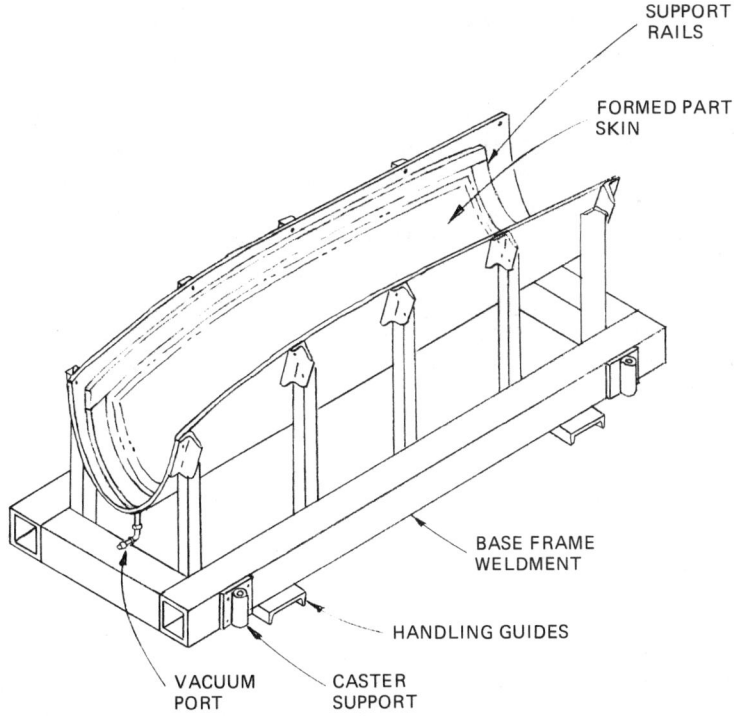

Fig. 3.22. Typical Sheet Metal Aluminum or Steel Mold or Tool.

Electroplated or Electroformed Nickel. Electroplated or electroformed nickel has a CTE close to that of steel. It also has a long series of distinct advantages over the machined, formed, or cast metals.

For instance, the porosity, as compared to aluminum in a casting, is nil. The cost per square foot is less than steel, and the detail of the finished production mold is so superior to other metal materials and processes that it becomes obvious that there is no other method to be used when forming certain metallic contoured molds.

Mandrels. In order to produce an electroplated or electroformed nickel mold, a mandrel must be made.

A mandrel is the formed or machined master contour upon which the plating is deposited to form the molding surface of the production mold or tool. Mandrels are sensitized to accept deposition from the plating tank and a complex undercut can be produced, if desired, by selectively bonding a "breakaway" onto the mandrel surface. These "breakaways" will release when the plated mold is complete and released from the plating mandrel. They can then be easily removed from the undercut cavity.

Disposable mandrels can also be used for forming complex contoured molds. Holes of various sizes, including threads, can be plated at selected locations along the surface of a mold form.

Mandrels or master patterns can be conventionally or automatically (N/C) machined. Most standard pattern materials can be used, as long as they can withstand the environment of the plating tank, which lasts about 10–20 days (.125 to .350 in. thick shell) at 100° to 115°F, and not absorb moisture.

Electroformed molds can be subsequently machined, if the nickel electroplate is backed with copper. This method is not highly recommended for production molds or tools, but is acceptable when machining after electroforming is a must.

With new and advanced mandrel-making methods of computer-assisted design and manufacturing (CAD/CAM), it is proven that, when practical, the electroformed nickel mold or tool is the best all-around choice as a high-production material and process.

3.6. MOLD AND TOOL REFURBISHMENT AND REPLACEMENT

Although the subject of refurbishment, maintenance, and repair is covered in greater detail later in this book, it is well to address the subject as an engineering topic now.

Mold complexity and production quantities and rates determine the frequency of refurbishment and determine the scheduling of mold or tool replacement.

One of the advantages in fabricating production molds and tools of plastic materials is that they can be easily modified, maintained, and repaired. Therefore, the proper selection of tooling materials is vital to the life expectancy, maintenance, and repair of the tool. Structural supports, changes in contour, and major modifications, if properly performed, will not distort the mold contour and will be strain free. Because metal mechanical fasteners, heat from welding, and internal metal stress is not present (as in metallic tools), contour and support frame distortion is eliminated.

Plastic frames, besides being easily modified, are light in weight (approximately half the weight of steel). Thus, they are easy to handle. No special hoists or machinery-moving equipment is necessary to transport the mold or tool. The plastic frame, besides providing an equalized CTE between the facing and the reinforcement substructure, also eliminates labor costs normally found in postprocessing welded steel frames.

If a mold or tool fabricated from nonmetallic materials is made in a controlled environment, using controlled materials, the resulting tool will last longer. Moisture, particularly, trapped in a mold or tool laminate will cause expansion and surface problems during tool use, and delamination of the tool surface.

The choice of hydrophobic hardened materials (which cure in the presence of moisture) will eliminate this problem somewhat. But ultimately, eliminating the introduction of moisture during the building and refurbishment of a mold will produce the best results.

In most cases, it is cheaper to modify a plastic mold or tool than to replace it. Modifying or changing the contour is sometimes simple. If a contour is to be changed, the supporting frame or substructure members must be installed before cutting away or removing the existing contour. Placing peripheral bagging (a strip bag around the mold edge) to hold the mold or tool down against the mold master prior to the modification of the structure will assure that the contour surface is in the proper position during frame modification.

If any additions are to be made to the mold or tool surface, these should be made only after the surface has been roughened, abraded and cleaned for bonding.

When repairing mold surfaces and substructures, the use of the same materials or formulated equivalents is highly recommended to avoid variations due to differences in each material's physical properties.

If the thermal coefficients of the materials are matched and the repair or modification material is chosen to cure at the use temperature (high-temperature curing materials), in most cases the mold will not distort and the bond of the modified area or repair will not fail. If a local modification or repair is necessary, and the entire mold or tool cannot be heated to cure the patch or addition, the localized application of low heat (up to 150°F) to the area of cure should be used. This area should not be greater than 6 in. X 6 in. and the local heat source — a lamp or blower — can be 12–14 in. from the surface to be heated.

Catalyzed and frozen ("B"-staged) materials, such as epoxy surface coats, and the prepreg tooling materials should be considered when possible, as these ma-

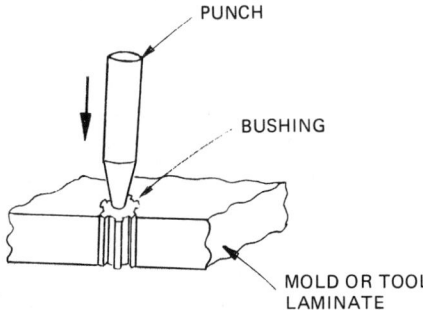

Fig. 3.23. Serrated Bushing for Easy Removal.

terials can be post-cured following the low-temperature gel or set of the material. Quick heating will cause shrinkage, which will result in warpage of the existing contour.

When replacing bushings or fittings, press fitting the hardware should never be done. Instead, these bushings or fittings must be put or encapsulated into place. If bushings are to be replaced frequently, the design of the original or replacement bushing should be considered. Imbedment bushings may be serrated along the length of the bushing if they are not to be replaced or removed. A removable bushing with serrations can be physically driven out of the laminate. If the bushing has serrations in the surface, it will assure that no twisting or spinning will occur during use (Fig. 3.23).

Adding metal braces to a mold or tool — a process that should be used only when absolutely necessary — should only be done when the metal is normalized before attachment. In addition, the nonmetallic material which contacts the brace should have the same, or very near the same, thermal coefficient as the metal. When long lengths of metal are used, a flexible attachment material is suggested and the cure temperature of the modification materials should be near to the mold's temperature of use.

Ultimately, only quality materials should be used when refurbishing or repairing molds, since the labor costs needed to effect the repair will far surpass the cost of materials if inferior compounds are chosen.

In all cases, it is suggested that the materials to be used for refurbishment, replacement, or repair be tested and proven adequate for work with the materials used to fabricate the original mold. It is thus highly recommended that test samples be made from these materials, when the time and materials are available.

4
Mold Materials

With the advent of plastic mold materials, certain casting and reinforcement processes were also developed. This chapter will deal with the materials used to cast, laminate, and otherwise reinforce plastic molds.

It may be well at this point to review the lamination and reinforcement process before proceeding to the materials themselves.

Laminates resemble a multilayered sandwich, with each of the various fillings bonded together by adhesive resins. These laminates, in addition may additionally be reinforced by the introduction of fibers, mixed into the resin. The fibers may be of various lengths, from long intermeshed to short chopped fibers, distributed at random throughout the entire plastic mass.

Although thermosetting resins are the predominate ones used as bonding agents, thermoplastic resins (PEEK–polyether ether ketone) may also be employed, particularly for structures that demand high flexibility. There is an advantage in this case of allowing objects to be formed from the laminate sheets by heating, pressing into shape, and cooling to set. However, in some cases, they are subject to heat deformation under stress at temperatures above 250°F.

The principal thermosetting resins used in laminated products are the following: phenolics, ureas, alkyds, melamines, polyesters, silicone, and epoxies. Sometimes these are used exclusively; at other times, combinations of resins plus reinforcing materials are used to construct the cast or laminate structure.

The first step in the manufacture of laminates consists of dissolving the resins in a suitable solvent to form the laminating solutions. The principal solvents used are alcohols, hydrocarbons, and ketone esters.

Next, the laminating stock is prepared in suitable widths, usually in rolls, and predried before it is impregnated with the resin solution.

At this point, the resin solution is placed in a V-trough and the stock is passed over guide rolls, then through the solution, then through a pair of squeeze rolls or doctor blades, and finally through a drying tower or tunnel where forced hot inert gases evaporate the solvent and leave the dried resin well-impregnated in the laminating stock. The amount and degree of resin impregnation will depend upon such factors as the viscosity of the resin solution, absorptivity of the laminating stock, temperature, and length of time of contact. Careful and correct control of these factors produces the correct amount of resin pickup, so that after drying, the resin content of the "prepreg" (preimpregnated stock) is within the range of usually 30 to 40% by weight.

Once the stock has been prepared, it is cut into the desired sizes and shapes by rotary cutters, guillotine knives, N/C controlled cutting devices, or cutting dies. Then, the cut sheet stock is sorted and assembled into stacks, one layer on top of another. The stacks of impregnated stock could be held between polished metal plates in a heated press or applied to the surface of a mold and compressed together under high pressures and temperatures, thus melting the resin and causing the sheets to form into a compact mass.

Finally, after curing (in the case of the thermosetting resins) or cooling (in the case of the thermoplastic resins), the mass sets or solidifies into a rigid laminated board, sheet, or form.

There are positive reasons for reinforcing plastics. The outstanding qualities that have made reinforced plastics valuable are: high strengths, resiliency with impact resistance, low thermal conductivity, a reduced coefficient of thermal expansion, excellent electrical resistance, good chemical and solvent resistance, and an ease of moldability. Contrarily, there are limiting qualities, which are: strength losses at about 350°F with medium heats of distortion, poor wear and abrasion resistance, creep under loads, especially at higher temperatures, and a low modulus of elasticity in bending and poor fatigue under repeated stress. Technological advances in recent years have, however, helped to overcome some of these disadvantages, and have allowed a resurgence of interest in and increased use of reinforced plastics.

Now that the process has been explained, and the inherent strengths and weaknesses chronicled, let us proceed to the mold materials themselves.

4.1. PLASTER AND GYPSUM COMPOUNDS

Plaster of paris has been used for pattern, model, and mold making for thousands of years, and until a few years ago was thought of as the only usable material.

Today, through research and consequent development, there exist a wide variety of specialized, scientifically formulated gypsum cements. The properties of these cements are designed to meet the specific patternmaking needs of the aerospace, automotive, foundry, and plastics industries, among others.

Patternmaking with gypsum cements can be classified into three basic methods of use: pouring, screeding, and splash casting. Combined, these methods offer the advantages of accuracy, dimensional stability, economy, and adaptability.

The basis of any gypsum cement is white or gray gypsum, a natural mineral. The gypsum is first finely ground and then calcined to form a uniform product.

Controlling this calcination, as well as further processing, makes it possible to obtain super-strength plasters which are termed "hard plasters" or "alpha gypsum cements."

Ordinary, or soft, plaster is formed in a fairly primitive, elementary way: Quarried gypsum rock is crushed, screened, and pulverized. Then, the flourlike gypsum is heated in kettles to remove three-fourths of its chemically combined water, dumped into pits, and cooled. Formulation follows. Before this sort of plaster, usually utilized in molding, dental work, or casting, can be poured, at least 65 pounds of water per 100 pounds of plaster must be mixed in.

Super-strength (or hard) plaster, which is four to five times as strong as soft plaster when it is formulated, is made from carefully selected gypsum blocks ranging from one-half to one and one-half inches in size. First calcinated under pressure in pressure vessels, the gypsum cement has the proper quantity of water removed, then is itself taken from the pressure vessels and cooled, pulverized, screened or air separated, and finally formulated as desired.

Gypsum made in this manner requires 30 to 40 pounds of water per 100 pounds of plaster to obtain a mix that can be poured, and it will produce casts that are not only four to five times as strong as those made with ordinary plaster, but that have a low coefficient of expansion and a wide range of densities.

The importance of recommended plaster to water ratios cannot be overemphasized. Termed "normal consistency," this amount of water required to be mixed, 100 parts of plaster by weight to a standard fluidity, has the power to change the compressive strength from 1,000 to 12,000 psi. The compressive strength of set gypsum casts is of maximum importance, because it correlates closely with resistance to breakage and the consequent length of useful life.

Obviously, a higher ratio of water to mix produces set plaster of a lesser density, hardness, strength, and durability. This occurs because the strength of gypsum is the result of the development, during setting, of numerous needlelike crystals that are tightly interlaced among each other. As more water is added to the original plaster, the final gypsum crystals are pushed farther and farther apart. Thus, the importance of the careful monitoring of water to gypsum ratios.

The future does not look cheerful for plaster masters and molds. Indications are that state-of-the-art and advanced composite mold making will cause a move away from these plaster masters and molds, and toward N/C machined models made from lightweight, thermally stable, low-cost materials. The labor-intensive plaster masters and molds of the past will be phased out to be replaced by accurate contours produced by automated equipment.

4.1.1. Plaster and Gypsum

Shop Equipment. Proper equipment is necessary to make gypsum cement patterns and models, and this equipment is especially easy to make, and essential for completing a good job.

Here is a basic list of tools and equipment necessary for making gypsum cement patterns and models:

Flat scrapers — One edge is smooth, and one edge is sawtoothed.
Kidney scrapers — Both smooth-edged and sawtoothed.
Spatulas.
Hand saws — Large-toothed and wide-set to enable the saw to clear itself.
Block planes.
Regular carving tools.
Cutting and filing equipment — For metal templates.
Angle plates — Plus all the other usual devices for patternmaking.
Water — Hot and cold water to provide any desired temperature, nearby for frequent washing of hands and mixing uses.
Work benches — The main body of the work should be in as convenient a height and position as possible. The entire surface should be flat and can be either marble, slate, polished plate glass, or treated hardboard. Iron benches can be used, but the rust resulting from moisture should be removed constantly. Granite surface plates are highly recommended as table tops. These plates are available in many sizes and provide an extremely accurate working surface, which can usually be accurate within 0.0001 in. All benches must have straight edges.

Storage of Bagged Gypsum Cement. Gypsum cements and plaster should always be stored in a warm dry room or building. Although it is packed in multi-walled bags with vapor-proof liners, the gypsum cement possesses a capacity to absorb moisture slowly out of the atmosphere. Thus, storage should be away from damp floors, heavy currents of air or steam, and should not be located where dripping or condensation from overhead piping can fall on it.

Rotation should be practiced. Under ordinary storage conditions, gypsum should not be kept longer than 90 days to insure normal working characteristics.

The Mixing Process. The best mix of gypsum cements comes about when the plaster and water are weighed. Volumetric measurements should never be used, since they cannot produce uniform results.

Once both components are weighed, the plaster *must* be added to the water (*never* the other way around). Strew the plaster into the water, and allow it to sit undisturbed for two to three minutes. Then, mix the plaster and water thoroughly until the slurry has creamed or thickened slightly (usually for from two to five minutes, depending upon the quantities being mixed).

As the size of the batches increases, more complex mixing methods must be employed: For batches of 5 to 25 pounds, hand mixing with the aid of a spatula is sufficient. For batch sizes from 25 to 50 pounds, a quarter-inch drill motor outfitted with a rubber disc is advised.

Larger batches require a direct drive 1760 rpm propeller mixer. Up to 50 lbs of slurry requires a 1/3 hp motor outfitted with a 3 in., three-blade 25° pitch propeller; from 50 to 100 pound batches require a 1/2 hp motor; batches over 100 pounds require a 3/4 hp motor. Each of these two larger mixers should be outfitted with a propeller measuring 4 to 5 in. in diameter. The rotation of the mixer should be such that the propeller forces the mix downward and the propeller clears the bottom of the bucket by 1 to 2 in. Best results are obtained when the propeller shaft is set at an angle of 15° with the vertical and is acting a little off the center of the bucket.

The setting time for cast plaster is affected by the plaster to water ratio. Normally, the cast gypsum plaster mold can be removed from the model within 35 to 60 minutes after pouring. The setting time can be accelerated by the use of warm water (130°-140°F) or by longer mixing. Cold water (50°-60°F) can be used to extend the fluidity or set of the gypsum cements without affecting the properties of the finished cast.

In contrast to other materials, super-strength gypsum cements do not shrink upon setting. On the contrary, they have a positive expansion. As the cast plaster sets, the mass begins to heat and expand. Through changes in the manufacturing process, the expansion can be controlled to insure consistent performance.

Pouring Gypsum Cements. In order to reproduce a solid pattern or mold, gypsum cements must be poured. A wood frame or flask is constructed around the properly sealed and lubricated model, and the gypsum cement is then poured immediately after mixing, while the slurry is in the fluid, or free-flowing, state. The fluid mix is usually poured into one of the corners so that it will flow across the model. Never pour directly onto the model, as the turbulence produced is apt to create streaks, flow marks, or (more seriously) numerous air voids on the working surface of the pattern or mold. Vibrating or joggling the work bench while pouring is recommended to insure reproduction of fine detail.

The pouring method is also used when making oversize patterns with medium high and high expansion gypsum cements. Unlike other gypsum cements, these materials require a lower consistency. Thus, they are mixed to a heavy, thick texture. They should be poured immediately after mixing and vibrated or tamped after the flask or mold is filled, to insure the reproduction of fine details.

Parting Compounds. Sometimes referred to as separators or releases, parting compounds are necessary to prevent adhesion of the set-up gypsum cement to a pattern or model, and in order to permit removal of the gypsum cement from the surface upon which it is set up.

A variety of methods, listed below, are used to apply these compounds.

Sealing with lacquer. When porous models, patterns, or molds are used, it is necessary to seal their surfaces before the separator is applied. Sealing is especially essentially for wood patterns in order to prevent dampness from causing the grain to swell and rise, with consequent release difficulty.

Applying a quick-drying lacquer by spray gun or brush provides an excellent seal for wood, plaster, and gypsum cement patterns and master models. Two coats of automotive light gray primer sealer lacquer or the equivalent has been found satisfactory in most cases. However, a clear lacquer is recommended where a transparent sealer is desired.

The lacquer can be applied immediately after the plaster or gypsum cement has set. It should be cut back enough with lacquer thinner so that the model or pattern readily absorbs the first coat. The second is applied after the first coat has thoroughly dried.

Other uses of separators. Separators must, in addition to preventing adhesion, provide a continuous, smooth, non-water soluble film upon the faces of the model, pattern or mold, possessing little or no frictional resistance to movement across its surface. A properly selected sealer-separator combination prevents penetration of moisture from the fluid gypsum cement slurry into pores of the pattern, model, or mold. It also allows free movement of the set gypsum mass as it expands.

In general, the requirements of a satisfactory separation medium are that:

- It spreads easily and uniformly in a thin, continuous, insoluble film.
- It will not react destructively with the plaster or gypsum cement surface, or with the pattern.
- It protects and lubricates the surface of the pattern.
- It prevents adhesion of the cast.

The use of powdered mica (one-half cup of powdered mica added to one quart of any plaster release lubricant) is highly recommended as an additive for all parting and separating compounds. Tests conducted with this powdered material showed improved parting qualities and a better mold surface.

Some common and widely used separators are listed below.

1. Petroleum jelly (vaseline): Cut back with approximately two parts of kerosene to one of jelly, and blended by careful heating and thorough stirring, then apply as a thin coat, well brushed out (Caution: fire hazard during blending).
2. Stearic acid and kerosene, or stearine: (Note: This separator is one of the most widely used of all parting compounds, and is particularly recommended for use with glue molds.) Formula: 1/4 lb stearic acid shaved to flakes: 1 pint kerosene. Melt the stearic acid by warming, remove

from the heat source, then add the kerosene with constant stirring until a uniform mixture is obtained. Apply to the pattern with a soft brush, preferably camel's hair or sable hair. If brush marks show, the mixture should be thinned with more kerosene or warmed slightly.
3. Light mineral oil: This is particularly effective as a fine fog spray on metal patterns.
4. Spirits of camphor: Used for extremely fine detail.
5. Olive oil or sperm oil: May be used if applied carefully while warm.
6. Hardened vegetable oils used as frying fats: Particularly recommended for use on lacquered wood.
7. Light lubrication oil: Effective if used very sparingly.

A final word of caution: If a plaster or gypsum cement mold is to be used for the casting of plastics, the manufacturer of the plastic should be consulted as to the proper sealer and separator, and the manufacturer's recommendations followed.

Utilizing the Period of Elasticity. The "creaming" or thickening stage is known as the period of elasticity. In order to be successful in making original models with plaster, the patternmaker should familiarize himself with all phases of this stage, since it is during this plastic period that the cement should be used to build up the models and to screed them to shape with templates. This technique eliminates the need for molds or mixing-in, which would be necessary if using it in its free-flowing state.

A little concentration and experience quickly shows how the progressive plasticity of the plaster can be utilized. Different areas of the pattern require different stages of plasticity, because as this plastic period progresses, the cement gains strength or body and can be built up to the contours required.

Although the usual period of plasticity is around 25 or 30 minutes, the length and character of this period will vary with different types of gypsum cement or plaster. Thus, the material should be selected to meet the requirements of the job.

Here is one example of the utility of the period of plasticity. Instead of making an entire pattern or model with one mix, successive mixes can easily be bonded together to build up larger shapes. This practice enables the craftsman to use each one at the proper stage.

Removal of the Set Gypsum Pattern or Model. When the cast heats, gypsum cement has set, and is now ready for removal. One or more of the following methods can be used:

1. Gently blowing with compressed air.
2. Suspension of the piece so that the weight of the mold aids the separation.

3. Wedging evenly around the surfaces with wedges or a sharp tool.
4. In extreme cases, soaking the cast and its pattern in warm water for a short time, followed by application of one or more of the previous techniques.

When molds or patterns of flexible modeling compounds are used to form the gypsum cement, no separator is necessary, and removal of the set object is simplified.

If, in moving or transporting the patterns from one spot to another, a corner or sharp edge is chipped or broken, these can be repaired by cementing with household cement, glue, or almost any of the vinyl-based wood adhesives.

Adding Reinforcements. When added strength is needed, reinforcements are added to the gypsum cement. The most universally used and versatile reinforcement is uncarded, long-fiber hemp, which may be obtained in any desired quantities from wholesale distributors, or pattern and plaster supply establishments. The hemp is matted in baling, so it must be picked apart to loosen the individual strands. A handful is then made into a flat bat and dipped into the gypsum cement mix to thoroughly saturate it. This bat can then be added to the underlayer of plaster, thus becoming an integral part of the pattern. The number of bats needed depends upon the size of the job.

In another variation, handfuls of the hemp can be dipped into the gypsum cement mix and formed into a rope to tie metal or wood reinforcements in place.

Other reinforcements are wire mesh, expanded metal, and metal rods. Wire can also be used as a reinforcement, but it must be totally encapsulated by the plaster, or must be wrapped with dipped hemp to furnish a mechanical bond between the wire and the gypsum cement.

Metal reinforcing rods, pipes, or tubing may be bolted or welded together to form a structural support. Steel pipes and tubing that have been bent and welded should be normalized to relieve strains or stress that may occur during fabrication. Metal reinforcements are tied to the pattern with fiber hemp bats impregnated with gypsum cement or plaster.

Wood bar reinforcements are used as supports for the finished pattern. The supports should always be tied onto the work with hemp fiber after the setting expansion has taken place to avoid warpage. Kiln dried wood is recommended, and should never be imbedded in the cement because the moisture in the gypsum cement will swell the wood and distort the pattern. All wooden reinforcements should be given at least three coats of lacquer before being tied to the pattern.

Storage of Patterns. Since gypsum cement patterns are dimensionally stable under high or low humidity conditions, any sheltered space in which the temperature does not exceed 125°F or dips below freezing, and which protects the pattern from weather, is satisfactory for permanent storage, provided the patterns are placed so that they cannot directly absorb water.

Since patterns which are not properly supported in storage may become distorted, flat slabs or patterns should be stored by standing them on end. If patterns must be stored lying flat, the surface upon which they rest must be flat. If this is not possible, wedges or blocks must be used at intervals to provide sufficient support. The shape and weight of the pattern will determine the type of support necessary to prevent distortion.

4.2. CEMENTITIOUS AND CERAMIC COMPOUNDS

To offset the expense and impractical nature of creating certain composite molds and tools for aircraft, a cementitious composite material is being developed.

This cementitious composite material will permit room temperature casting of very high strength, thin-walled, complex-shaped tools for composite parts for aircraft.

Besides containing an awesome compressive strength factor of 43,000 lbs/sq in., this material does not act as a thermal barrier, thus making it practical for use in an autoclave. Containing a thermal diffusity comparable to that of nickel, at a fraction of nickel's cost, this material has the further virtue of having vacuum integrity (see Fig. 4.1).

In contrast to cementitious materials, castable ceramic tooling compounds have a low tensile strength and low heat conductivity. However, they are relatively inexpensive, are easily repaired, and can, through various methods, be made to last the lifetime of a program — be it prototype (5 units or less) or large volume (over 100 units) production.

Furthermore, ceramic tooling compounds have the least detrimental effect on composite stress-strain properties because of their low CTE.

When selecting ceramic tooling compounds, the following properties should be considered.

Method of cure. Ceramic tooling compounds are cured by a chemical mechanism called "hydraulic" setting, which introduces water as an integral part of the ceramic structure. At room temperature, a saliceous and/or aluminous cement gel is formed, which coats and binds the ceramic filler. Then, temperatures between 200°–500°F are introduced to evaporate excess water and convert crystal/amorphous bodies to a physically stronger phase.

Thermal properties. Ceramic compounds are noted for their low heat conductivity, low CTE and large heat capacity.

Tensile strength. Tensile strength of ceramic tooling compounds is generally low — about equal to their compressive values.

Finally, it is important to acknowledge certain advantages, disadvantages, and limitations when designing specific ceramic tools. Although the geometry and

82 ADVANCED COMPOSITE MOLD MAKING

Fig. 4.1. Cementitious (Stainless Steel Filled) Composite Tool. *(Courtesy of Cemcom Research Associates)*

size of the composite part will be the primary consideration in the tooling design, the following must also be taken into account:

1. Low maintenance and repair costs.
2. A long, useful life.
3. Ease of management in the shop.
4. Sealed surfaces for vacuum maintenance and release application.
5. Hard, damage-resistant surfaces that still permit template scoring and layout.
6. Directional reinforcement if tensile shear loads are expected to exceed 500 psi.
7. Integral heating to improve heat transfer, which in turn will match the composite cure profile.

All of the above considerations can be met by ceramic tooling compounds.

4.3. ELASTOMERS

Elastomeric materials, as sheets and castable compounds, have been adapted for use in the formation of composites by taking advantage of their CTE, flexibility, and — in some instances — built-in release qualities.

Recent discoveries have even adapted low-temperature elastomers for use as molding intensifiers, flexible caul plates, and pressure pads. Such compounds as latex, neoprene, ethylene propylene, polyacrylic, and fluoroelastomers have been adapted, modified, and reinforced for mold making and for tooling use.

A major advantage of their existence lies in the use of the newer silicones as sheeting for permanent reusable vacuum bags. These materials can be laid up and stacked to form conformal bags, thereby reducing and eliminating the bridging found in conventional film bagging (Figs. 4.2 and 4.3).

Lower-temperature materials such as the EPR, butyrate, acrylate and, fluoroelastomers have increased molding improvements by their ability to be formed

Fig. 4.2. The Processing of Reusable Vacuum Bag Materials. *(Courtesy of the Keene Laminates Div. of Keene Corp.)*

84 ADVANCED COMPOSITE MOLD MAKING

Fig. 4.3. The Compounding and Manufacture of Reversion-Resistant Silicone Rubber. *(Courtesy of the Keene Laminates Div. of Keene Corp.)*

into corners and points beneath the forming bag. Their ability to withstand temperatures up to 350°F and exposure to hydrocarbon oils, solvents, and amine curing agents make them a preferable material for use as intensifiers.

4.3.1. Silicones

The silicones that form elastomers have quickly gained widespread use in the mold-making sector of composites.

Both one-part and two-part systems are being used as tooling materials. The one-part systems cure with a reaction to atmospheric moisture, and the two-part systems cure in any thickness at heats from room temperature to the temperatures found in conventional press molding or injection molding machines.

The linear thermal coefficient of most silicone rubbers that have been measured fall into the range of 1 to 2.1×10^{-5}. This range is consistent over a 75 to 480°F temperature range. The rubbers are said to have a linear expansion of approximately 17 times that of carbon steel, which is why they are used to mold composites by thermal expansion molding techniques.

4.3.2. Polysulfides

Soft and rubbery, polysulfides have the single disadvantage of exhibiting shrinkage when exposed to prolonged temperatures over 180°F.

On the other hand, they cure quickly at room temperature with no shrinkage, give an excellent reproducibility of detail, and have a short-time resistance to very high temperatures — for example, they can withstand a temperature of 5000°F for up to one minute.

Polysulfides have the added advantage of being cheaper than either silicones or polyurethanes.

4.3.3. Flexible Epoxies

The flexibility of epoxy resin systems is varied in two ways: 1) by adding a flexibilizer to a resin-hardener mix, or 2) by the simple use of a flexible hardener.

When epoxies are at the upper hardness range, they are tough and strong, but highly flexible epoxies tend to cold flow. In a general sense, these epoxies are relatively slow curing, and are prone to crack or shatter under sharp impact.

4.4. WOOD

High quality pattern lumber, though sometimes difficult to obtain and expensive to purchase, is nevertheless an important material to the patternmaker. As expensive as high quality lumber is, wood is an ideal material, because it carves readily and requires little labor to eliminate the defects which could adversely affect a pattern.

On the other hand, judicious buying must be practiced, since inferior, unworkable lumber is easily obtainable — and in this case, money saved in purchasing will result in higher expenses later in the pattern-making process.

Recent developments in stabilizing wood through autoclave impregnation of the cells using thermoset materials such as polyurethanes have improved the physical characteristics of wood, but synthetic materials are still proving to be better than wood for patterns.

Wood is composed of approximately 60% cellulose, 28% lignin, plus an assortment of minor quantities of other materials. The cells of wood are similar to straw in minute form, and cellulose forms the framework of the cell wall. Lignin has two functions: it is the cementing material that binds the cellulose together and it is mixed with cellulose in the cell wall proper. Color, odor, and natural resistance to decay are credited to approximately 2% of the components other than cellulose or lignin.

Both soft and hardwoods are utilized by patternmakers. Softwoods are primarily used in master patterns and for short-run production patterns. Hardwood patterns are used for long-run production patterns in which wear is an important consideration.

4.4.1. Softwoods

Ranking from most used to least, softwoods employed by patternmakers are: 1) sugar pine; 2) northern white pine; 3) Idaho pine; and 4) Spanish cedar.

Following are characteristics, recommendations for usage and a survey of the grades available:

Sugar Pine. The most commonly used lumber in the pattern industry today, sugar pine is distinguished by its conspicuous resin ducts and softness. A light wood which weighs approximately 2.3 lbs per board ft, it maintains its shape well when properly seasoned. A type of wood whose supply is unfortunately diminishing, it can be found in its ideal form in the high altitudes of California.

Select grades. Select grades are the top quality pieces from the log. They are either clear or characterized by minor blemishes such as small pitch streaks or pockets, pink knots, minor season checks or other minor blemishes too insignificant to prevent utilization of the stock as finish material.

Selects are regularly available in 4/4 (1 in.), 5/4 (1¼ in.), 6/4 (1½ in.), 8/4 (2 in.), 10/4 (2½ in.), 12/4 (3 in.), and 16/4 (4 in.) thicknesses. All thicknesses are usually shipped in random widths and length with a relatively small proportion of short lengths and narrow widths.

1 and 2 clear sugar pine (B and better select). This is the top grade sugar pine, the finest product of the tree. Most of it is entirely clear and thus can be used with almost no waste. At the lowline segment of the grade, and occasional piece may be found with some pink knots, a small smooth knot, a small dry pitch pocket, light, localized pitch, a small area of light stain, or a fine season check. Both fronts and backs of this grade are of extremely high quality.

Because of its characteristics and quality, this grade is suited for the most accurate patterns of practically every type and description.

C select sugar pine. Resembling grade 1 and 2 clear in appearance, this grade contains slightly more numerous occurrences of the above-listed flaws, usually on the backs. C select pieces may also contain a few pin knots or small tight knots.

Allowable in this grade are also small season checks, one or two small pitch pockets, small streaks of medium pitch and medium stain over not more than 1/3 of its arc, *if* all of the above do not occur in serious combination.

C grade is also used in high grade accurate patterns. Because of its minor flaws, it contains the added advantage of being more available and a trifle less expensive.

D select sugar pine. The lowest grade of finishing lumber, this grade resembles C select, but has larger or more numerous admissable flaws. A type often found in D select is a piece showing a clear or nearly clear face with numerous or rather serious imperfections on the back, or one defect requiring a cut to eliminate.

D select is not particularly suited for patterns on which much carving is to be done, since the defects would interfere and increase production costs.

Special pattern grade. This grade consists of low line pieces and cast-outs from the D select grade and extreme top line of the number 2 and better common grade.

It is widely used where little carving is required, such as segment work or large patterns where cost is a factor.

Common grades (number 2 and better). This grade is used on the large and medium class of patterns where the price of the lumber is a factor and little carving is required.

Shop grades. There are 3 shop grades commonly used by patternmakers: Factory selects (number 3 clear), number 1 shop, and number 2 shop. They are known as cutting grades and are characterized by knots which can be large and loose. These knots are generally cut out of the board, leaving clear lumber in various lengths and widths.

Percentages of clear lumber are:

Factory selects (number 3 clear). Minimum of 70% of number 1 door cuttings.
Number 1 shop. 50% to 70% of number 1 door cuttings.
Number 2 shop.
   Either: 25% number 1 cuttings, or 33½% mixed number 1 and number 2 cuttings, or 40% number 2 cuttings.

Northern White Pine. This, the finest of the soft pattern lumbers, is found in Northeastern sections of the United States, the Appalachians, and in most of the states and provinces bordering on the Great Lakes.

Its characteristics are strength, straight and close graining, and a high degree of dimensional stability. This grade can be carved readily across the grain.

It weighs approximately 2.4 lbs per board ft.

Its serious drawback is its relative scarcity, and therefore high expense. There is very little virgin growth left and the lumber that is presently being cut is very narrow, and since it is second growth, somewhat pitchy.

Idaho Pine. Although this pine is native to Idaho, some wood comes from Washington and Montana. Resembling northern white pine, it is not quite as dimensionally stable, slightly more difficult to work, harder, heavier, and more brittle.

Spanish Cedar. Graded, oddly enough, by hardwood grading rules, Spanish cedar is, to anyone who uses it, clearly a softwood.

Often overlooked by the average pattern shop, this wood is available at a moderate price in usable lengths and widths. It is characterized by its straight grain, ability to stay straight even in thin sections (when properly cured), and freedom from major defects. It carves exceptionally well and is easily glued.

Caution must be observed to buy lumber from Spanish cedar trees grown in specific localities known for their excellence, texture, and ability to accept proper seasoning.

The finest grade of Spanish cedar is characterized by wide boards of good length with practically no defects. Thicknesses most commonly stocked are 4/4, 5/4, 6/4, 8/4, 10/4 and 12/4.

4.4.2. Hardwoods

When wear and accuracy are prime factors, hardwoods are the answer. Thus, they are ideal for long-run production patterns. In descending order of usage, the following hardwoods are those most employed by the patternmaker: 1) mahogany; 2) hard maple; 3) cherry (Currently scarce and expensive, and so of little value to patternmakers); and 4) walnut. The characteristics and grades of mahogany and hard maple follow.

Mahogany. The best-known and best-suited of all hardwoods for pattern use, it is the standard hardwood for the pattern industry.

The principle types of mahogany are: Mexican, Honduras, Peruvian, Cuban, Phillipine, and African. Since mills in certain selected areas of Mexico, Honduras, Peruvian, Cuban, Phillipine, and African. Since mills in certain selected areas of Mexico, Honduras, and Peru produce a superior lumber of fine grain, texture, and quality, only Mexican and/or Honduras and Peruvian varieties of mahogany are suited to the requirements of the pattern industry.

All mahogany lumber should be thoroughly seasoned and kiln dried.

Random widths and lengths of this lumber are available in: 4/4 (1 in.), 5/4 (1¼ in.), 6/4 (1½ in.), 8/4 (2 in.), 10/4 (2½ in.), 12/4 (3 in.), and 16/4 (4 in.) thicknesses.

Mexican and/or Honduras mahogany. So similar that they can be included in the same category, these two mahoganies have an exceptional ability to resist heat, cold, and extremes of dryness and moisture and still maintain their dimensional stability. Furthermore, they withstand wear and abrasion remarkably well. The grain tends to be straight and close and the wood carves easily. The wood weight is 3.1 lbs per board ft.

Peruvian mahogany. Containing all of the characteristics listed for Mexican/Honduras mahogany, Peruvian mahogany also is a closer-grained, firmer-textured lumber that is slightly harder than its Mexican and Honduras counterparts.

Weighing 3.5 lbs per board ft, this mahogany is recommended to patternmakers who want the ultimate in a hardwood pattern that will stay straight and true and wear exceptionally well.

Mahogany is graded on 5 levels:

- Firsts and seconds. Containing not less than 35% of firsts, this top graded mahogany can contain reverse side defects of such a minor nature that they can be disregarded. Firsts shall be 91 2/3% clear on the face and the seconds 83 1/3% clear on the face.

  Recommended for the most accurate patterns and models of all descriptions, this lumber comes in dimensions ranging from 6 in. and wider and 6 to 16 ft in length. No worm holes are permitted in this grade.

- Selects. Selects grade the same as firsts and seconds on the better face and not below number 1 common on the reverse side. In this grade, there is a tendency to develop more scrap due to defects. Dimensions extend from 6 in. and wider and from 6 ft and longer.

- A grade (also termed A wormy). Scattered pin worm holes are permitted in this grade and are not considered a defect. These holes are easily filled with a wood filler. Recommended uses are for almost every pattern purpose, but particularly in middle-class patterns where price becomes a factor. Widths range from 6 in. and wider and length from 6 ft and longer.

- N grade (also termed N wormy). Comparable to a number 1 common, this grade permits worm holes and other minor defects. It is recommended for use where price is a factor and appearance is not. Slightly more scrap is developed in this grade than the preceding one. In this grade, there is a tendency to develop more scrap due to defects.

Hard Maple. Maple has lately been disappearing from the scene as a lumber for patterns and models. Surviving uses are in big foundries where patterns undergo considerable wear and abuse, and in the automotive industry. Because of its strength and hardness, maple can be used as a die for forming metal where only a few pieces are required.

Available only in top graded (first and seconds and selects), thicknesses are as follows: 4/4, 5/4, 6/4, 8/4, 10/4, and 12/4.

4.4.3. Seasoning

Softwoods are rarely seasoned.

Hardwoods that are produced in the tropics are usually both water- and air-cured. Transported by water, the logs are relieved of strains, mellowed, and lightened in resin content. Piling in certain patterns also air-cures the wood and further mellows it, lessening kiln time and lowering the ultimate cost.

4.4.4. Kiln Drying

Softwoods are dried to a moisture content ranging from 10% to 12%. Dried to a lesser percentage, these woods would take on moisture and create a dust problem when machined.

Since they tend to expand and contract at a much lower rate than softwoods, hardwoods are dried to a moisture content of from 8% to 10%.

Boards will take on moisture and expand and lose moisture and contract, depending on their percentage of moisture to relation to the environmental humidity.

4.4.5. Purchasing

Pattern lumber should be purchased far enough in advance to permit storing within the plant for a reasonable length of time before it is used. Stored so that air freely circulates around each board, this wood will acclimate itself to the conditions under which it must be worked in that particular shop.

Generally speaking, the highest grade of lumber compatible with the work should be purchased. A superior pattern is obtained with the higher grades, and there is less scrap and considerably less labor involved in cutting and carving to shape.

The following rules should be followed for maximum results:

1. Insist on a recognized grade of lumber.
2. Insist that each board be marked as to grade.
3. Insist that each board be marked as to the footage.
4. Tally each load for footage and piece count.
5. Insist that the supplier certify the footage and grade on either an invoice or shipping slip.
6. Obtain a carbon copy of the original tally.
7. Accept no deliveries in the rain except in enclosed areas. (When lumber dries, it shrinks, and checking results.)

4.4.6. Storage

Since the humidity ranges above 12% in most large industrial cities in the summer, spring, and fall, and since both soft and hard lumber is kiln dried to a moisture content varying from 8% to 12%, lumbers will pick up moisture.

Thus, "dead piling," to minimize this pickup is necessary. Dead piling is piling one row of boards directly on the other with no sticks or strips separating the layers.

In a heated shop, where the humidity may be as low as 6%, lumber will tend to dry to 6% and to shrink in the process. But, if it is allowed to stand for a period of time, it will acclimate itself to the humidity of the shop and will then be dimensionally stable and can be worked to close tolerances. If it is not allowed to acclimate itself, the lumber, particularly after planing and jointing, will shrink, causing problems.

Thus, it is recommended that humidifiers and dehumidifiers be utilized in lumber storage places, and that the lumber rack or the benches where work is being done are not in the path of overhead heaters.

Humidity can be checked by a wet and dry bulb instrument and the moisture content of individual boards can be tested by a moisture meter.

The consequences of not observing humidity conditions are obvious: When wood patterns and coreboxes, built accurately and dried out to a 6% moisture content are then taken to a foundry and surrounded by wet sand, core fit and centerlines might not check, if acclimation has not occurred.

Patternmakers sometimes talk about splitting sixty-fourths of an inch on wood patterns, and using verniers and occasionally micrometers. A 12 in. flat-grained pine board will shrink 1/16th of an inch when the moisture content is reduced from 12% to 7%. Wood shrinks most in the direction of the annual growth ring tangentially, commonly called flat grain, and about one-half to two-thirds as much across these rings radially, called edge grain. Practically no shrinkage or expansion takes place longitudinally, except when a board is excessively cross grained, and lengthwise shrinkage is a combination of crosswise and longitudinal change.

4.4.7. Tallying Lumber

A lumber rule or tally stick that is graduated so that it measures the number of surface board feet in lengths from 8 to 16 feet is an invaluable tool for patternmakers.

Boards are measured in thickness in quarter-inch increments. Thus, a 1 in. board is termed 4/4, and a 2 in. board 8/4, etc. If you measure a 4/4, or 1 in. board 12 in. wide and 12 ft long with a tally stick, it will indicate that there are

12 surface board feet in that piece. If you measure an 8/4, or 2 in. board, the tally stick will indicate 12 surface board feet, but you then double the surface board foot measurement, thus indicating that the piece contains 24 board feet.

Other thicknesses are more complex: In the case of 5/4, you would add one quarter of the surface board feet to the actual surface board feet to get the total and in the case of a 6/4 piece, you would add one half of the same.

In tallying lumber, a material measured with a board rule in actual widths, pieces measured to the even half foot shall be alternately counted as of the next higher and lower foot count. Fractions below the half foot shall be dropped and fractions above counted as of the next higher foot.

Tally sticks can also be used to check incoming lumber, and its presence can often pay handsome dividends.

4.5. REINFORCED PLASTICS FABRICATED BY LAY-UP AND CASTING

4.5.1. Resins

The materials used to fabricate plastic tools are, in most cases, specially formulated epoxy resin systems. These systems employ various reinforcements, fillers, thixotropic agents, and wetting agents and/or diluents to provide specific physical properties and/or handling characteristics.

Epoxy. Among all of the resins, epoxy resins tend to rank high as matrix materials for many fiber composites.

The reasons for this are manifold: First, epoxy resins and curing agents are available in a wide variety of viscosities and cure and use temperature ranges, and thus they can be formulated to give a broad range of properties after curing, *and* to meet a diverse spectrum of processing requirements. Second, epoxy resins adhere well to a wide variety of reinforcing agents, fillers, and substrates. Third, because the chemical reaction between epoxy resins and a curing agent does not release any volatiles of water, the shrinkage after cure is generally lower than that for phenolic or polyester resins.

Applications. The above-mentioned qualities make epoxy resins particularly usable in various composites and structural molds and parts, in adhesives, in mold-making and tooling compounds, and as potting and encapsulating compounds.

Since no volatiles are given off during the cure of epoxy resins, materials can be processed into void-free cured products. Furthermore, since epoxies cure with considerably less shrinkage than that encountered in vinyl polymerizations, reduced stress is seen in the cured product.

Thus, epoxy resins are particularly useful as coatings, and almost half of them are sold as such.

Noncoating applications are the following: casting and molding, reinforced composites, and adhesives.

The first commercial epoxy resins were the reaction products of bisphenol acetone and epichlorohydrin, which gives the diglycidyl ether of bisphenol A(DGEBA) and higher homologs. These basic resin characteristics and the control of its molecular weight, make it the most widely used of the resins.

The next useful class developed were the novolac epoxy resins. These were formed by reacting phenol or substituted phenols with formaldehyde, which is subsequently reacted with epichlorohydrin. The particular qualities of novolac epoxy resins have made them useful in areas that require lower and lower amounts of ionic species, total halogen, and alkali metal impurities.

Later research produced alicylic epoxy resin, by reacting hydrogenated bisphenol acetone with epichlorohydrin.

Epoxy resins are also produced from cycloaliphatics such as cyclopentadiene via peracetic acid oxidation, and from a variety of polyolefinic compounds such as polyethers, polyesters, unsaturated animal and vegetable oils, and butadiene derivatives.

For applications that demand even higher thermal capabilities and good mechanical properties in hot/wet and hot/dry conditions, epoxy resins have been developed which work well in high temperature adhesives, composites, and electronic encapsulation.

Finally, resins have been produced which contain 15% to 50% bromine. Their principal use is for addition to other epoxy resins for flame retardance.

Properties. The molecular weight range of the most widely used epoxy resin (the reaction products of bisphenol acetone and epichlorohydrin, which gives the diglycidyl ether of bisphenol A(DGEBA) and higher homologs) can be controlled by the operating conditions and by varying the ratio of the reactants, epichlorohydrin and bisphenol A.

In a general sense, the resins with the molecular weight value "n" being equal to 0 to 1 are liquids.

When "n" is 2 or greater, the resins become solids at room temperature.

For special applications, ultrahigh molecular weight epoxy resins with "n" values of 50 to greater than 150 are made and sold.

In novalac epoxy resins, the "n" values range from 0 to 100, which produces resins at room temperature that are liquids when "n" is 0 to 2 and are semisolids to hard solids as "n" is increased to greater than 2.

Resins based upon the use of phenol are presently the industry standard, while novalac epoxies, based upon 0–cresol, result in resins with increased thermal properties.

All resins must be reacted with a hardener, crosslinking agent, or catalyst to become a thermoset resin and develop desirable physical properties. Since the choice of hardener will determine the final physical properties of the cured thermoset epoxy resin, extreme care must be given to this choice.

In general, resins with a high percent of epoxy functionality are cured through the epoxy group, while the higher molecular weight epoxy resins are cured via the hydroxyl groups along the resin backbone.

The largest volume hardeners are the polyamides. Their ease of handling, long working life, good adhesion, and water resistance, and the large amounts that must be added to the resin to effect the cure have earned them this distinction.

Next are the urea/formaldehyde-type curing agents. Their chief value is in their ability to crosslink the epoxy backbone via the hydroxyl groups.

Anhydrides, applied as curing agents either as low viscosity liquids or solids, are noted for their production of resins with a high thermal stability and good chemical resistance.

Composites and structural molds and parts. By functioning as matrix materials in laminated and filament-wound composites for structural molds and parts, epoxy resins have served in a large number of aircraft, space, military, and industrial applications, in printed circuit boards in the electrical industry, and as a material in composite storage tanks and pipes in the chemical and petroleum industries.

Since they are adaptable, epoxy resins can be formulated for a multiplicity of processes, such as wet filament winding or wet lay-up and dry winding or dry lay-up with prepregs.

Although epoxies are more expensive than most other resins, their superior performance often makes them ultimately more economical.

Adhesives. Since epoxy resins provide the highest adhesive strength of any known polymeric material, and, since they can thoroughly wet a wide variety of substrates with minimal shrinkage, they can be used to join many dissimilar materials. Their ability to be formulated to cure at many different temperatures and times becomes an important consideration when producing adhesives commercially.

Mold making and tooling compounds. Epoxy resins have proven useful in tooling as a substitute for metals. They are considerably more economical to make, can be modified quickly and inexpensively, and have high dimensional stability, good mechanical properties, and low shrinkage. These last three qualities allow the production of replicate molds and parts within extremely close tolerances.

Curing epoxy resins. There are a variety of techniques available to determine the best cure cycle within the parameters of short processing time, good properties, and low thermal stresses.

The degree of cure is defined in terms of any of the properties that change during the curing reaction and reach a constant value at the end of the cure. Physical and mechanical measurements should be correlated to chemical changes so that the resin properties and the molecular structure can become the basis for the choice of cure cycle.

Since the critical crosslinking reactions occur in the resin system, many composite cure-monitoring methods have been developed for the resin alone. This is a valid method, since the total composite curing process is similar enough to the cure process of the pure resin to make the results for the resin transferable to the composite.

Chemical analysis. Wet chemical or physical analysis methods such as solvent swell or titration are often used to directly measure the chemical reaction during the cure. Infrared spectroscopy is also employed, with wide success. Less widely used is the process of nuclear magnetic resonance to track functional group disappearance.

Thermal analysis. Thermal conductivity, specific heat, and heat content depend on the extent of cure, and so can be used to assess cure advancement. Differential scanning calorimetry is the method usually employed.

Dielectrical measurements. The mobility of molecular segments, and therefore the extent of cure, is inferred by the response of molecular dipoles to an oscillating electric field. Thus, this easy to use and simple to interpret means of measuring cure in prepregs is gaining widespread acceptance.

Monitoring of shrinkage. This can enable the selection of a cure cycle that minimizes the residual stresses in the final product.

Estimation of mechanical properties. Available methods of doing this range from simple hardness tests to more complex static mechanical tests or sensitive dynamic mechanical tests.

If adequate quality control of the starting material and processing procedures is exercised during production, extensive monitoring procedures need only be used once, to select a cure cycle.

Characteristics of cured epoxy resin systems. Once an epoxy resin has been cured, it acquires specific characteristics that are determined by its chemical structure.

A higher crosslink density can yield an improved resistance to chemical attack.

A lower crosslink density can result in reduced shrinkage during the curing process, and it can also improve toughness by permitting greater elongation before breakage.

The greater the number of aromatic rings a cured epoxy resin contains, the greater its thermal stability and chemical resistance.

Epoxy resins that are cured with aromatic curing agents must be watched closely. They are likely to be more rigid and often make a stronger cured mold or product than those cured with aliphatic curing agents. But, they require higher cure temperatures because their very rigidity reduces the molecular mobility needed to properly position two reactive end groups for reaction.

Processing parameters. Finally, resin systems — epoxy or otherwise — must be formulated to meet processing requirements.

Wet lay-up and wet filament winding methods require a low viscosity to the uncured resin system to permit good wetting of the fiber phase and good resin distribution in the composite. Thus, reactive diluents are often added to reduce viscosity.

Aliphatic amine curing agents react faster with epoxy resins than do aromatic amines. However, when an accelerator is used, the working lives of epoxy-anhydride systems can become as short as a few hours, depending upon the amount and type of accelerator used.

Since mold-making and structural part prepregs are usually stored at low temperatures to inhibit further reaction of the partially cured resin until desired, a controllable and reproducible cure advancement that allows both sufficient working life as well as shaping and arrangement of the prepreg is critically important.

Potential. Since epoxies are synonymous with adhesion, liquids can easily be formulated for myriad applications. Formulated with the appropriate hardeners to give an appropriately slow or fast curing system, and with the added characteristics of complete cure (with no volatiles during the cure), minimal shrinkage, and good bonding, their potential as excellent adhesives to glass, graphite, kevlar, metal, and stone make their use in this capacity widespread.

Since marine and maintenance coatings are usually ambient-cured, solvent-based systems, there is a general trend for these systems to move toward higher-solid (less-solvent) and/or water-based systems.

Finally, in commercial and military aircraft and aerospace mold and part applications, the use of carbon fiber and glass fiber epoxy-reinforced composites, providing high strength-to-weight ratios and retention of physical properties at high temperature and high humidity conditions, is increasing rapidly.

Phenolic. Phenolics, or phenol formaldehyde resins, are produced from the reaction of phenol (carbolic acid) with formaldehyde in the presence of a catalyst. This produces a polymer that assumes a variety of forms: as powder or flakes for molding compounds (which are available only in dark colors), or as a liquid for casting, bonding, coating, and impregnating.

The major negative characteristic of phenolics is their brittleness, which necessitates the presence of fillers.

This, however, is greatly offset by their many positive characteristics: hardness, rigidity, cost savings, dimensional stability, possession of excellent insulating properties, heat resistance up to 500°F, low moisture absorption, and possession of a chemical inertness to most common solvents and weak acids. Due to these properties, phenolics are ideal for use in the design of precision devices.

Phenolics are used as liquids in the lamination of fabrics, paper, and veneers, and in coating and adhesive applications. Thus, they are found in distributor caps and coil tops, telephones, tool housings, brake linings, and home appliance handles and parts.

Impregnated wood fibers are molded into such products as salad bowls, and reinforced laminated sheet is used for automobile body parts, roof panels, and large ductwork. Phenolic bonding agents are used in the foundry industry for cores and shell molding, and the construction industry uses a large volume of phenolics as an adhesive in bonding plywood.

For mold- and tool-fabricating materials used to form automotive, marine, and aircraft–aerospace nonmetallic parts, paper and fabrics are impregnated with liquid phenolic resins, laminated into thick slabs, bonded together, and subsequently machined into finished contours as the final mold-making procedure.

Polyesters. Best known as fiber glass (because it is combined with chopped fibers or glass mat), polyester is produced from the polymerization of certain alcohols and acids. It is generally found in two forms: as a resin to which a catalyst is added to complete the curing during molding, or as a completely polymerized polyester film. The polymer is a clear, colorless liquid that is then colored, filled, and reinforced to meet a multitude of specifications.

Premixed molding compounds with fillers are available for immediate use, although these compounds will react if not used quickly or kept cold. More practically, polyesters are available as liquid resins with accompanying catalysts which are mixed at the time of use.

Matched die, bag, or hand lay-up molding is used to produce such items as boat hulls, automobile, and aircraft body components, luggage, wash tubs, and roofing sheets. Premix molding by the compression process produces such products as hammer handles and automotive ductwork.

In addition, polyesters are used as patching compounds, and coatings on masonry, wood, and metal. Cast polyesters are used for embedding decorative jewelry, scientific specimens, and cast-in electrical parts. High temperature polyesters lend themselves to being used in tool fairing or surface smoothing compounds and in expensive prototype parts and molds.

Heat Resistant Resins. There are certain composites that employ high-temperature resins as matrix materials. These high-temperature resins are linear or cross-linked aromatic/heterocyclic polymers that have a high glass transition temperature and can withstand continuous exposure in air at temperatures above 600°F, without exhibiting a significant loss of structural integrity.

Although these polymers do undergo thermo-oxidative degradation during elevated temperature exposure in air, they degrade at relatively slow rates. Furthermore, it is speculated that these polymers degrade into stable residues, thus increasing their service life at elevated temperatures.

The key to creating high-temperature resins is the synthesis of polymers that contain a multiplicity of aromatic/heterocyclic molecular structural units. These units, which contain a minimum number of oxidizable hydrogen atoms, are able to absorb thermal energy.

But there is a drawback: The structural units that are responsible for the thermal and oxidative stability of these polymers are also responsible for their intractability, which makes it extremely difficult to process them into useful articles.

Because of these characteristics, attempts to utilize these polymers as matrix resins for advanced composites either failed or were not considered to represent economically feasible methodology, and their future looked bleak indeed.

However, developments within the class of high temperature polymers known as polyimides from 1972-1974 not only rekindled interest and experiment, but also made it possible to realize the potential of high temperature polymer matrix composites.

At the present, polyimide advanced-fiber composites are being used or considered for use in a variety of 600°F structural applications. The polyimide matrix materials most widely used by prepreg manufacturers are either the NR-150B series of condensation-type polyimides or the monomeric reactant, addition-type polyimide PMR-15 developed at NASA's Lewis Research Center.

Condensation-type aromatic polyimides can be prepared by reacting aromatic diamines with aromatic dianhydrides, aromatic tetracarboxylic acids, or dialkyl esters of aromatic tetracarboxylic acids.

Generally, the diamine-dianhydride reaction is preferred for preparing polyimide films and coatings, whereas the other two combinations of reactants are preferred for polyimide matrix resins.

Polyimide prepreg, or precursor, solutions are prepared by dissolving the appropriate reactants in a high boiling point aprotic solvent or in a solvent mixture containing an aprotic solvent.

4.5.2. Curing Agents

Epoxies are used in coatings, adhesives, tooling, civil engineering, electrical insulation, composites, and construction materials.

They are most often used as thermosetting materials which crosslink to form a three-dimensional nonmelting matrix. Where epoxies are used as thermosetting materials, generally a coreactant is used to achieve crosslinking (curing).

In room-temperature-curing, two-component epoxy systems, the coreactant which serves as the crosslinker is generally an amine such as diethylenetriamine or triethylenetetramine.

Where an elevated temperature cure is possible, a number of different curing agents can be utilized. These include aromatic amines, anhydrides, carboxylic acids, phenol novolacs, and amino resins (such as substituted melamines and urea formaldehyde).

Other hardeners will catalyze homopolymerization of the epoxy. An example is a Lewis acid.

The chemical reaction between epoxy resins and a curing agent does not release any volatiles or water. Thus, the shrinkage after cure is customarily lower than that for phenolic or polyester resins.

The wide variety of available epoxy resins and curing agents can be formulated to give a broad range of properties after cure and to meet a diverse spectrum of processing requirements.

Applications. Generally, epoxy resins are used in various composites and in many structural parts, as potting and encapsulating compounds, adhesives, tooling compounds, and molding powders.

Since epoxy resins are very resistant to bases, acids, and humidity, they are ideal for potting and encapsulating. A high heat distortion temperature can be attained, and these resins also exhibit little shrinkage and high volume resistivity, enabling replication of parts to extremely close tolerances.

Epoxy resins provide the highest adhesive strength of any known polymeric material. They can thoroughly set a wide variety of substrata with minimal shrinkage. Thus, epoxy resins can be used to join many dissimilar materials. They can also be formulated to cure at many different temperatures and times, an important consideration when producing adhesives commercially.

Epoxy resins have been used since World War II in tooling. Particulate or fibrous reinforcing agents are easily incorporated to reduce cost and improve dimensional stability. Epoxy resins are used instead of metals for two principal reasons: They are more economical to make and they can be modified quickly and inexpensively.

Finally, one of the most important uses of epoxy resins as matrix materials is in laminated and filament wound composites for structural parts. These composite structural parts have been used in aerospace, military, and industrial applications. Laminates are also used for printed circuit boards in the electrical industry, and in composite storage tanks and pipes in the chemical and petroleum industries.

Epoxy resins are adaptable and can be formulated for different processes: wet filament winding or wet lay-up and dry winding or lay-up with preimpregnated fiber strands, fabrics, or tapes (prepregs).

In general, epoxies are more expensive than most other resins, but their superior performance often makes them more economical in the long run. Caveats before matching curing agents and epoxy resin systems:

- Epoxy resins cured with aromatic curing agents are likely to be more rigid and often make a stronger cured product than those cured with aliphatic curing agents. However, these epoxy resin systems require higher cure temperatures because their very rigidity reduces the molecular mobility needed to properly position two reactive end groups for reaction.
- The greater the number of aromatic rings that a cured epoxy resin contains, the greater its thermal stability and chemical resistance will be.
- A lower crosslink density can improve toughness (if strength is not significantly lowered) by permitting greater elongation before breakage.
- A lower crosslink density can also result in reduced shrinkage during cure.
- A higher crosslink density leads to an increase in the heat distortion temperature (and glass transition temperature), but too high a crosslink density lowers the strain-to-failure (increased brittleness).
- A higher crosslink density also yields an improved resistance to chemical attack.
- The performance of anydride-cured systems is better in an acidic medium than in a basic medium.

Curing. The epoxy group can bond chemically with other molecules, forming a large three-dimensional network. This process, called curing, changes a liquid resin into a solid. A major area of epoxy technology is devoted to the study of curing-agent properties.

A CAUTIONARY NOTE: Before we proceed with the defining of choices of curatives for epoxies, we must address recent findings about curing agents and resins that contain suspected carcinogens. As an example, 4.4'-Methylenedianiline (MDA) and VCHD (Vinyl Cyclohexene Dioxide) are among those suspect materials. Information on this subject is available from federal agencies such as OSHA, NIOSH, and others.

Amine Curing Agents. Aliphatic amines are commonly used curing agents which permit the formation of a crosslink between epoxy molecules. For thorough crosslinking, the hydrogens of the primary and secondary amines should be matched one-to-one with the epoxy groups.

The reaction between aliphatic amines and epoxy groups will usually proceed at room temperature. However, heat is required when rigid aromatic amines are used.

The carbon-nitrogen bond formed in crosslinking is stable against most inorganic acids and bases, but it is less stable against organic acids than linkages formed by other curing agents. In addition, the electrical insulating capabilities of amine-cured epoxy resins are not as high as those obtained with other curing agents, probably because of the polarities formed by the hydroxy groups during the cure.

Tertiary amines (no hydrogen on the nitrogen) are Lewis bases that cure epoxy resins in an entirely different manner. They are added to an epoxy resin in small nonstoichiometric amounts that have been empirically determined to give the best properties.

The curing agent operates as a true catalyst by initiating a self-perpetuating anionic polymerization. This homopolymerization of one epoxy molecule to another molecule results in a polyether. The ether linkages are fairly stable against most organic and inorganic acids and bases. Like ester linkages, they are more thermally stable than the carbon-nitrogen linkages formed by an amine cure.

Of the two most popular Lewis bases, EMI is the most efficient, producing the highest degree of crosslinking and the highest heat distortion temperature.

The two following curing agents deserve special comment.

Aminopolyamide, a versatile fatty amino polyamide, is a flexible polymer that contains primary and secondary amine functional groups as well as amide functional groups. Although referred to in the trade as an amide, this curing agent, which also acts as a flexibilizer, cures epoxy resins mainly through its amine groups.

It produces increased flexibility and impact strength in cured products, and it has less skin-irritation potential than the regular amine curing agents. However, an aminopolyamide-cured system is less resistant to chemicals and solvents than are the other epoxy resin systems.

Dicyandiamide (DICY) is usually employed with solid epoxy resins in prepreg laminates because it is a latent curing agent — that is, little or no reaction with the epoxy resin occurs at room temperature, but full cure is rapidly initiated at elevated temperatures. Dicyandiamide causes especially adverse skin reactions in some people and thus should be handled with extreme care.

The following is a list of amine curing agents:

- Triethylenetetramine (TETA)
- Diethylenetriamine (DETA)
- Diethylaminepropylamine (DEAPA)
- 2.6-Diaminopyridine (DAP)
- 40% MDA–60% Diethy MDA
- 33.3% MPDA–33.3% MDA–33.3% Isopropyl MPDA
- 40% MPDA–60% MDA

- Aminopolyamide
- 4.4'-Diaminodiphenylsulfone (DDS)
- m-Phenylenediamine (MPDA)
- 4.4'-Methylenedianiline (MDA)
- Dicyandiamide (DICY)
- Aliphatic Polyether Triamine (APTA)
- Tetraethylenepentamine
- 2-Ethyl-4-methylimidasole (EMI)
- 2.4.6-Tris(dimethylaminomethyl)phenol

Anhydride Curing Agents. Anhydride curing agents react with epoxy resins to form esters. They usually require the application of heat in order to initiate full cure. In fact, the epoxy-anhydride reaction can be so sluggish that a small amount of accelerator will often be added to speed up the curing process.

Anhydride curing agents usually require careful storage to prevent degradation as a result of moisture absorption.

The following is a list of anhydride curing agents:

- Trimellitic Anhydride (TMA)
- 3.3',4.4'-Benzophenone-tetracarboxylic Dianhydride (BTDA)
- Methyltetrahydrophtalic Anhydride
- Succinic Anhydride (SA)
- Maleic Anhydride (MA)
- Phtalic Anhydride (PA)
- Nadic Methyl Anhydride (NMA)/Maleic Anhydride Adduct of Methyl Cyclopentadiene
- Hexahydrophtalic Anhydride (HHPA)
- Dodecenyl Succinic Anhydride (DDSA)

The following is a list of accelerators for anhydride curing agents:

- Boron Trifluoride-Monoethylene Amine (BF3MEA)
- Benzyldimethylamine (BDMA)
- Dicyandiamide (DICY)
- 2.4.6-Tris(dimethylaminomethyl)pehnol
- 2-Ethyl-4-Methylimidazole (EMI)

Lewis acid catalytic curing agents. Boron trifluoride is the only Lewis acid that has achieved great popularity as a curing agent for epoxy resins in composites. Added in small amounts to the epoxy resin alone, it functions as a catalyst by cationically homopolymerizing the epoxy molecules into a polyether.

For prepregs, which are often stored for weeks before fabrication into a component, the use of a latent curing agent such as this is an absolute necessity. Epoxy resin systems containing BF3MEA are popular for use in potting, tooling, laminating, and filament winding. It has been found, however, that prepregs and cured systems containing BF3MEA generally have a poor resistance to humidity.

Epoxy reactive diluents. The following is a list of reactive diluents:

- Cresyl Glycidyl Ether
- Butyl Glycidyl Ether
- Phenyl Clycidyl Ether (PGE)
- Heptyl Glycidyl Ether
- p-t-Butyl Phenyl Glycidyl Ether
- Octyl Glycidyl Ether
- Allyl Glycidyl Ether

Epoxy resins. The following is a list of epoxy resins:

- Polyglycidyl Ether of Phenol-Formaldehyde Novolac
- Diglycidyl Ether of Bisphenol F (DGEBF)
- Diglycidyl Ether of Bisphenol A (DGEBA)

The Monitoring of the Cure. The degree of cure can be defined in terms of any of the properties that change during the curing reaction and reach a constant value at the end of the cure.

The most often used monitoring tests are listed here:

- Wet chemical or physical analysis methods.
- Infrared spectroscopy.
- Nuclear magnetic resonance (NMR).
- Heat content measurement by differential scanning calorimetry (DSC).
- Dielectric measurement.
- Monitoring of shrinkage during cure.
- Measuring of mechanical properties of the cured resin.

It must be emphasized that once a cure cycle is selected, extensive monitoring procedures need not be used again, provided that adequate quality control of the starting material and processing procedures is exercised during production.

4.5.3. Reinforcements and Fillers

Reinforcements and fillers are added to and used with plastics to control viscosity, weight, and thermal properties; reduce shrinkage, exothermic heat, and cost; and to increase pot life, strength, and wear resistance. They are also used to add color and texture. Fillers are usually inert and do not react with the resin. Proportions and types of fillers are controlled by the plastic formulators to meet specific requirements.

The weight of plastics can be adjusted by the use of various types of fillers. The heavyweight fillers used include metal powder and particles, barite, sand, and tar. Lightweight plastics such as foams, cast resins, and paste resins can be poured or applied over or into contours to make lightweight cores or space fillers.

The designing of tools made of fiberglass and graphite-reinforced epoxy involves the same basic considerations as those made of metals and other familiar materials. With glass- and graphite-reinforced epoxy, however, the designer has a wider choice of control factors. This choice is available in part because of the directional control of fiber strength, and partly because several reinforcing media having different characteristics can be incorporated into the individual tool. Ply orientation and stacking of plies (balanced lay-up) also affect strength control.

The highest strength properties are found in laminates employing thin glass or graphite cloths, at least on the outer stressed surface. The overall strength decreases as the fabric thickness increases. The diminishing effect of increased thickness is greater in compression. Impact strength, however, varies directly with the thickness (i.e., thicker fabrics have higher impact strengths).

Fibrous Reinforcements. Reinforcements, aligned in various manners within a matrix, yield a composite that exhibits relatively uniform, if low, mechanical strength in all directions.

The chief advantage of a reinforcement is its excellent resistance to attack by water, and, in the case of fiberglass E-glass fibers, by acids and bases. Its disadvantage is its relatively low mechanical strength — a prime consideration in choosing this reinforcement for specific tasks.

History. The knowledge that molten glass can be stretched into fine lengths probably goes back thousands of years. Until the nineteenth century, these gossamer-like threads were attached to ornamental glass objects as fine and delicate decoration.

Late in the 1800s, however, glass fibers were interwoven with silk fibers, producing exquisite and expensive gowns for French and American ladies of wealth and position.

The multimillion dollar industry of textile fiberglass began in 1939, with the formation of the Owens-Corning Fiberglass Corporation. Since that time, this has been overshadowed by the production of fiberglass insulation products.

Properties. Moisture resistance — Fibrous reinforcement (F.R.) simply does not absorb moisture. As a result, it resists rot, and neither swells, stretches, nor disintegrates, and retains its full strength in the most humid of environments.
Chemical resistance — F.R. resists degrading by most chemicals, by fungus, bacteria, and insects.
Heat and fire resistance — Since F.R. is inorganic, it neither burns nor supports combustion. It also has a low coefficient of thermal conductivity. Thus, it becomes eminently usable in high-temperature environments, or conditions where fire resistance or retention are desired.
High tensile strength — In some applications, the F.R. strength-to-weight ratio makes it stronger than steel wire.
In addition to the above, most types of F.R. do not conduct electricity, thus making them ideal for use as electrical insulation.

F.R. use in composites. Since fibrous-reinforced thermosetting and thermoplastic resins are engineering materials, the process by which they are combined and the contribution of each component to the overall composition determine the properties and the performance critera of the end product.

The wide use of F.R. plastics as an engineering material in the last twenty years, and the recent introduction of advanced composites — those resins reinforced with fiber forms other than fiberglass — have led to an increased understanding of the use of fiberglass reinforcing agents in composites.

It has been found that the type, form, amount, and alignment of the fibers have a strong bearing on the physical and mechanical properties of a composite.

Furthermore, although the thermal and flammability properties of a composite are primarily determined by the resin matrix, the chemical and/or electrical properties may be influenced by the resin matrix and/or the selected reinforcing agent.

Here are some of the methods of the fibrous reinforcing of composites:

Roving is a series of parallel, continuous strands or filaments. This is further broken down into two distinct methods: *Conventional roving,* in which a number of single strands are wound together to achieve the required yield, and *single-strand roving,* which consists of a single strand of filament drawn from a bushing containing the proper number of orifices to result in the single strand having the required yield.

Woven roving is a process by which many rovings are woven into a heavy, coarse-weave fabric designed for applications that require rapid thickness buildup over large areas, such as the manufacture of fiberglass boats, marine products, and many types of tooling.

Matting is divided into three forms: 1) *Continuous-strand mat,* which consists of underchopped continuous strands deposited and interlocked in a spiral fashion. The characteristic of this mat is its openness and springiness and the fact that, because of its mechanical interlocking, it does not require much binder for adequate handling strength; 2) *Surfacing mat (or veil),* which is an almost transparently thin mat of single, continuous filaments used as a decorative surface reinforcing layer in hand lay-up or press-molding processes (see Fig. 4.4). In a practical sense, these veils minimalize transfer of the texture of the primary reinforcement through to the finished surface of a component. This material is particularly useful in the new surface coats of high temperature tools; and 3) *Chopped-strand mat* (Fig. 4.4), which is nonwoven material, in which the strands from roving are chopped into one- to two-inch lengths, are evenly distributed at random onto a horizontal plane, and then are bound together with an appropriate chemical binder.

It must be emphasized that the type, form, amount, and alignment of the fibers have a strong bearing on the physical and mechanical properties of the composite. This is further discussed in the following paragraphs.

Handling characteristics of a fabric or composite are determined by the amount of yarn and the weave pattern that holds the yarns together. If this weave pattern is too tight, the fabric will resist conforming to certain contours, and will not accept resin. Thus, the composite will be weak.

If, on the other hand, the weave pattern suffers from openness or looseness, there will not be sufficient fiber in the fabric to develop its maximum possible strength. Furthermore, it will distort, which will result in an inability to align the fibers with preferred strength axes.

In general, the longer the float (i.e., the portion of a warp or filling yarn that extends unbound over two or more yarns lying at a 90° angle to it), the higher the composite strength.

Further strength characteristics lie in the reinforcement-to-resin ratio. The higher the reinforcement-to-resin ratio, the greater the mechanical strength of the composite.

In filament winding, parallel strands of reinforcement are wound around a

Fig. 4.4. Types of Mat.

cylindrical mandrel. As a consequence, the round filament cross section allows close packing, which in turn results to a very low resin-content capability.

A general rule is this: reinforcement loadings should contain no more than 38% nor less than 20% resin for maximum mechanical strength. Resin, it must be remembered, contributes little to overall mechanical strength.

Composites reinforced with strands aligned parallel to each other have their maximum mechanical strength and stiffness in the direction of strand alignment.

When half of the strands are laid at right angles to the rest, the mechanical strength at either angle is less than that of the parallel alignment. Also, as the distribution of the strands varies between the 0 and 90° angles, the mechanical strength varies accordingly.

This can be offset by rotating alternate layers of reinforcement or varying the yarn distribution between the 0 and 90° directions. A balanced fabric with equal yarn distribution between the warp and filling directions will have comparable (but not necessarily equal) laminate properties in those directions.

Composites reinforced with unidirectional fabrics have their maximum mechanical strength in the direction corresponding to the greatest concentration of yarn.

Since this is true, it would seem logical to align the reinforcement in a random manner within the matrix (as in a chopped-strand mat), since this would yield a composite that exhibits relatively uniform mechanical strength in all directions. Ironically, this does not happen. Instead, it leads to a composite of relatively low mechanical strength in all directions.

The type of weave selected for a tool is important, since the reinforcing material is an integral part of a laminate. There are three general types of fiberglass or graphite cloth that can be considered:

Plain cloths. These are cloths in which each warp and fill yarn passes over one yarn and under the next (see Fig. 4.5). Plain weave cloths are intended for

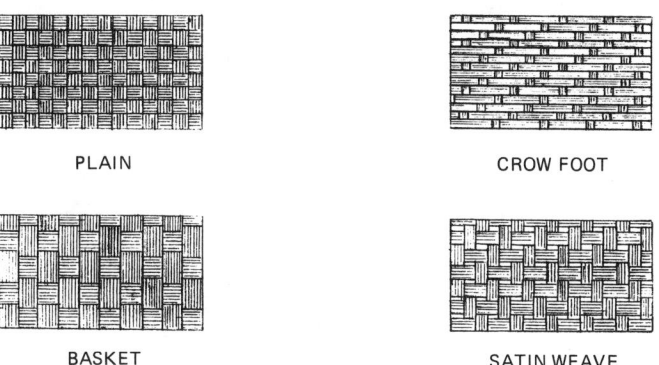

Fig. 4.5. Types of Woven Fabrics.

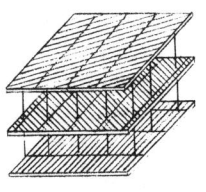
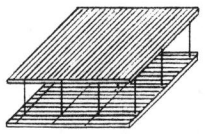
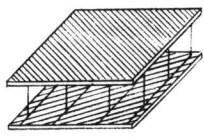

TRIAXIAL   BIAXIAL

Fig. 4.6. Unidirectional Knitted Fabrics.

use where design requirements call for a uniformity of strength pattern through 180°, or where a fast laminate thickness build is required.

Unidirectional fabrics. These are made with strong warp yarns and relatively few weaker fill yarns. They furnish maximum strength in one direction, thus their name (see Fig. 4.6). Unidirectional fabrics are intended for use a) where the maximum strength is required in one direction, b) where the load pattern can be definitely established and the material can be oriented to meet the known requirements, c) where local areas require additional stiffness or impact resistance, and d) where high strength with minimum weight is required.

An economic caution: Because these unidirectional fabrics must be cross-laminated to a specific pattern, the cost of fabrication is increased by the hand labor involved.

Satin fabrics. These are designed as eight shaft satins, in which each warp and fill yarn passes over seven yarns and under one.

These eight shaft satins have been found most satisfactory in heavier fabrics, but they also have their uses in laminates. In laminates, they give higher physical strengths than plain weave fabrics of comparable thickness.

They are thus best suited 1) where a part or mold, requiring high strength in all directions combined with light weight, must be made with a minimum of lay-up labor, 2) where smooth, decorative finishes are desired, and 3) where high-strength laminates are desired at low material cost. This is accomplished by employing thick, long shaft satins in place of thin, square woven cloths.

4.5.4. Prepregs

Growing percentages of advanced composites are utilized in the current design of high performance sailboats, aircraft, and missiles. These composites consist of a matrix (usually a resin) which may be one of an imposing array of new high temperature plastics, reinforced with graphite, Kevlar, or S-glass fibers.

Occasionally, processing requirements include processing in excess of 700°F, but the usual high temperature in this context is in the 250°-500°F performance area.

Aircraft engineers have found that the production of large graphite laminates requires new approaches in tooling. A traditional material such as aluminum expands ten times as much as a graphite/epoxy composite, and this can result in a part that is trapped or even destroyed in a female mold, or oversized beyond tolerances on a male mold.

The solution is to match the coefficient of expansion (CE) of the tool to that of the part. There are surprisingly few candidate materials available.

Graphite tools for advanced composite use are usually 0.635 cm (1/4 in.) thick, so they are light in weight. The laminate is supported and made rigid with a network graphite faced/aluminum honeycomb or solid laminate panels, called eggcrate. Some mold makers are rigidizing the laminates with integral stiffeners molded in the back surface of the tool during lay-up of the laminate.

High quality laminates have been made from prepreg (i.e., pre-impregnated fabric) since the 1950s. This controlled application of resin to woven or unidirectional reinforcement has made possible our space hardware, modern aircraft, and printed circuit boards. Prepreg for tooling was introduced around 1975, with no commercial success. By 1980, advanced composites had become a production reality, and the tooling community aroused itself. There are now thousands of graphite/epoxy tools in service, which are actually carbon cloth tools, since it is neither economic nor necessary to graphitize the carbon fiber for tooling use (see Fig. 4.7).

Fig. 4.7. The Manufacture of Tooling Prepreg. *(Courtesy of the Toolrite Div. of Fiberite Corp.).*

110 ADVANCED COMPOSITE MOLD MAKING

The great care taken to fabricate a graphite tool for advanced composite use is in sharp contrast to the procedures followed years ago for wet lay-up, fiberglass/epoxy tools. Today's prepregged graphite tool should last for hundreds of 350°F autoclave cycles. This should eliminate the main objection to soft tooling for composites. The key is the use of long-proven technology, plus adequate quality control procedures.

It is important to emphasize throughout the importance of a low coefficient of expansion in these materials, because in most cases one goes directly to a high-temperature tool lay-up. This is only one of many factors governing the choice of a material, and it may be that one or two properties are more significant than all the others combined, for a particular need. Thus, there can be no single choice for all tooling needs, and each material has its own limitations and advantages.

Prepreg is often produced by the hot melt method. This procedure involves no solvent and no tower to partially react or cure (B-stage) the resin. The catalyzed resin is calendered on one side of the fabric (Fig. 4.8). This offers advantages for tooling use.

Thicker lay-ups can be quickly debulked under vacuum, because there is a ready path for air removal. There are no volatiles trapped in thick laminates, since the prepreg is solvent-free.

Labor costs are reduced when the thickness is in the range of .025 to .040 in. per ply. The expensive 3K graphite filament can be supplemented or replaced with a more economical 6K and 12K filament from Hercules, Celanese, or Union Carbide.

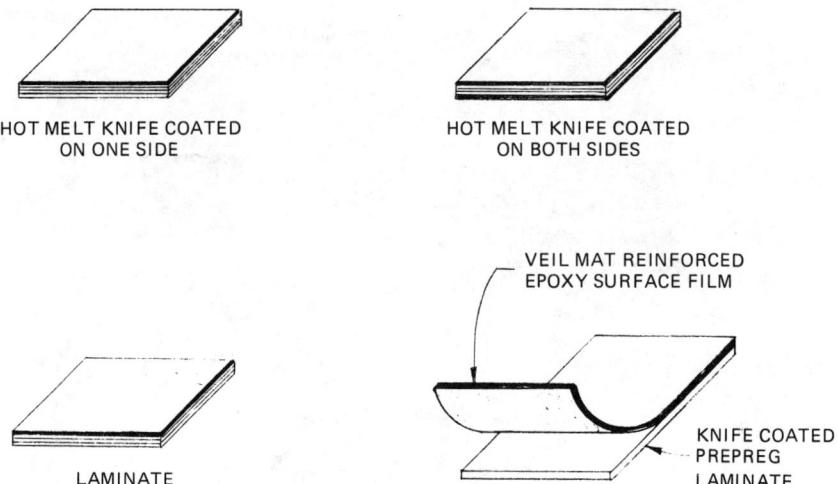

Fig. 4.8. Hot Melt Prepreg and Veil Reinforced Epoxy Surface Film.

There are many excellent laminating systems from which thousands of autoclave tools have been made. Graphite tooling should always be considered, since neither aluminum tools nor epoxy/fiberglass are suitable for some contours.

4.5.5. Reinforced and Nonreinforced Materials for Nonmetallic and Electroformed Patterns and Molds

As a result of new technology in model and master making, the need for new model and master pattern material has become apparent. To match coefficients of thermal expansion, master model and mold materials have become reinforced with matching reinforcements, such as graphite and fiberglass fibers. Slab materials of various forms have been developed, based upon machinability, cost, and availability of raw materials. Laminated paper reinforcements for high-temperature slab stock, mass-cast light weight master material, and even solid graphite block (Fig. 4.9) has been adapted for this use.

These materials have been used as N/C (numerical controlled) machining masters. N/C milling machines are specialized to produce repetitive contours after careful programming of the operations to be performed. The coded instructions are usually comprised largely of numbers on N/C tapes, which can be stored in compact locations.

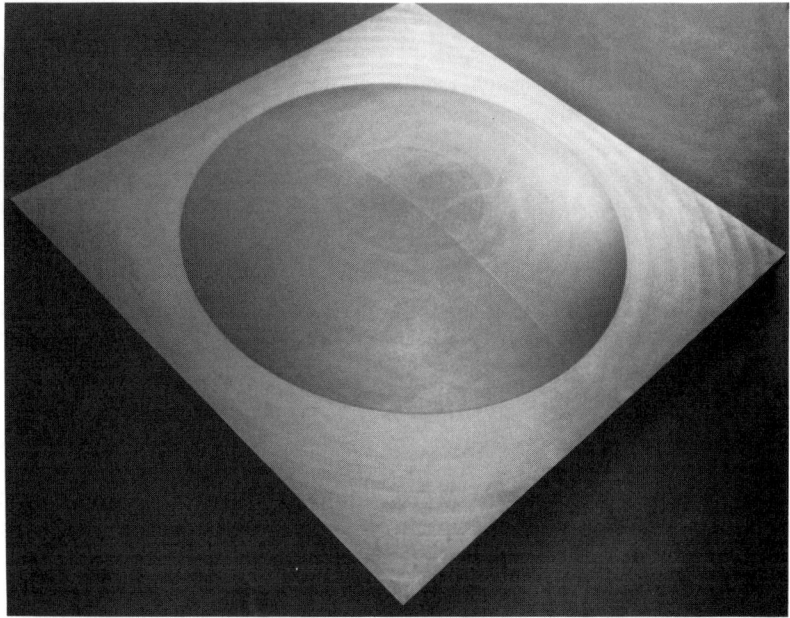

Fig. 4.9. N/C Machined Master Molds. *(Courtesy of Airco Carbon Co.)*

Fiberglass and Graphite Board as Machine Masters. Various forms of extruded and cast epoxy are modified, using reinforcements of graphite and fiberglass, to form slabs of modeling stock, which in turn are bonded into stacks using both room-temperature and high-temperature adhesives.

Since graphite fabric is strong, lightweight, dimensionally stable, and highly workable, it has proven to be extremely versatile in use.

The temperature of use depends upon the final application for the machined model. In the form of a tooling lay-up master, the temperature might be as high as 350°F. On the other hand, as a plating mandrel, the temperature might only approach 110°F.

Another form of graphite model master material is the monolithic graphite block, chosen because graphite has proven to be an exceptional tooling material for the fabrication of the reinforced plastic composites used in aerospace applications.

Basically, monolithic graphite consists of a homogeneous, compacted calcined petroleum coke that is combined with a pitch binder. Once formed to approximate shape by either extruding, molding, or isostatic pressing, the material is heated first to its carbonizing temperature and then to its graphitization temperature (over 2500°F) to provide a uniformly constructed product.

Unlike graphite fabric-reinforced epoxy tooling (which produces anisotropic properties) monolithic graphite is extremely uniform, and therefore provides consistent properties in all directions.

Monolithic graphite tooling is engineered as a complete system regardless of size or complexity. It is segmentable for producing large composite shapes. In either case, tooling systems are manufactured to provide surface characteristics that are not only vacuum tight, but compatible with standard release agents.

Monolithic graphite tooling is also designed to provide maximum design and application versatility when complex contouring is required (see Fig. 4.10).

Because of its low CTE, monolithic graphite tooling will not expand as most other materials tend to do. (During a typical curing temperature of 350°F, aluminum will expand nearly seven times more than graphite; steel will expand more than three times as much.) In fact, when subjected to heat, graphite remains stable and provides low thermal stress. This results in better dimensional control of fiber-reinforced plastic composites, with less stress due to part expansion.

Monolithic graphite is easily machined into precision tools with a one-step operation. Since the forces required to machine graphite are significantly less than metal, it can be machined considerably faster. For instance, in typical milling operations, tool steel can generally be machined at 20 ft/min and aluminum at 100 ft/min. Graphite can be machined at speeds up to 200 ft/min. Thus, since most equipment used to machine graphite is also designed for metal, a

MOLD MATERIALS 113

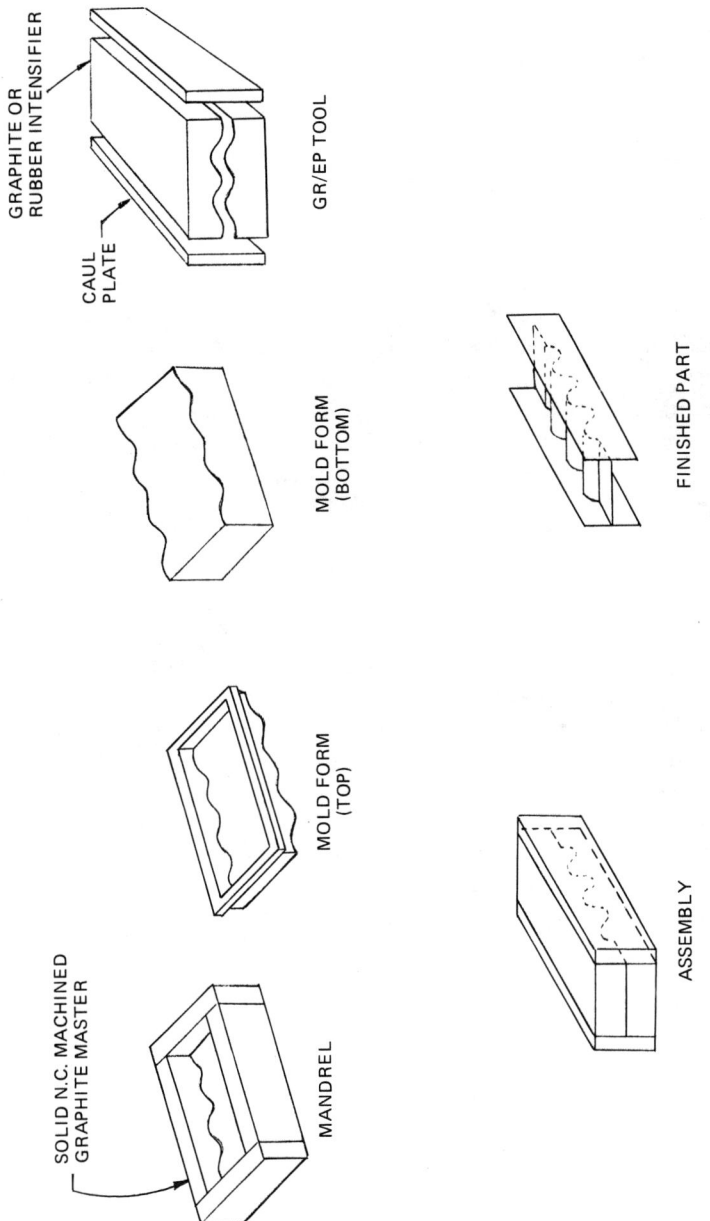

Fig. 4.10. Graphite Block Molding Sequence.

good surface finish and closer tolerances over longer lengths can be virtually assured.

Monolithic graphite also provides comparable thermal conductivity and requires less energy to cycle composites through the various temperature ranges with maximum consistency than either aluminum or steel.

At a typical curing temperature of 350°F, steel requires almost three times as much energy per cubic foot as graphite, while aluminum requires more than twice as much energy per cubic foot. This means, obviously, that graphite can provide savings in time and energy, provided it is used correctly and efficiently.

Laminated Phenolic. This all-purpose, high-strength paper-based phenolic sheet material is used in pattern shops, machine shops, and industrial plants, along with other materials. It can replace high-cost metals for many applications, and is especially practical where a light-weight, corrosion- and abrasion-resistant mold material is required.

This type of sheet material weighs only half as much as aluminum. This quality is important in saving freight costs. Furthermore, this relative lightness reduces the equipment and personnel needed for handling it as a tool mold or fixture.

When used as a base material to machine plating mandrels, the phenolic base resists ordinary acid, solvent, and bacterial action.

This hard-wearing product will give exceptionally long service under the toughest conditions of use. The material comes in sheets of various thicknesses. The thickness and overall size of the sheets can be increased by bonding additional material to the original sheet. Moisture absorbtion should be considered when choosing a particular type of sheet material, as most forms are hydroscopic.

Syntactic Foams for Cast-in-Place Net Molds and Machining Masters. Classically, foams have been produced by generating gas cells within a material. Desired densities have been obtained by controlling the sizes and number of gas cells. But syntactic foam is a different material entirely, produced by introducing hollow microspheres into a resin matrix that acts after mixing. The term syntactic implies an orderly arrangement of spherical filler particles. Thus, as seen in Fig. 4.11, syntactic foam is a composite material in which small preformed bubbles of glass, plastic, or ceramic are mixed with a binder, usually a thermosetting resin, and castable ceramics.

These foams are a bit heavier and a great deal stronger than gas-blown foams. In fact, they can withstand five to ten times greater hydrostatic pressure than any type of plastic foam.

The eye can detect the difference between these two: cells in gas-blown foam are usually large and easily discernable; cells in syntactic foam are microscopic, giving the material a homogeneous appearance.

MOLD MATERIALS

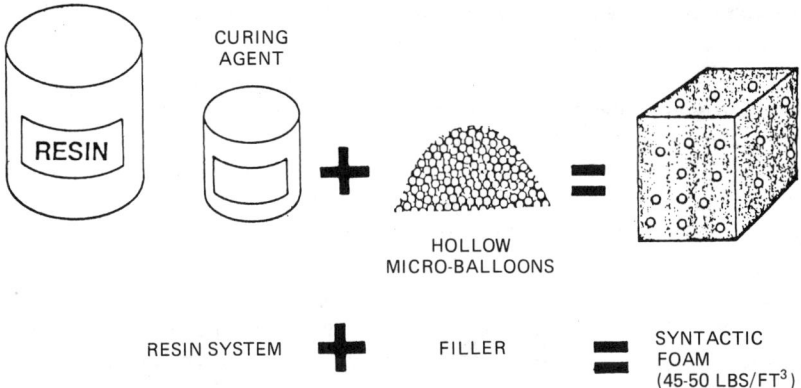

Fig. 4.11. Syntactic Foam System Components and Manufacturing Sequence.

In the making of tools, the fillers in syntactic foams are largely in the form of glass microballoons. These are used with casting resins to reduce both density and cost. For example, polyester resin filled with hollow glass spheres has become an extremely popular form of synthetic wood, closely duplicating the weight and strength of wood, and thus ideal for reproducing intricate furniture patterns.

In a parallel use, the spheres are added to SMC and BMC materials to lower the weight and cost of plastic moldings for automotive applications.

Since microspheres extend loading, a reduction in cost is possible in many formulations. Most often, it is the optimum balance of performance and cost that gives syntactic foam its edge over other systems. In time, because of its performance and economical advantages, syntactic foam should widen its applications range.

Specific properties. In terms of solid resins as well as foams, hollow microspheres seem to significantly improve the base material's compressive, flexural, and shear moduli, as well as its shear and compressive strengths.

Compared to other available fillers and fibrous reinforcements, the modification of matrix materials by microspheres tends to be more isotropic, which results in more even stress distribution under load. Moreover, spherical particles introduce less stress concentration than do oriented or irregular fibers. Thus, impact and fatigue strengths are also improved.

Since hollow microspheres act as path barriers for electrical and thermal flow, syntactic foams offer good insulation both electrically and thermally. Furthermore, since most matrix materials are polymeric and thus usually excellent electrical insulators, syntactic foam that employs conductive or coated microspheres can be formulated to be conductive electrically and thermally. To in-

116 ADVANCED COMPOSITE MOLD MAKING

crease the electrical and thermal conductivity, other types of conductive fillers may be added.

It is well to note that the CTE in syntactic foams is low, due to the reduced amounts of matrix material and the structural independence of spherical fillers against the expanding matrix material.

Mechanical properties of syntactic foams vary with the types of microspheres and resin systems used. Organic microspheres such as phenolic and epoxy are limited to the lower end of the temperature spectrum. Glass microspheres are mechanically sound, which means that the syntactic foam would be safe in any organic matrix formulation.

The specific gravity of syntactic foams ranges from 0.2 to 1.5, depending upon the type and size of the microspheres, as well as the type of resin and curing agents used. Where applications require high mechanical values, additional modification may sacrifice lower density values.

Ceramic microspheres can maintain structural integrity up to a temperature of 3000°F. The only suitable matrix material for such high-temperature devices would be ceramic itself, since ceramics are the strongest of all microspheres, thus making them ideal for situations in which high stress occurs.

It must be stressed that flow characteristics are important in casting thermosetting resins such as epoxy, polyester, phenolic, polyurethane, and polyimide (Fig. 4.12). Since hollow microspheres offer the lowest ratio of surface area to

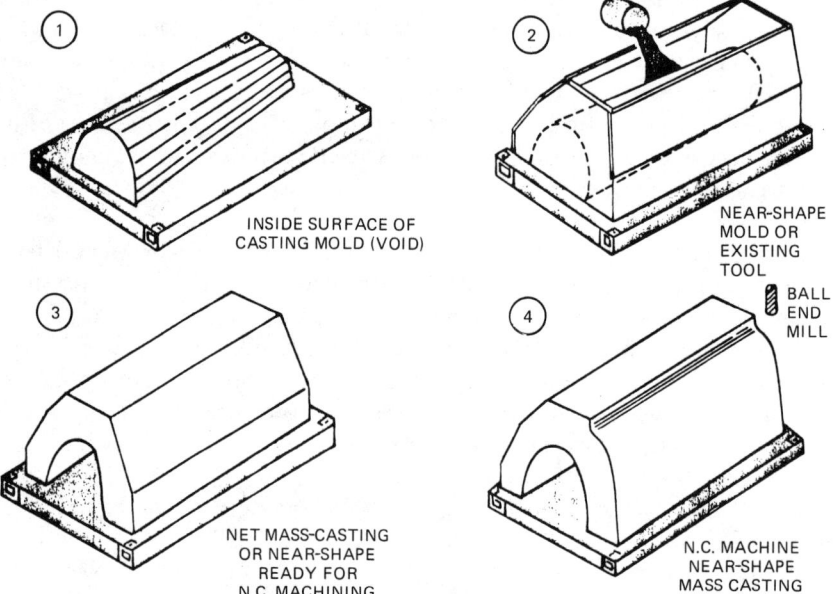

Fig. 4.12. Syntactic Foam Casting Sequence.

volume, they allow the highest volume loading with minimum surface area interference.

In terms of strength: New glass and ceramic hollow microspheres are four to five times stronger than earlier ones. In selecting materials for their specific strengths, shear, tensile, compressive, and flexural strengths should all be evaluated, in addition to the material's primary mechanical properties.

Specific uses. Aerospace. Since no other industry considers strength-to-weight ratios more critical, syntactic foam finds its widest use in the aerospace industry. Here, it is used to bond honeycomb and PVC/PUR (polyvinyl chloride/polyurethane) foam cores to aluminum and composite skins. It is also used as an embedment compound for threaded inserts in core materials such as PUR and PVC foams and balsa, as well as honeycomb.

Further uses in the aerospace industry are as a sandwich core, as a core-splicing compound, used to fill in between the pieces of core that are used as the sandwich fill structure in a part, and in the ablative shields of reentry vehicles such as space capsules and warheads.

The most significant present-day use of syntactic foam in the aerospace industry is the casting of it (in block form or net) to a contour to form cast-to-shape or N/C machined mold masters to form composites (Fig. 4.13).

A

Fig. 4.13. A, B, and C. Epoxy Syntactic Foam Used as Lightweight Laminate Tool Back Fill. *(Courtesy of Grumman Aircraft Systems)*

118 ADVANCED COMPOSITE MOLD MAKING

B

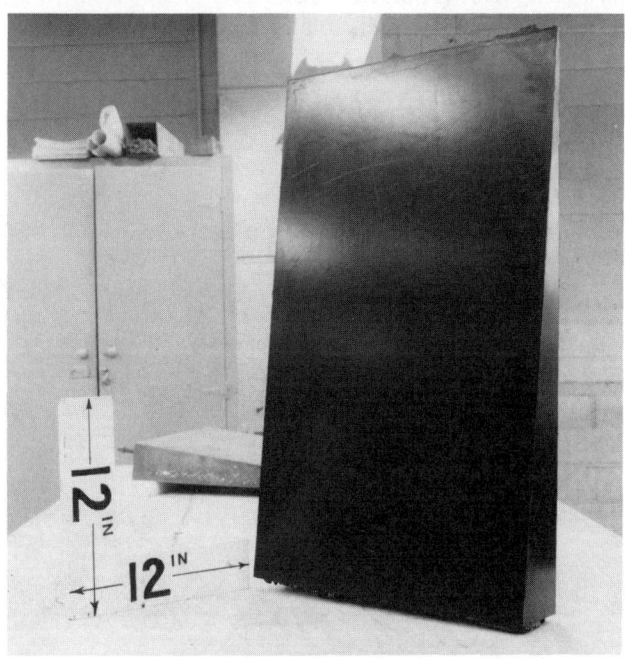

C

Fig. 4.13 Continued

Marine. In much the same way as it is employed in the aerospace industry, syntactic foam is used here for tools, mold masters, core lamination, core splicing, filleting, insert embedment, and as sandwich core. It has become a component in ship hulls, rudders, decks, and masts, as well as being utilized as structural flotation in hulls, buoys, cable floats, bulkheads, gunwales, and launch platforms.

A dramatic application of syntactic foam is in deep-submergence ocean vehicles. Since well-formulated foams can withstand hydrostatic pressures up to 131 MPa (19,000 psi), oceanic submersibles can probe to great depths with little concern for the collapse of on-board flotation for buoyancy control when fabricated from glass fiber and epoxy syntactic foam. Thus, hulls formulated in this way have been able to withstand the pressures of a submergence to 20,000 feet.

Medical. When syntactic foam is combined with fibrous laminates, it produces a product that is strong, stiff, and lightweight. Thus, artificial limbs, bone sculpturing, and instrument cases employ this material in their manufacture. In an ancillary use, medical helicopters are formed from these lightweight and strong products.

Automotive. The design of more lighter-weight, higher-strength automobiles, with stressed skins and unibody construction techniques, demands high-performance body fillers. Thus, epoxy syntactic foam, with its reduced density and increased toughness, has significantly improved the performance of body putty. Epoxy also bonds better to polyester/glass panels than conventional body filler. Since more cars today are fitted with weight-reducing composite body parts, it is increasingly important to employ a body filler that is similar in density and mechanical properties.

Future uses (of which development currently under way) of syntactic foam will include employment as an acoustical firewall, acoustical/vibration energy pads, structural core filler for panels, structural adhesives, and composite steering wheel cores.

Industrial syntactic foams are used for the production of molds for composites (Fig. 4.14), vacuum forming and compression molding, flame barriers of firewalls, metal foam using ceramic hollow microspheres and damming putty for welding. Cost reduction is achieved by using thermoset coatings to finish mold surfaces heavily filled with microspheres, since they have been shown to maintain a glossy finish due to the smooth finish of the microspheres.

Chemical Foams. Patterns and molds made from polyurethane foam have been used for some time for aircraft and aerospace applications. In this respect, checking fixtures, preform lay-up contour molds, structural parts of high temperature tools, and support substructures are some of the applications for chemical foams.

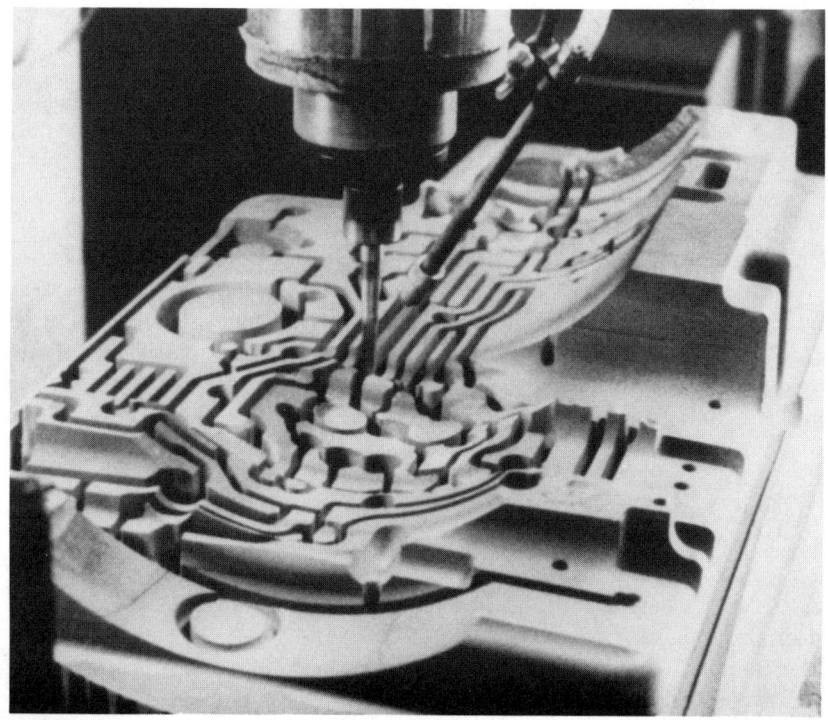

Fig. 4.14. A and B. Machined Syntactic Foam Tool. *(Courtesy of Ren Plastics Div. of Ciba Geigy Corp.)*

Recently, high temperature polyimide foams have been used as integral stiffeners on the back of laminate molds and tools.

In addition, rigid foam and the closely related isocyanurate foam are among the most efficient insulation materials used commercially, while high-density foams are typically used in structural parts.

The term polyurethane refers to the chemical compound resulting from the reaction of isocyanate and an alcohol. Polyurethanes fall into a number of basic chemical categories.

Urethane foams are prepared by the reactions of di- and multi-functional isocyanates with polyols and water in the presence of a catalyst. Other components necessary to bring about the desired properties include silicone surfactants, and frequently a blowing agent such as chlorotrifluoromethane or methylene chloride, which supplement the blowing reaction to enhance foam properties. Flame retardants and fillers may also be present.

Rigid, isocyanate-based foams include isocyanurate as well as urethane products. The processing, applications, physical properties, and insulation efficiency are similar for both types. Rigid urethane is specified for use in applications involving temperatures up to 200°F. Isocyanurate foams, often called trimer foams, can withstand use temperatures of up to 300°F and have greater inherent flame resistance than urethane.

Most rigid foam is made from PMDI (polymethylene polyphenyl isocyanate), though much urethane insulation foam is still made from a crude TDI (toluene diisocyanate) that is not distilled after manufacture.

In rigid foam systems, catalyst combinations of amines and organotins provide the desired reaction profiles. Recent catalyst development for molded systems has been in the area of delayed and heat-activated catalysts for improved flow characteristics.

Isocyanurate-modified rigid urethane foams are designed for improved flame retardancy.

Where structural and screw-holding strength is required, high density foams are water-blown. However, when high quality, decorative surfaces or insulation foams are needed, they are halocarbon-blown. If water is added to halocarbon-blown formulations, a compromise between these two aims can be achieved.

Catalysts for reaction injection molding (RIM) need to be fast for short reaction and demold times. Highly active amines such as triethylene diamine and low levels of an organotin are used in RIM systems.

Recent advances in chemical foams include improvements in MDI (diphenylmethane—4,4'diisocyonate)-based prepolymers to achieve easier processibility. These new materials offer a viable alternative to current TDI MOCA (methylene orthochloroaniline) systems, which have presented toxicity problems.

4.5.6. Adhesives

Although many plastics materials adhere to each other by the mere application of heat to the surfaces to be united, the use of adhesives for plastic-to-plastic, plastic-to-metal (or wood), metal-to-metal, and metal-to-wood attachment has largely replaced the older, time-consuming metal fasteners.

Because practically every type of metal and some types of plastics require a specific adhesive for bonding them to themselves or another material, adhesives have grown through a wide and expanding contemporary usage.

These adhesives are available in a multitude of forms. They can appear as solutions of film-forming resinous materials, as meltable stocks such as films or rods, as two-component packs of resin compounds and curing agents, or as powders which may be melted or dissolved in various solvents.

Since the chemical structure of an adhesive bears a direct relationship to its possible applications, it becomes logical to classify adhesives according to their chemical structures.

In general, it is convenient to break the classification of adhesives into two specific areas: flexible (or thermoplastic) adhesives and rigid (or thermoset) adhesives. The two categories, with their specific components are listed here.

Flexible (or Thermoplastic) Resin Adhesives

    Acrylic resins
    Isobutane resins
    Rubber-based adhesives
    Protein adhesives
    Cellulose acetate
    Cellulose nitrate
    Carboxylmethyl cellulose
    Ethyl cellulose
    Hydroxyethyl cellulose
    Silicone elastomers
    Polyamides
    Styrene resins
    Vinylidene copolymers
    Vinyl ethers
    Vinyl chloride copolymers
    Vinyl alcohol
    Vinyl acetate
    Vinyl acetals

Rigid (or Thermosetting Resin) Adhesives

    Epoxies
    Silicone resins

Alyds and polyesters
Phenolics
Polyurethanes
Ureas and melamines
Furanes

For thermosetting adhesives, the bonding of metal to wood requires two compatible adhesives — one applied to the metal surface and one to the wood. Once this is accomplished, pressure and heat are introduced to produce a strong, permanent bond.

There are five classes of thermoset adhesives: epoxies, polyesters, phenolics, polyimides, and silicones. The structural adhesives used are primarily of the epoxy type, although various blends of the above classes are often compounded to provide a compromise utilizing the valuable properties of each. Also, some patching, fairing, and trowelling compounds are composed of high temperature pastes.

Once cured, the resins are cross-linked into three-dimensional molecular networks that are insoluble and nonmeltable. This is why the materials most commonly used in the structural adhesive bonding of composite molds and tools are thermosetting resins. They are added to adjust the shear moduli of thermoset adhesives. In addition, they are miscible and co-react during the curing cycle to form inert impact resistant polymers.

For bonding wood and paper, phenolics, ureas, vinyl acetate, and casein are highly recommended. Phenolics are sometimes applied as a solution by a spreader before uniting the parts. Alternately, in powder form they are sometimes applied to the preheated surfaces to be joined. The resorcinol type of phenolics are particularly effective in bonding wood products such as marine veneers and weatherproof plywoods. In some of the higher-grade plywoods and veneers, the adhesive is applied as a film in a sandwich-style construction. In any case, heat and pressure are required to set or cure the phenolic adhesive, which creates a strong, rigid, water-resistant bond.

Packagers currently favor contact cements, which are usually a rubber-based adhesive in a solvent solution. Applied to the surfaces to be joined and allowed to dry before being pressed together, this adhesive not only forms a tight, flexible bond, but it can be used as part of a mechanized, fast, automatic sealing process.

The newest method of bonding, however, is achieved through the use of hot-melt adhesives. These are applied hot from rollers or contact brushes, after which the areas of application are pressed together to give an instantaneous bonding when the cemented area chills. A distinct advantage of this method is its elimination of the hazardous fumes of the solvents in contact cements.

As stated in the beginning of this section, the self-sealing qualities of some plastics eliminates the need for adhesives.

Vinyl, polyethylene, polystyrene, and nylon films are readily heat-sealed when aided by the application of a small amount of solvent on the surfaces to be sealed and then applying heat and pressure, followed immediately by cooling. To accomplish this in the packaging of plastic film and the joining of plastic pipe, elaborate semi- and fully automatic electrical heat-sealing equipment has been recently designed and is now widely utilized.

Furthermore, electronic or high-frequency heat, developed some time ago and currently in wide use to seal vinyls and polyethylene, remains the best method for rapidly producing a strong integral bond. The types of electronic sealing equipment currently in use are a bar sealer for semicontinuous operation, and a roller sealer for continuous operation.

Adhesives as Film Surface Coats. The structural adhesives used in the aircraft industry are now being used in the facing of prepreg composite tools. As a result of this discovery, tool material manufacturers have developed reinforced structural films to be used specifically as film surface coats.

In general, the adhesives are in the form of a semisolid precured B stage form. In this form, the adhesive is precatalyzed and cast or extruded on carrier scrims, fabrics, or veils, or onto a release film and aged to produce an artificial solid. The solid B stage material is rolled, interleaved with the release film, and stored on rolls at temperatures below $40°F$. When ready for use, these materials are thawed, applied to the molding surface, and cured by heat activation and pressure.

When the adhesive or film material is applied under the prepreg tooling material, it turns into a liquid and cures in front of the fabric-reinforced prepreg laminate to an insoluble epoxy surface.

Some adhesives are used as paint on surface coats, and as such can be used to repair molds and tool surfaces. The adhesives can be formulated as high-viscosity compounds, precatalyzed for use as a single component adhesive, or supplied with an appropriate curing agent. They are used to bond vertical tooling members and are adaptable to secondary bonding. It is preferable to use single-component adhesives where applicable, so that they cure with the entire tool structure. Two-component systems lend themselves to intermediate tool components and temporary molds.

Epoxy and modified epoxy adhesives are most frequently used in composite bonded structures. These adhesives are the most versatile and provide the best balance of properties required for joining composite tool components.

The reasons are precise: epoxies are easy to process, exhibit low shrinkage on cure, possess excellent wetability on many types of adherends, and provide high load-carrying capacity. The generally recommended glue-line thickness is 0.004–0.008 in. for maximum joint efficiency, although adhesive thicknesses of up to .125 in. have performed satisfactorily. Epoxy adhesives are somewhat degraded in strength at temperatures above 350°F. The maximum long-time service temperature of existing epoxy-phenolic adhesives is 400°F, or at temperatures below 260°F when saturated with moisture (approximately 2% by weight).

To prepare for bonding to a cured laminate or precured cast surface on a mold or tool surface, three methods have proven successful. The first is the peel ply method, where heat-sealed and scoured nylon peel ply material is applied to the B stage laminate in the area to be bonded. This material does not wrinkle easily and produces a suitable surface. The peel ply can be removed, providing a surface ready for bonding. The second method is grit-blasting with a fine aluminum oxide. The third method utilizes a hand-sanding operation followed by a solvent wipe and an air dry step.

4.5.7. Honeycomb Panels

Prefabricated stiffeners are used in conjunction with fiberglass epoxy laminates and epoxy casting materials for structural support purposes on molds and tools (Fig. 4.15).

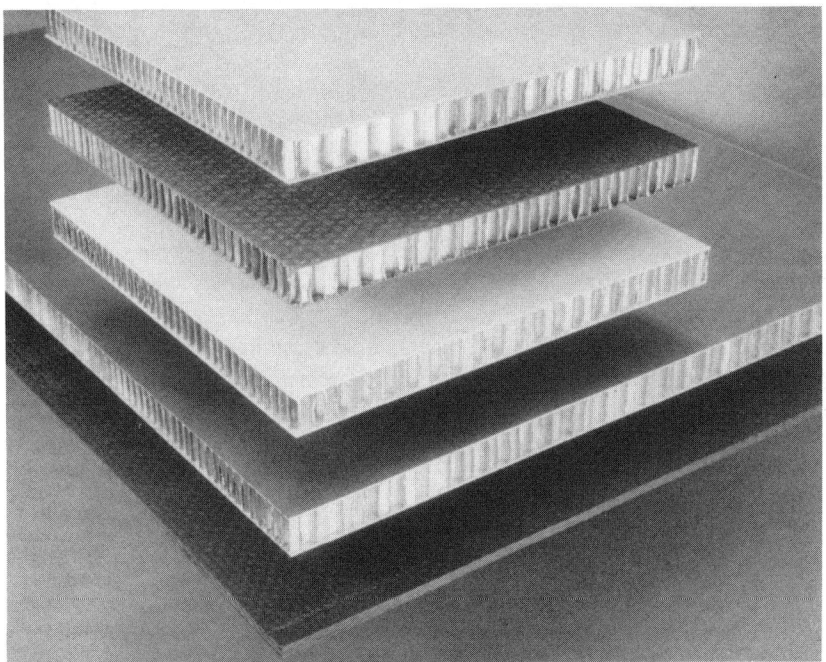

Fig. 4.15. Honeycomb Panel for Tooling Substructure. *(Courtesy of Hexcel Corp.)*

One of the most widely used of these is the honeycomb panel. Used for eggcrate back-up boards, these stiffeners are made from honeycomb flat stock consisting of three layers of fiberglass or graphite cloth for the skin with a perforated aluminum honeycomb as the core material.

The panels may either be purchased from outside manufacturers who use both prepreg or wet lay-up skins or fabricated in-house if the presses are available.

There are three basic kinds of aluminum honeycomb:

1. Fiberglass/epoxy faced with an aluminum honeycomb core.
2. Graphite/epoxy faced with an aluminum honeycomb core.
3. Aluminum faced with an aluminum honeycomb core.

In addition to honeycomb reinforcements for back-up substructures on molds and tools, recent advancements have shown that solid laminates of 1/4 in. thick fiberglass or graphite fabric-reinforced boards make a more positive structural support for the tool.

Fiberglass/Epoxy-Faced Honeycomb Reinforcement. These panels are commercially available and are normally constructed of 3/8 in. cell — a 0.003 in. expanded aluminum honeycomb core faced on both sides with two plies of type 1500 fiberglass cloth that has been impregnated with epoxy laminating resin. The panels are customarily sold in 4 × 8 ft sections in 1/2, 1, and 2 in. core thicknesses.

The main advantage of this type of sandwich panel is its high strength-to-weight ratio. Large laminating tools, where tool weight is an important consideration, demand this type of reinforcement.

There is great ease of use in fiberglass sandwich panels. They can easily be cut on a band saw, fitted, and attached to the back of a laminated tool in much the same manner as that used to apply solid laminated sheet reinforcements. No special tools are required under normal circumstances. Good adhesion can be obtained without other means of surface preparation, such as sanding the skins.

These panels are used extensively in the construction of master models built by the N/C machining method. Used as the supporting structure under the extruded low-density tooling compounds, they have proven to be highly dimensionally stable while providing strength and lightweight characteristics.

Graphite Sandwich Board. Graphite sandwich board is an engineered tooling back-up system. The core material used is aluminum commercial grade honeycomb, with a 3/8 in. cell size and a density of 3.6 lbs/ft. The facings are graphite with a minimum thickness of .015 in. The graphite facings are designed with a peel ply on both surfaces. After the peel ply has been removed, the surface of the board has an appropriate texture for subsequent bonding. The graphite facings are bonded to the aluminum core with a structural grade adhesive on a fiberglass scrim cloth which forms a barrier between the aluminum and the graphite. Also commercially available, the panels come in 4 × 8 ft sheets with an overall thickness of 1 in.

Also easily prepared, these panels can be cut to size with a band or skill saw fitted with a high-speed metal blade, carbide-tipped blade, or with a plywood blade. The chief features of graphite sandwich board are the following: long-term stability at ambient and elevated temperatures, excellent thermal conductivity and a low thermal coefficient of expansion, rigidity, strength, and light weight. There is, in addition, a strong convenience factor: A firm surface requires only the removal of the peel ply for the subsequent production of high-quality bonds.

Its major application is in situations demanding a flat, stiff panel at minimum weight. Thus, graphite sandwich board is used in back-up bonding tools, back-up systems for reinforced plastic check fixtures, master die model duplicates, custom back-up tooling structures, and multipurpose use on hydroplanes and yachts.

Aluminum-faced with an aluminum honeycomb core. Commercially available, these sandwich boards consist of a high-strength aluminum honeycomb core under aluminum alloy facings. The core material is aluminum commercial grade honeycomb with a 3/8 in. cell size and 3.6 lbs/ft density. The facings are 5052 aluminum with a thickness of either .020 or .025 in. The panels have an overall thickness of 1 in. and are available in standard 4 × 8 sheets.

As in graphite panels, cutting to size can be done with a band or skill saw outfitted with a high-speed metal blade, carbide-tipped blade, or a plywood blade. Aluminum-faced panels are rigid and strong with minimum weight, possess good thermal conductivity, and a long-term stability at ambient and elevated temperatures. They possess a unique quality in that they have an excellent bonding surface that requires only solvent wiping for subsequent production of high-quality bonds.

Aluminum honeycombs are generally used in applications requiring a flat, stiff panel of minimum weight. Since it has excellent dimensional stability with light weight, it is used in back-up systems for reinforced plastic check fixtures, master die model duplicates, and custom back-up tooling structures. It has also been used for wall panels in portable structures such as stages, shelters, and simulator enclosures.

Solid fiberglass or graphite fabric reinforced laminates. The most recently employed form of eggcrate substructure is a solid laminate board composed of epoxy resin in systems reinforced with fiberglass or graphite fabric and press

128 ADVANCED COMPOSITE MOLD MAKING

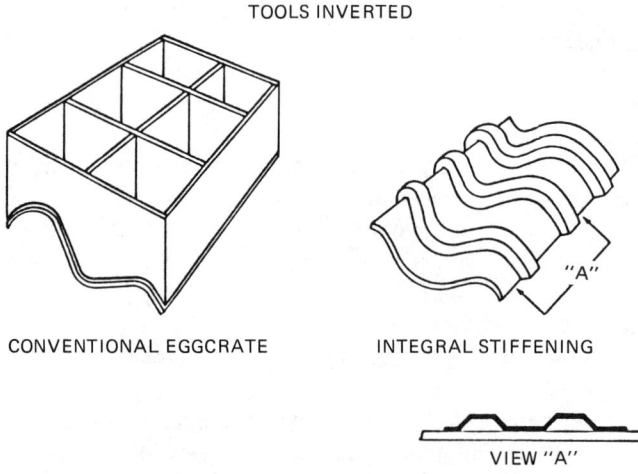

Fig. 4.16. Elimination of Eggcrate Structure.

molded into 1/4 in. thick 4 × 8 ft panels. These panels, made by wet lay-up or prepreg materials, match the physical properties of the face laminate of the mold or tool, and are therefore considered the best choice of a material for back-up substructures. This material is stronger than any of the honeycomb forms, and is more durable and long lasting. The cost is greater, of course, but by design, inclusion of lightening areas, and reduction of stiffeners, the material becomes the best of all possible choices.

In conclusion, although all of the choices demonstrated seem to indicate that one should use an eggcrate back-up structure, the recent trend is towards elimination of honeycomb and substructure totally. Integrally molded stiffeners, formed in the last stages of mold or tool lay-up, will be the main reason for the future elimination of the back-up structure (Fig. 4.16).

5
Preparation of Masters

Masters, models, and patterns are the names used to identify the totality of the shape, form, contour, and configuration of an object that must be created or duplicated. The methods used today to form these masters, models, or patterns can be classified into two general categories, 1) manual design and fabrication, and 2) computer-assisted design and manufacturing.

In either case, the master model maker, whether that person is a hands-on craftsman or computer designer, must conceive the same overall shape or form in his or her mind as the structural part designer who created that shape or form. This configuration could be the master pattern necessary to form a mold copy, or it could be the actual mold contour itself, which is used directly to form the finished part.

Since it may be difficult to produce the actual mold contour directly using the first method (manual design and fabrication), a model or pattern is usually required. Since the second method (computer-assisted design and manufacturing) provides automated means for designing and producing a mold contour directly, models and patterns are superfluous.

It then becomes obvious that the first method is quickly becoming outmoded. (The advantages of automated design and mold/tool fabrication lend themselves, in fact, to producing some of the necessary component parts of the manually fabricated model or pattern structure required for the first method). So, bearing this in mind, this chapter will review the technology of the first method, but will concentrate on the advanced state of the art by expanding and exploring the entire scope of the second method.

More must be stated about the visualization by the model maker of graphic presentations created by the designer of the object.

Since this understanding and visualization must often be accomplished in a three-dimensional form, the model maker is truly an artist — or at least a creator. From the basic line drawing of the manual design, the model maker must visualize, in space, a mold master that must be created manually, first as a concept, and then, depending upon the mold making process and finished part manufacturing method, as an object adjusted for normal compensating factors such as shrinkage, distortion, and shape change.

The mold maker must then imagine and realize how the formation of the mold form from the master's surface is used to form a cast, laminatated, or other type of metallic or nonmetallic finished part.

In addition, the model maker must be aware of all of the fabrication methods required to produce a particular kind of finished part.

A tall order for a human being, then — to be adept in machine practice, materials, and processes, and the resulting mold or tool usage requirements, and to have the imagination to visualize the mold, master, or pattern as the part.

In order to advance the state of the art for mold and tool makers, a shifting of talent from the engineering sector to the manufacturing sector has created a new and highly technical position in manufacturing engineering design; that of the mold and tool designer of today.

Once merely a skilled draftsperson, with a creative mind and hands-on experience, this individual has now evolved into the sophisticated, CAD-CAM mold and tool designer of the present and of the future.

Trial and error, once an accepted part of the master model manufacturing process, has been virtually eliminated. By the use of CAD and CAM, transformations and translations of engineering data, the choice of mold materials and processes, N/C methodizing, and even the proof of the manufacturing process before production, can be accomplished. In fact, by using automated mold and tool design, lead time and cost savings have become extensive.

Of course, major changes have occurred not only in the design technique area. Materials for master models have evolved as well. Plaster, the original material used for shaping, and high density wood, used for machining master patterns, are being replaced in every industry sector by machinable plastics, which have been precast in slabs and bonded together, or mass-cast in place.

These machinable plastics take on various forms. Some are light-weight chemical foams formed by expansion of the plastic resin system, upon curing, to reduce density. Some are formed by filling a resin system with light-weight fillers, such as hollow microballoons or bubbles. Others exist that need only to be poured near-shape and machined, or cast exactly to the finished contour. These state-of-the-art materials will now be transformed by the manufacturing processes as we proceed through each phase of master pattern preparation, fabrication, and pattern or mold finishing.

5.1. PREPARING MASTER PATTERNS AND LOFTING TEMPLATES

Master patterns are sometimes prepared from lofting templates, which are flat sheet metal sections of the finished contour that have been shaped around the outer edge either by N/C machining or by hand.

Although some shops still form their templates by handcutting and finishing of the edges, most templates are fashioned today by automation.

PREPARATION OF MASTERS 131

The automated method, similar to that used to automatically cut fabric part patterns, can also be used to lay out the lofting base so that the lofting template can be fit accurately in the base slots. Manual templates to transfer location data to the base can also be used.

If automated equipment or the base template is not available, the template base can be "laid out" by scribing the lines and locations directly into the base surface.

The holes for the attachment of wire screen and spacers can be drilled automatically or manually, with the templates placed in stacks. Holes should be spaced approximately 6 in. apart and approximately 1 to 1 1/2 in. from the template edge. Good judgment with respect to their location, number, and placement will result in a secure and rigid support structure. The following figure (Fig. 5.1) will suggest the form of the master pattern frame assembly.

The lofting template structure can easily be inspected while it is open and prior to the application of the wire screen. If inspection dictates changes, adjustments can be made at this point.

Once these adjustments have been made, the entire structure can be "tacked" in place, piece to piece, fastener to pieces, using a quick setting, thixotropic, polyester resin system paste material.

The wire screen, usually about 1/4 in. square mesh size, is now cut to fit just inside the templates, and outside of the threaded support rods. Usually, it is spaced off the support rods, using 1/2 in. thick blocks, cut to suit. The blocks,

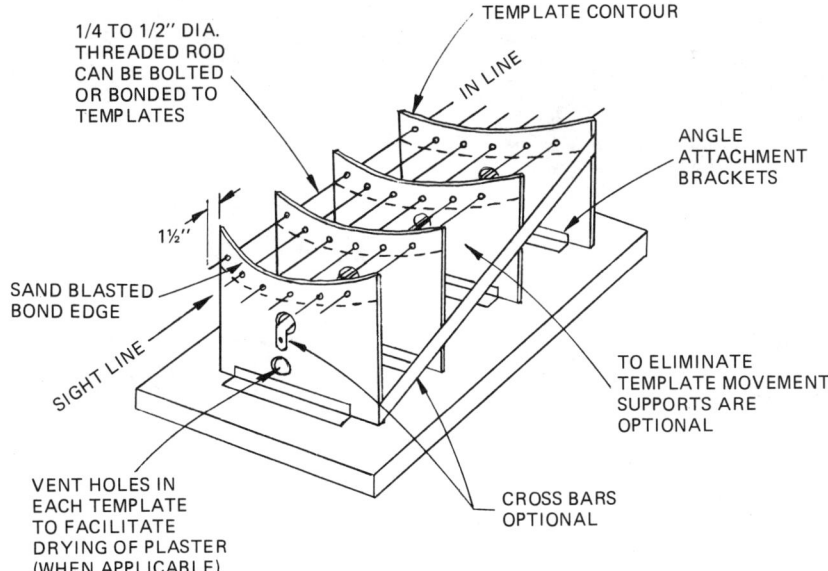

Fig. 5.1. Master Pattern Frame Assembly.

made of any nonmetallic substance (usually plaster) are wired between the screen and the rod.

Then, the wire screen is attached as required, using prefabricated spring clips or soft draw steel wire. Attachment locations will vary, but good judgment is the rule for location, starting about 2 in. inboard of the lofting templates.

The splining materials used to fill the space over the wire screen will vary according to personal or company preference. Some companies prefer a plaster combination composed of a rough coat, with or without plaster/bats, with a superfine coat above. Others prefer mixed plaster with plastic, achieved by blending mixed plaster with an epoxy resin system that contains a hardener (hydrophobic) which cures in the presence of moisture. Still others have applied plaster and splined with plastic. Finally, others prefer the plastic splined over plastic. (Plaster and other materials referenced in this chapter are covered in greater detail in Chapter 4).

Two coats (rough and fine), whether plaster or plastic, are usual. The amount of plastic or plaster that will be applied in the first coat will vary. A good rule to follow is to fill about three-quarters of the distance from the wire screen to the final spline edge on the first coat.

Once the application of the rough coat or coats has been accomplished, the spline step is performed. Using a splining tool fabricated from a piece of thin spring steel, the finished coat is applied to one section — and only one section — at a time. At least four lofting templates are picked up as guides during the spline pass (Fig. 5.2). This will assure that the space being filled will not be flat

Fig. 5.2. Plaster Being Splined or Screeded Over Lofting Template/Screen Substructure. *(Courtesy of Grumman Aircraft Systems)*

PREPARATION OF MASTERS 133

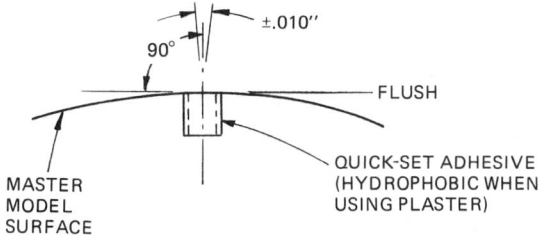

Fig. 5.3. Bonding a Bushing into a Master Pattern.

in contour. The reason for this is one of economy: more than one space can be filled at the same time only if the contour is mild or near flat.

After the application of splining material is set, dry, or cured, the reference and trim lines are scribed in the required locations. Hole patterns and any other information are transferred to the finished surface at this time. If the pattern has been made of plaster, the finished surface should be coated with several coats of sanding sealer or lacquer sealer.

The above procedures are suggested as a guide. There are numerous other methods that can be followed that will result in the same final master pattern finish.

When required, bushings that will assist in locating the "splash" or copy on the master pattern surface may be bonded into the surface of the master pattern. If this method is employed, a slightly oversized hole should be drilled and the bushing bonded in place with a quick-setting adhesive. Care should be taken to bond the bushing perpendicular to the pattern surface (within .010 in.) and at a predetermined distance from a scribed edge and centerline. Figure 5.3 illustrates the bushing placement.

5.2. MASTERS FABRICATED BY N/C MACHINING

Most of the lay-up molds used for prototypes or small orders are made from the plaster masters described in the previous section. Copies of these masters, called plastic-faced plasters (PFPs) must be made to provide the contour that will withstand a temperature high enough to cure the mold laminate resin system being made from the surface. These plastic faced plasters, which are castings of plaster over the shape to be reproduced, are usually made from a plaster "splash," which is a copy of the master model (Fig. 5.4).

The numerical controlled (N/C) master, on the other hand, eliminates a number of steps so that one can go directly to the laminate mold. An N/C machined master may be used to produce a room temperature (R.T.)/high temperature (H.T.) or 200°F precured, 350°F free-stand post-cured laminate mold (Fig. 5.5). The concept of N/C machined master models is, however, an innovation in model making, where master dimension data is available.

134 ADVANCED COMPOSITE MOLD MAKING

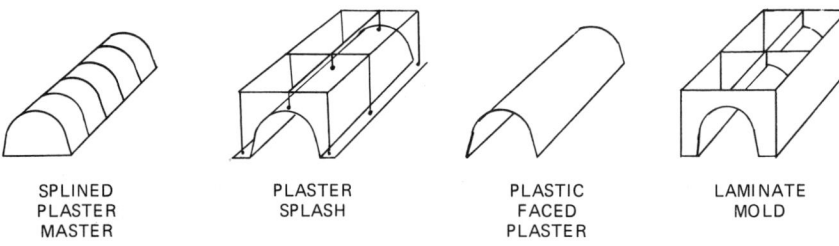

SPLINED PLASTER MASTER   PLASTER SPLASH   PLASTIC FACED PLASTER   LAMINATE MOLD

Fig. 5.4. Master Pattern and Mold Fabrication Sequence.

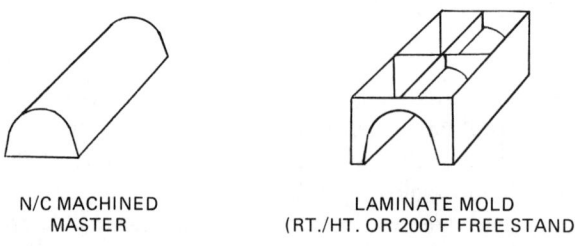

N/C MACHINED MASTER   LAMINATE MOLD (RT./HT. OR 200°F FREE STAND)

Fig. 5.5. One Step Master Pattern Method.

Undercuts, when necessary, can be formed by using wash-away plaster on plaster masters and molds. N/C machined master model undercuts can be formed by machining a breakaway section that will fall off with the removal of the laminate mold. The breakaway section can be pinned to the master with fragile plastic or wood pins (Fig. 5.6).

N/C machined masters have other advantages besides the one-step procedure that leads to a finished laminate mold. Plastic materials for precast slabs and

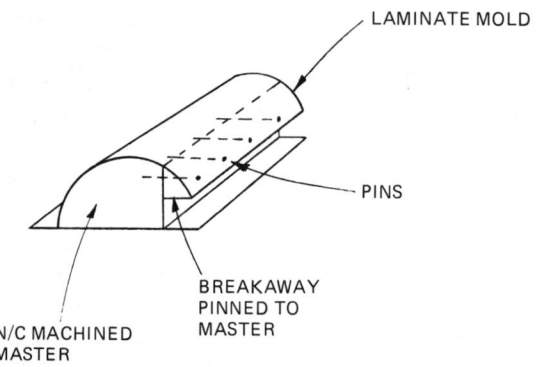

Fig. 5.6. Breakaway Master Pattern.

PREPARATION OF MASTERS   135

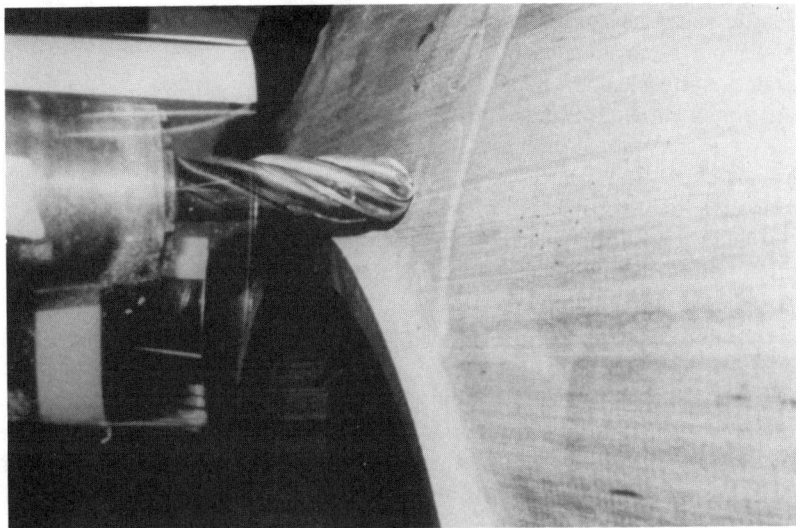

Fig. 5.7. Precast Machinable Bonded Model Stock as Automobile Master Pattern. *(Courtesy of Ren Plastics Div. of Ciba Geigy Corp.)*

near-shape mass cast master models can be machined at high speeds (60 to 200 in./min). Some materials can be machined to the finished dimension in one shot, cutting as much as 3–4 in. of material in the one pass. Figures 5.7 and 5.8 show a master model in the machining process, and the finished model.

Fig. 5.8. N/C Machined Precast and Bonded Model Stock. *(Courtesy of Ren Plastics Div. of Ciba Geigy Corp.)*

136 ADVANCED COMPOSITE MOLD MAKING

Plastic materials can be cut with no lubricant, and this is the preferred procedure. Properly formulated plastic master slab materials produce a curl rather than dust waste when machined.

As stated earlier, the innovation of N/C machined master models depends upon the availability of dimension data. In the case, when the coordinate data does not exist, CAD and CAM can be used to change a design from manual plaster to automated machining and to check or evaluate the design as it develops and evolves. Automated digitizing, the automatic recording of coordinate contour points is one such method of accomplishing this task.

CAD/CAM not only allows for ease of design, it increases accuracy by expediting a multiplicity of functions. Designs are proven in the computer, so time spent in prototyping and development of the master is reduced. Besides developing the loft data in the computer, the loft template data can be used to create a three-dimensional wireframe skeleton of the contoured master model. Even the machine path route of the cutter can be viewed, monitored, and readjusted prior to the production of the master. Tooling holes, substructure patterns, trim lines, surface data, and any other mold or tool information is automatically transferred to the surface of the master model. In the final analysis, improved use of materials and reduced waste become immediately apparent in the planning of the precast slab model stock layout.

Another major advantage of N/C machined master models is the fact that an electroformed plating mandrel can be formed in one step. Although the topic will be covered in detail in Chapter 7, the one-step forming of electroformed plating mandrels is directly associated with the machining of the master contour, and thus deserves to be noted now. Machining the contour of the master model is the same operation that is performed to produce the nickel electroform mandrel master for the production of a nickel-plated tool. To facilitate simple master model preparation, a mass-cast, near-shape, master model can be prepared more quickly than a bonded one. Simple plywood near-shape molds are used to form the mass casting (Fig. 5.9).

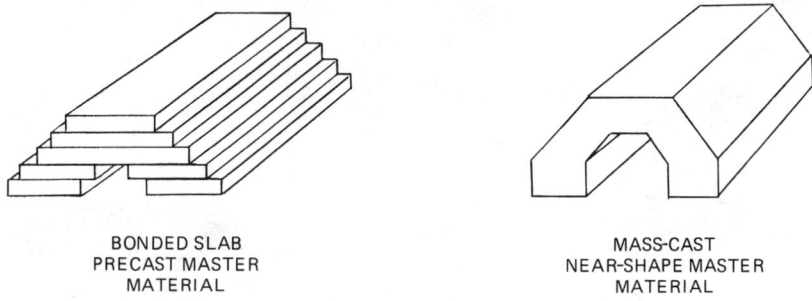

BONDED SLAB
PRECAST MASTER
MATERIAL

MASS-CAST
NEAR-SHAPE MASTER
MATERIAL

Fig. 5.9. Near-Shape Master Pattern.

A mass-cast master model can even be eliminated in some instances by properly formulating the mass-casting material for use as a forming tool, such as in metal stamping, a stretch-draw forming die, etc., and the contour that is then machined becomes the finished tool surface in one step.

The choice of the precast slab or mass-casting resin system should be a careful one. A system must be chosen that:

- has low shrinkage
- can operate as a master model at 200°F
- is not abrasive and is easily machined, drilled, tapped, cut, or sanded
- has good compressive strength (7,000–9,000 PSI is acceptable)
- has low water absorption
- has a hardness rating of approximately Durometer Shore "D" 65–75
- can be easily bonded or coated with finishes
- has a density of approximately 40 lbs/ft^3
- has a tensile and flexural strength of 4,000 PSI or more
- has a thermal coefficient of expansion of less than 20×10^{-6}

Various methods are used to support precast or mass-cast master models. Important considerations are that the plastic master should be supported by a very stable frame, and a copy of the master (another casting, lay-up, etc.) should be made immediately after the finished machining and surfacing finish operations have been completed.

Steel frames, plastic frames of fiberglass epoxy tube, fiberglass, and graphite epoxy skins with aluminum honeycomb, as well as various combinations of the foregoing, are constructed to support the precast slab or mass-cast master.

There is a difference of opinion regarding the construction techniques to be used when bonding together precast slab stock. The present bonding style appears to have evolved from the formerly prevalent use of wood as master model material. The impregnated wood was classically precut and bonded, as shown below (Fig. 5.10). The new, suggested method is shown on the right.

When bonding precast slab material to itself and to the base materials, it is important to try to choose a material that has a thermal coefficient that matches the slab material being bonded. If a urethane slab material is chosen, then a similar class of adhesive should be used. An epoxy slab material, for instance, would require an epoxy adhesive. Figures 5.11 and 5.12 show machined precast and bonded models.

At the time of bonding, precast slab materials sometimes require the squaring of bonding sides, sanding of bond surfaces, solvent cleaning of surfaces to be bonded, and the imbedding of mechanical fasteners between layers.

In addition, the use of surface fillers or fairing compounds to smooth and finish machined surfaces is sometimes necessary to provide an improved surface

138 ADVANCED COMPOSITE MOLD MAKING

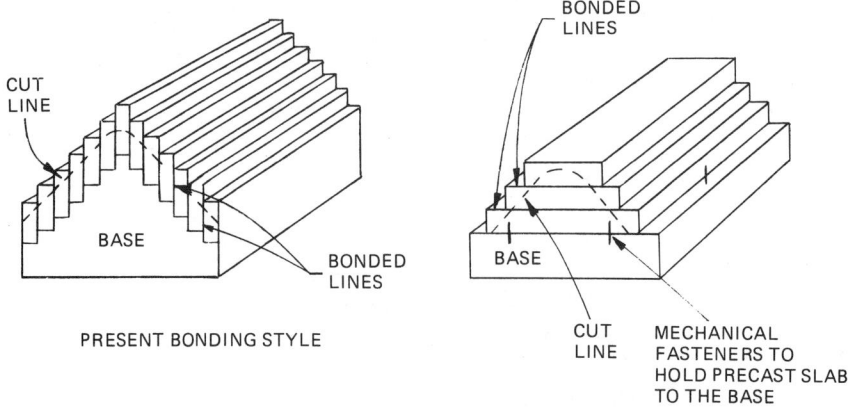

Fig. 5.10. Bonding Precast Machinable Plastic Model Stock Material.

Fig. 5.11. N/C Machined Precast and Bonded Model Stock. *(Courtesy of Ren Plastics Div. of Ciba Geigy Corp.)*

Fig. 5.12. N/C Machined Precast and Bonded Model Stock. *(Courtesy of Ren Plastics Div. of Ciba Geigy Corp.)*

upon which the intermediate laminate mold, cast copy, or plated reproduction can be made (Figs. 5.13, 5.14, and 5.15). These fillers are usually wet sanded, using 400 to 600 grit papers, following cure. It is important to try to choose casting compounds, adhesives, and coatings that are formulated to blend one to one by weight or volume, so that proportioning of the resin and curing agent is made easier for the mold or tool maker.

High density chemical foams in the density range of from 12 to 20 lbs/ft^3 are used to prove N/C machining tape programs and to show physical representations of CAD/CAM designed contours. The foams, which are usually foamed polyurethane, are used to form models and masters for stylists in various industries, such as the automotive, other transportation, marine, aerospace, and aircraft industries. This foam may also be used to form the near-shape contours against which mass-cast masters can be formed or an actual N/C machined master model itself (see Figs. 5.16, 5.17, and 5.18).

This foam is probably the most inexpensive material for master models available, and can be cast in place or machined from slab or bun stock. If cast, the material can be poured into wood, metal, or various nonmetallic casting or laminating materials, as well as various elastomeric molds. It is imperative to

Fig. 5.13. Finishing and Fairing of the Surfaces on N/C Machined Precast and Bonded Model Stock. *(Courtesy of Ren Plastics Div. of Ciba Geigy Corp.)*

Fig. 5.14. Finishing and Fairing of the Surfaces on N/C Machined Precast and Bonded Model Stock. *(Courtesy of Ren Plastics Div. of Ciba Geigy Corp.)*

PREPARATION OF MASTERS 141

Fig. 5.15. Finishing and Fairing of the Surfaces on N/C Machined Precast and Bonded Model Stock. *(Courtesy of Ren Plastics Div. of Ciba Geigy Corp.)*

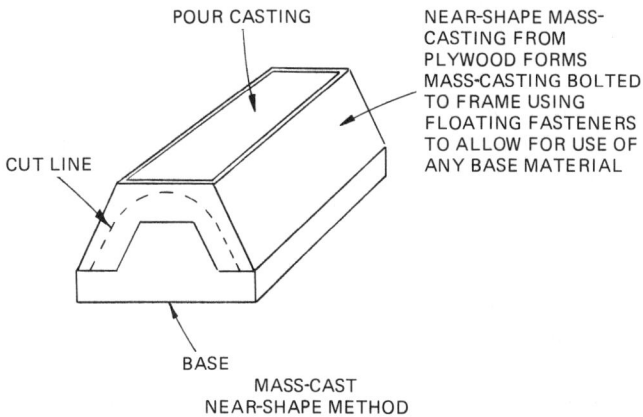

Fig. 5.16. Mass-Cast Near-Shape Master Model.

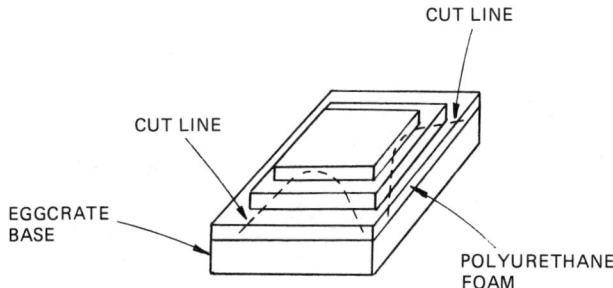

Fig. 5.17. Foam Master Model.

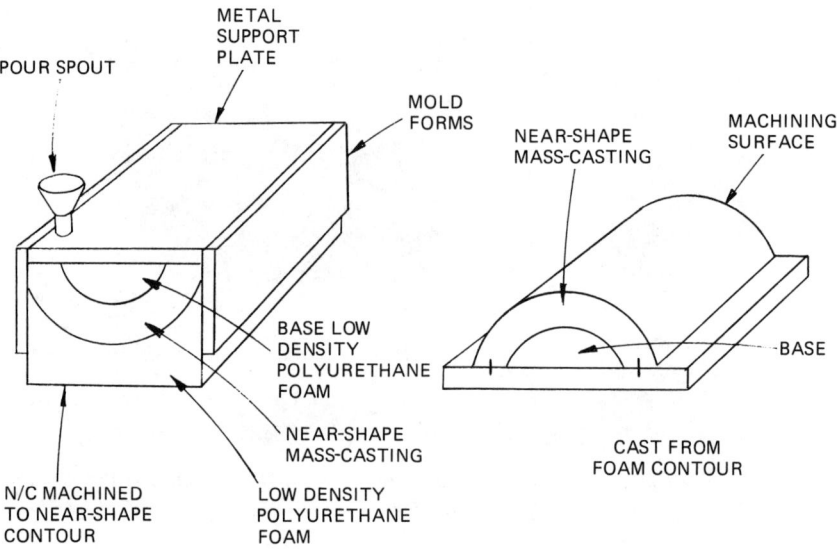

Fig. 5.18. Low Cost Mass-Casting Method.

choose a foam system that is safe for the operator or mold maker to mix and apply, since recent toxicity regulations indicate that some of the foam systems are unsafe unless properly handled. A foam system should also be safe for machining.

Another recently introduced approach to forming master molds involves the use of machinable wax as a master model material. Similar to machinable precast and mass-cast slab plastic material and high-density foam, the machinable wax can be used to proof CNC/NC tooling programs or the contour of any surface that should be examined before the production use of the generated machining program. One distinct feature of the wax material is that it can be melted and reused. In addition, it possesses the ability to be glued or machined at any speed, and, as a 100% petroleum product, it is relatively nontoxic, nonabrasive, and self-lubricating.

5.3. MASTER MODEL, MOLD TREATMENTS, AND RELEASES

Master models, as well as molds and tools, must be treated and sometimes coated with various sealants and releases to provide the necessary functional surface upon which laminates, castings, and duplications can be made. Besides assuring positive release of the master pattern and copy, the properly chosen release also eliminates mold cleaning, and produces parts ready to finish.

In addition, various coatings, films, releases, or barriers act as separators between the master pattern, mold or part. The reasons for the release to occur is that there has been an interruption or lessening of the surface polarity.

Releases come in many forms: solid, liquid, wax, and in aerosol compositions containing carnauba, silanes, silicones, plastic films, and paste waxes. The choice of the proper release system should be based upon quality reliability, technical service, unique qualities of the system, and cost. The systems that will be described here are not only necessary, but products for the manufacture of master patterns, molds, and parts could not exist if the release materials described were not available. Materials such as tapes, prepregs, film adhesives, etc. all rely upon release coatings.

It is important not to choose an inexpensive release agent just because it is inexpensive. The cost of releases, whether film or liquid, are insignificant when compared to the cost of labor to apply, maintain, and remove release materials or replace damaged molds. Removal — if possible — of a stuck part could be such a costly task that the mold, tool, part, or all three could be lost.

Releases should be chosen that, whenever practical, stay on the master model, mold, or tool forming the part. Releases that stay on the mold will not contaminate the part either internally as a laminate or casting or impair finishing of the part surface. A positive release of a mold, tool, or part can be affected by applying more than one release agent over another, forming a shear plane such as: sealer, coating release, 2–5 wax releases. Releases that transfer to parts could require strong solvents that might attack the parts during removal of the release. Solvents also play an important role in the cleaning of the mold or tool surface — a process that will be covered in detail on page 148.

To dramatize this: One of the basic ingredients in some epoxy strippers is methylene chloride. You can well imagine what this might do to a mold surface if improperly used. In addition, this type of solvent is also considered unsafe for the user.

General purpose waxes, petroleum jelly mixtures, and stearates are used for lower temperature master pattern release applications. The general purpose

144 ADVANCED COMPOSITE MOLD MAKING

releases, some of which contain fluoroethylene propylene (FEP) should be used below 350°F.

It must be pointed out at the very beginning that these solid or liquid coatings, as treatments and releases, affect the surface of the master, model, mold, or tool, and prevent adherence of the formed copy to the surface. Some releases are waxes, and as such, will not adhere to the material being molded.

Furthermore, it is important to note that transfer of a treatment or release may be cause for a failure, since the polymer mixture of the material being molded may be inhibited. Transference will, if the coating is a mold release, affect finishing the manufactured part. Thus, the surface of the master or model must be treated before the application of a release.

A good grade of quick-drying lacquer is suggested to treat various wood, plaster, or porous master or model surfaces. Automobile-type primers work well, and are also supplied in clear form when required. At least two coats are required before the appropriate release is applied. The first coat is usually thinned, so that it penetrates into the surface. In the case of gypsum or plaster, this first coat may be applied soon after the material has set. In the case of wood, the lacquer is meant to seal in and/or out moisture so that the wood grain will not swell and rise above the contour surface. Solid treatments like sheet wax or sheet putty are used to compensate for the thickness of the laminate or part materials. The sheet material should be applied to the thickness of the part being formed. The sheet material will then compensate for the space needed in the mold to contain the part material during the forming process. Sheet materials are applied over the sealed master.

The master pattern is mold-released on the exposed surfaces, around the sheet wax, only after the wax is applied. Sheet wax is sometimes used to define trim lines, locations outside of the waxed area or other reference information. The wax may be trimmed about .250 in. inside a reference line, after which the reference line can be transferred to the rigid laminate mold or tool as shown in Fig. 5.19.

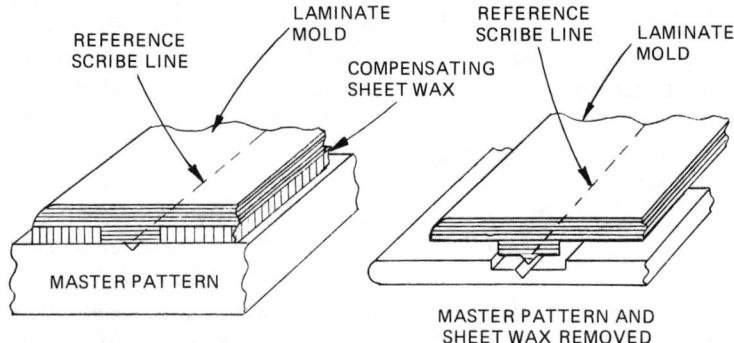

Fig. 5.19. Installing Scribe Lines in a Master Pattern.

PREPARATION OF MASTERS   145

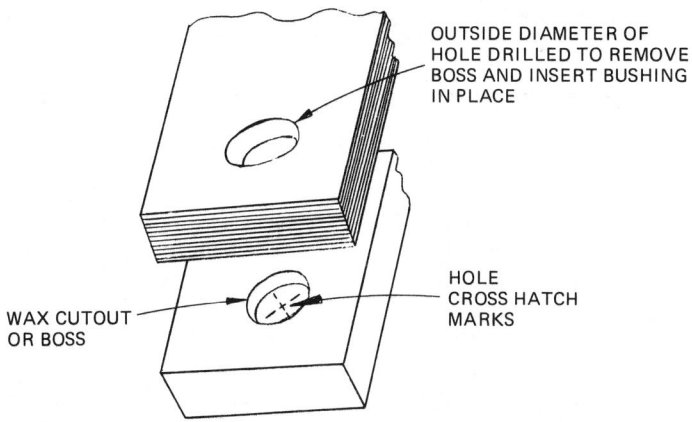

Fig. 5.20. Spotting a Bushing Location.

This step may be removed during the finishing of mold and tool edges. Variations of the above approach have also been used successfully.

An advantage of this method is the ease with which it allows for the spotting of holes to locate bushings that are not transferred by pins. The solid treatment wax sheet is cut out so that the laminate material can pick up the crosshatch marks of the hole location. If there is a hole already in the master pattern, it can easily be filled with an appropriate sealer, such as putty or clay. It is important to always make the cutout (boss) smaller than the hole to be drilled, so that the boss disappears during drilling (Fig. 5.20).

Before proceeding to specific releases, it is important to note that releases vary according to the material being used to duplicate the master pattern, and when the material of the master changes. The Frekote* subsidiary of the Dexter Corporation, Hysol Division, offers a handy chart as a reference, shown in Fig. 5.21.

As noted before, general purpose releases that contain fluoroethylene propylene (FEP) should be used below 350°F.

Polytetrafluoroethylene (PTFE) releases, on the other hand, work well to temperatures of 500°F and slightly above. Because of the fact that the dispersions are merely sprayed or coated onto the exterior surface, care should be taken to remove releases containing PTFE from the master pattern copies or parts being molded.

Silicone mold releases are useful to temperatures approaching 400°–500°F. Care must be taken in choosing such a release, since manufacturers' formulations differ from one another.

---

*Frekote is the reg. trademark of the Dexter Corp., Hysol, Div.

PRODUCT	PLASTIC: COMPRESSION/TRANSFER	PLASTIC: LAMINATING	PLASTIC: REINFORCED PLASTIC	PLASTIC: POLYESTER RESIN INJECTION (RTM)	PLASTIC: ROTATIONAL	MOLDING AEROSPACE COMPOSITES: GRAPHITE AND BORON COMPOSITES	MOLDING AEROSPACE COMPOSITES: FIBERGLASS LAMINATES	MOLDING AEROSPACE COMPOSITES: ADHESIVE BONDING	RUBBER: INJECTION/ COMPRESSION	RUBBER: TRANSFER	RUBBER: TIRE MANUFACTURE	RUBBER: RTV SILICONE RUBBER	OTHERS: POLYURETHANE SHOE SOLES, UNITSOLES HEELS, SKI BOOTS	OTHERS: EPOXY POTTING ENCAPSULATION, FILAMENT WINDING	OTHERS: DRY LUBRICANT REQUIREMENTS	OTHERS: URETHANE ELASTOMERS	OTHERS: FURNITURE/RECREATIONAL PRODUCTS FROM EPOXY RESINS, HIGH DENSITY POLYURETHANE SKINNED FOAM	OTHERS: ELECTRONIC AND ELECTRICAL	OTHERS: FOUNDRY
FREKOTE FRP			X	X															
FREKOTE 44	X	X	X	X	X	X	X	X	X	X	X							X	X
FREKOTE HMT	X	X							X	X	X							X	X
FREKOTE 33	X	X			X	X	X	X	X	X	X							X	
FREKOTE 700	X	X			X				X	X	X							X	X
FREKOTE 31					X														
FREKOTE 1711									X	X		X	X	X		X	X	X	
FREKOTE LIFFT					X				X	X			X			X	X		
FREKOTE EXITT					X				X	X			X			X	X		
FREKOTE NO. 1									X	X				X	X				

Fig. 5.21. Frekote Mold Release Selection Chart.

Silanes, which are non-silicones, are stable to 900°F. This allows this resin-type mold release to be used on cast ceramic, monolithic graphite, and other high-temperature master patterns. These machined or cast master patterns or molds can then be used to form advanced thermoplastic composites. Combinations of some of the above, together, have also worked.

Polyvinyl alcohol (PVA) (as a liquid coating or film), polyvinyl chloride (PVC) (as a liquid coating or film), polyethylene, nylon, polyester, and other thermoplastic films also act as releases.

Release films, papers, fabrics, and peel plies are another form of material that can be applied to master patterns or molds prior to producing a cast or laminated surface. It is sometimes wise to vacuum bag the release material itself down to the contour being duplicated, and then adjust out the pleats that form during the bagging step.

Peel plies are usually coated fabrics. Thus, the fabric style must be chosen so it conforms to various contours. Peel plies will, on the rear of a master pattern, laminate, or copy, provide a surface that need not be sanded for subsequent bonding, the ease of removal of this release fabric leaves the impression of the fabric in the molded laminate as it is removed, thereby forming a rough surface (Fig. 5.22).

Care must also be taken when choosing these fabrics, since some materials are not specifically designed for these applications, and so will not adhere properly to the master pattern laminate copy surface during pattern duplication. Some of the materials that *can* be used include fiberglass, nylon, polyesters, and other thermoplastic fabrics.

A release dispersed in the mold or part polymer is called an *internal release*. Internal releases are *not* recommended for the manufacture of master patterns

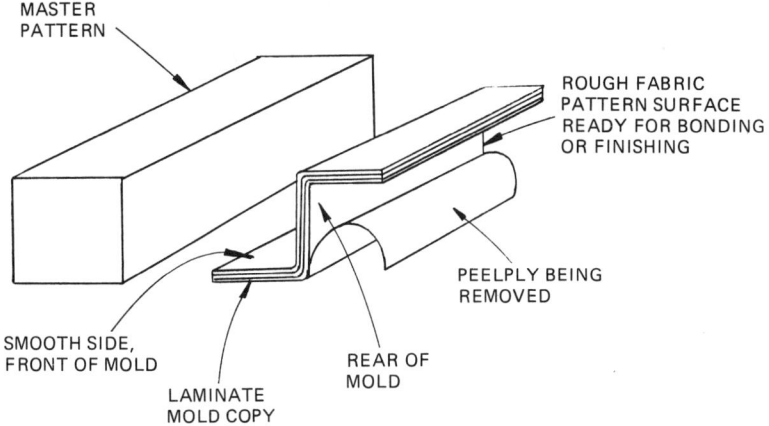

Fig. 5.22. Application of a Peel Ply.

or the laminate duplicates formed from their surfaces for a number of reasons, which includes the fact that subsequent repair/modification would be impossible.

In the use of metal master patterns and molds, fusible PTFE and mixed dispersions or blends have been used. Permanent releases applied to the mold surface are selected only in special applications, where the coating thickness will not interfere with the overall dimensions of the contour or shape being duplicated.

The external release, preferred by almost all mold and part fabricators, can also be a curable coating that may be applied by painting, brushing, or dipping. As with many other coatings, the safety of the user as well as those around the user is important. Government regulations affect the users of mass quantities of treatments, coatings, and releases containing solvents, so the solvent's or solid's content, amount of applications, amount to be applied, and other requirements relating to the evaporation of solvent carriers should be considered when choosing the release. Harmful solvents in the hands of the operator should be avoided at all costs.

It should be noted that routine and proper maintenance of the mold surface is critical to the life of the mold, as well as assuring that the parts being molded will not hang up and bond to the mold surface, thereby destroying the mold and part.

Cleaning of the mold surface after the application of the release is a primary maintenance function. Solvents, such as toluene, work well as releases for various oils. It is thus imperative to consult your release company technical service group to recommend the proper solvent for the release chosen and the surface to which it will be applied. General purpose detergents and other commercial and industrial cleaners work well on metallic mold surfaces. Again, the release to be removed will govern what cleaner will be necessary to be used to remove the release, or to prepare the mold for releasing.

Whether the mold is metallic or nonmetallic, cleaning can also be accomplished by using various abrasives such as plastic, glass, and natural blasting media. In fact, the surface of a mold or tool can be cleaned by blasting in an open shop area, using a piece of equipment that collects the blasting media while it is dispensing it (Fig. 5.23).

Cleaned molds, as well as new molds, sometimes require sealers in addition to releases. The sealers must also contain solvents that do not affect mold surfaces, since the sealer must sometimes fill microscopic voids in the mold surface and might also require an oven bake if the sealer is a coating that cures. Entrapment of an undesirable solvent and a subsequent oven bake might affect the mold surface.

Many of the composite reinforcements and fillers in resin systems *abrade* the mold surface in certain areas, removing the mold releases. These select areas may be touched up while the mold or tool is in service, if the proper release is chosen. Hot molds require the proper choice of solvent carrier type and a

PREPARATION OF MASTERS 149

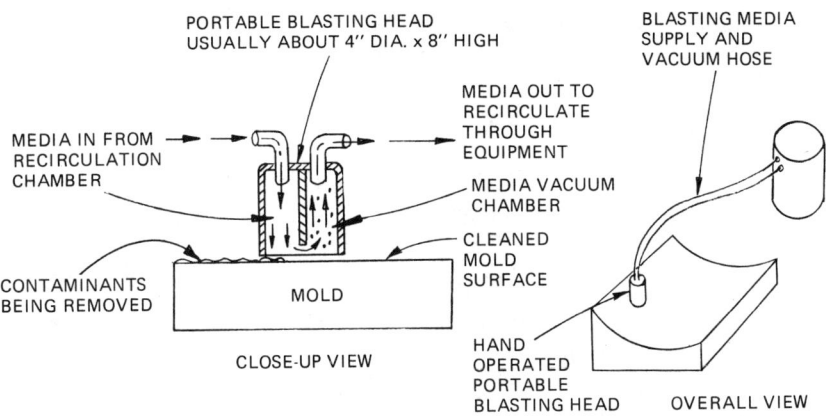

Fig. 5.23. Portable Vacuum Sand or Grit Blaster.

heated mold is sometimes required for special process cycles. An external release that is applied to a sealed mold and cures as a permanent coating to the mold surface is the best release choice.

Permanent release coatings do not transfer or leave transferable residues on a fabricated mold. For part forming, a permanent release applied to a properly prepared mold surface will provide many trouble-free releases before having to be reapplied.

Now, with the advent of high-temperature part materials such as the polyimides, bismaleimides, and the advanced thermoplastic resins and blends, temperatures of mold use in many instances reach 750°F. The thermoset molds and metal molds must be coated with nontransferable releases as well, so that the fabricated part will not be contaminated or adhere to the mold.

Because parts bonded to a mold must be physically removed, the original plastic mold may sometimes emerge extensively marked, gouged, or damaged. It is therefore imperative that the proper type of mold release, as well as a reliable manufacturer, be chosen to supply the required treatment, sealer, or release.

Some of the treatment and release systems offered by Frekote can provide that safety margin. For example, solvent solutions that contain extremely stable reactive materials which polymerize and bond intimately to a clean mold are available. Polymerization is usually completed after the release is applied to the mold surface.

Another characteristic of releases that is becoming increasingly important is thermal stability. Frekote high temperature mold releases are reported to be the only releases capable of use at 900°F. This makes them ideal for use over high-temperature cast and laminated ceramic and monolithic graphite (carbon) molds. Since these castable, machinable, and laminating-type ceramic mold materials are used at temperatures up to 3000°F and will eventually require releases in excess of 900°F, various industries will eventually require releases which work at temperatures in excess of 900°F.

The releasing of a cast or laminated ceramic surface will ultimately depend upon the material that will be formed over the ceramic mold or the type of process for which the ceramic mold will be used, such as the forming of thermoplastic composites, RIM polyurethanes, injection and/or rotational molding of thermoplastics, and various other high-temperature thermoset materials.

A listing of the most widely used releases available at this time is given below.

- Silanes as proprietary high-temperature curable coatings.
- Silicones as greases, resins, and fluids.
- Fluorocarbons as dispersions, sheets, films, and combinations with fabrics. (Note: some of these coatings are permanent.)
- Waxes as synthetic and natural blends of materials to mixtures of carnauba, paraffin, animal and vegetable bases, and microcrystalline origin.
- Metal salts as stearic acid (a fatty acid) and derivatives such as calcium, lead, magnesium, aluminum, and zinc salts. Lead is the least recommended, for reasons of toxicity.
- Inorganic compounds as powders in the form of crystalline flakes such as mica or talcum.
- Polyethylene as film.
- Polyamines as films, extruded or cast, which are resistant to solvents and heat.
- Polyvinyl alcohol (PVA) as a film or liquid, and as a liquid, usually in a water solution.

Although not release materials, the following methods fall into the category of releasing molds and/or parts, and are thereby applicable at this point. These methods are recommended to remove cast or laminate molds from a mold being duplicated or off the surface of a master pattern.

Release films. Release films of various types, according to the temperature of mold use, are applied along the edge or flange of a mold, but outside the area where vacuum bag sealant tape might be applied (Fig. 5.24). Release films can be applied to selected areas along the mold periphery, especially in complicated contour areas, because the purpose of the film is to act as a quick-release for breaking the vacuum between the surface being duplicated and the mold being formed. The release film can be applied to the selected area, which might be a tab that can be removed during the trim operation (Fig. 5.25).

Releasing wedges. Wedges formed from thermoplastic materials such as FEP, PTFE, and polyacetal material can be used to pry a laminate loose (Fig. 5.26).

PREPARATION OF MASTERS 151

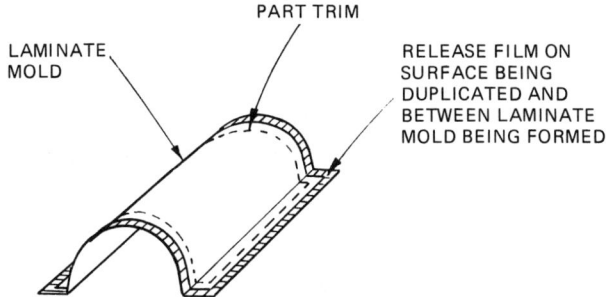

Fig. 5.24. Releasing a Laminate Using a Release Film.

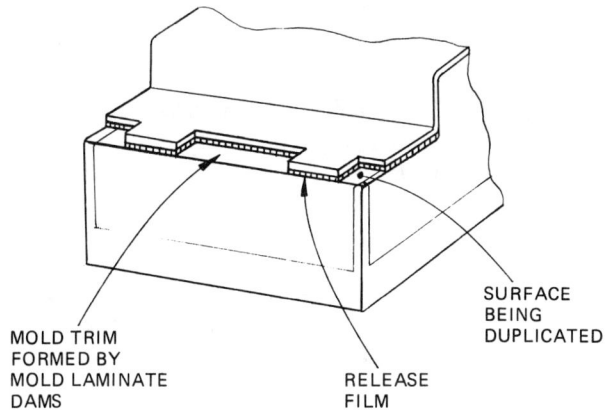

Fig. 5.25. Releasing a Laminate Using a Release Film and Tabs.

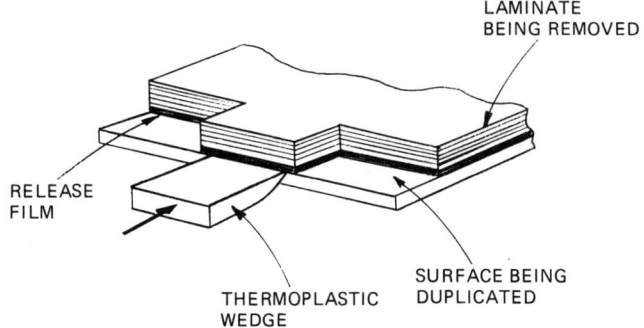

Fig. 5.26. Releasing a Laminate Using Wedges.

152   ADVANCED COMPOSITE MOLD MAKING

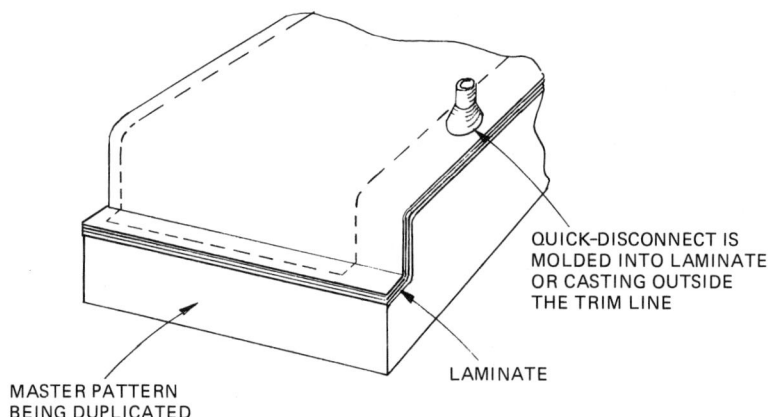

Fig. 5.27. Releasing a Laminate Using an Air Fitting.

Compressed Air Fittings. The same method that is used to mold in or laminate in a bushing can be used to mold in a compressed air quick disconnect fitting, so that low pressure air (less than 10 psi) can be used to release the vacuum, formed at the surface being duplicated (Fig. 5.27). This quick disconnect fitting can also act as a through-the-mold vacuum fitting for drawing vacuum on parts being fabricated.

Threaded mechanical lift. A bolt or large cap screw is laminated or cast into the mold surface (Fig. 5.28). Then, after the laminate or casting is cured, this bolt or screw is turned and used to break the vacuum or lift the mold off the surface being duplicated.

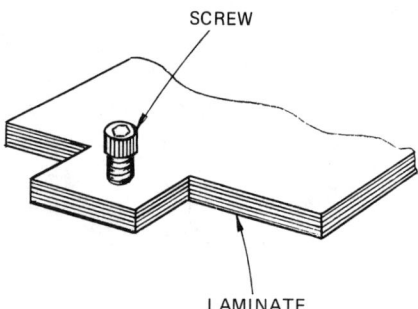

Fig. 5.28. Releasing a Laminate Using a Molded-in Bolt.

6
Fabricating Master Molds and Nonmetallic Molds and Tools

Now that we have investigated the materials available to fabricate masters, molds, and tools and discussed the processing of master patterns and models, it is time to review, establish, and project the state of the art in the fabrication of the master or intermediate molds and nonmetallic production molds and tools.

In some cases, the intermediate mold must be expended during the production of the nonmetallic production mold. It is for this reason that less expensive materials and processes are employed to produce these copies. Plastic faced plaster (PFP) master molds with the front face of plastic (epoxy), reinforced or nonreinforced with fabric, is one form of inexpensive mold. This mold copy is similar to the plaster splash, associated with plaster masters, and will be discussed in detail in this chapter.

Other forms that will be discussed include versions in which the plaster is mixed with plastic (epoxy) and is applied as a backing to the mold or tool. Polyurethane and epoxy syntactic foams, formed by filling a resin system with light-weight hollowsphere fillers are also available precatalyzed and frozen or prepackaged for easy application and cure at room temperature.

Another aspect of state-of-the-art fabrication involves in-house processing of impregnated mold materials that are normally purchased as prepregs.

Elimination of some of the usual materials, like surface and gel coats, elimination of process equipment through unique material formulations, and even the fabrication of production molds directly from the master model will be demonstrated.

Mold and tool support structures, sometimes defined as "back-up structures" will be compared by cost reduction techniques to the elimination and support of a mold or tool laminate by integral stiffening.

Sometimes, substructures are essential, and therefore must be attached to support the laminate. Thus, the technique of preparing and attaching these substructures will be presented.

Back-up materials such as honeycomb panels, solid laminate panels, and even combinations of metallic and nonmetallic tube frames will be studied.

Adjustable molds and tools using the substructure as an adjustment support frame and elastomeric "bag side tools" will be an interesting topic for those who design molds in which the finished contour cannot be projected, or both sides of a molded laminate must be finished after molding.

Finally, the discussion of metal-faced plastic laminate molds and tools will complete this section.

One very significant mold-making technique which should be considered when choosing the mold-making material and process is the use of extremely high temperature ceramic mold materials that can be cast directly from the master model, be formed into a laminate using ceramic woven fabrics, and even be constructed to be the master model and production mold all in one. Carbon-carbon (graphite monolithic) block is also another form of high-temperature material, but it has its limitations in that the cost is high, it cannot be cast to shape, and it usually requires machining by experienced carbon processing companies. Its one major advantage appears to be that it will produce a mold exactly to the dimensions machined into its surface.

It might be helpful to note at this point that this is the first time that a grouping, in concise format, of all of the materials and processes about to be presented has been offered. This particular presentation, then, is designed not only to allow the designer, fabricator, procurement individual, administrator, or employee to review, compare, and choose an advanced or state-of-the-art mold-making material or process, but also to assure quality results and extensive cost reductions in the choice of that material and process.

6.1. MASTER MOLDS

Master molds are formed through the application of a number of materials and processes. The choice is one that involves cost in almost every case, although sometimes cost is overruled by the need for extreme accuracy or some other technical design consideration.

As of this date, the basic master mold is still the plaster-faced plaster (PFP) mold. Soon, however, the N/C machined master model, a mass-cast mold or laminate, when temperature resistance allows, will become the basic master mold. The PFP formation process involves some of the following basic materials:

- Plaster or gypsum cement
- Uncarded long-fiber hemp, in bales or bats
- Fiberglass fabric reinforcement (in some cases)
- Epoxy coating to form a plastic face
- Frame and/or back-up materials, such as wood or metal

The PFP or PPP (Plastic, backed with plaster mixed with plastic-epoxy) mold is the copy of the master model and splash mold. As such, the PFP or PPP mold acts like a master mold made from master models formed from plaster, N/C machined model stock, or even an existing mold contour.

These molds, which have a plastic face, or are made from plaster alone, are used to form duplicator or Keller dies and production laminate molds. Duplicator or Keller dies are the master patterns from which dies of various kinds (foundry, autoclave, etc.) are made. The surface of the Keller copy is automatically traced on equipment such as a pantograph machine or trace mill.

The PFP mold is also the master mold used to produce a high-temperature laminate mold or the intermediate laminate mold that will form the production mold or tool.

One recent method used very successfully to prepare accurate nonmetallic mold masters is the preparation of a near-shape laminate (such as glass or graphite epoxy) contour, vacuum bagged or autoclave processed, room temperature or 200°F cured prepreg. The near-shape, supported laminate is then machined accurately to the finished master dimensions, and usually the "rate" molds or production molds, in the form of a laminate, are made from the laminate surface.

Mass-cast, N/C machined master models, close to the dimensions to be machined (usually .125 to .150 in. away from the finished machined dimensions) are suggested for the preparation of the master model.

The master model material chosen should be inexpensive, easy to machine, have a known and consistent CTE, and be capable of being cast into large volumes with little shrinkage.

The steps in this process are illustrated in Fig. 6.1.

When a simple contour must be formed, rolled metal approximately .125 to .150 in. thick can be used to form the near-shape surface upon which the near-shape laminate will be formed, prior to machining. This is illustrated in Fig. 6.2.

Cost and economic factors will be discussed shortly, but more significant in the choice of mold and tool materials and processes is the fact that the new thermoplastic composite materials are formed at such high temperatures that machinable or castable ceramic materials are by far the wisest choice for the master and, in fact, the production mold. In this case, specifically, ceramic's low CTE and part accuracy surpass all the other mold materials available.

6.1.1. Economic Factors

The cost of producing master molds involves both metallic and nonmetallic materials. In the formation of graphite epoxy laminate parts and bonded assemblies, the master mold can, in some cases, be used to form the finished part.

A comparison of the weight per cubic foot in all materials choices is a logical and easy way of relating weight, density, cost, and other factors. For instance, steel, the material used for many composite molding applications, approaches 500 lbs/ft^3. Thus, not only is steel heavy and difficult to machine, its cost

156 ADVANCED COMPOSITE MOLD MAKING

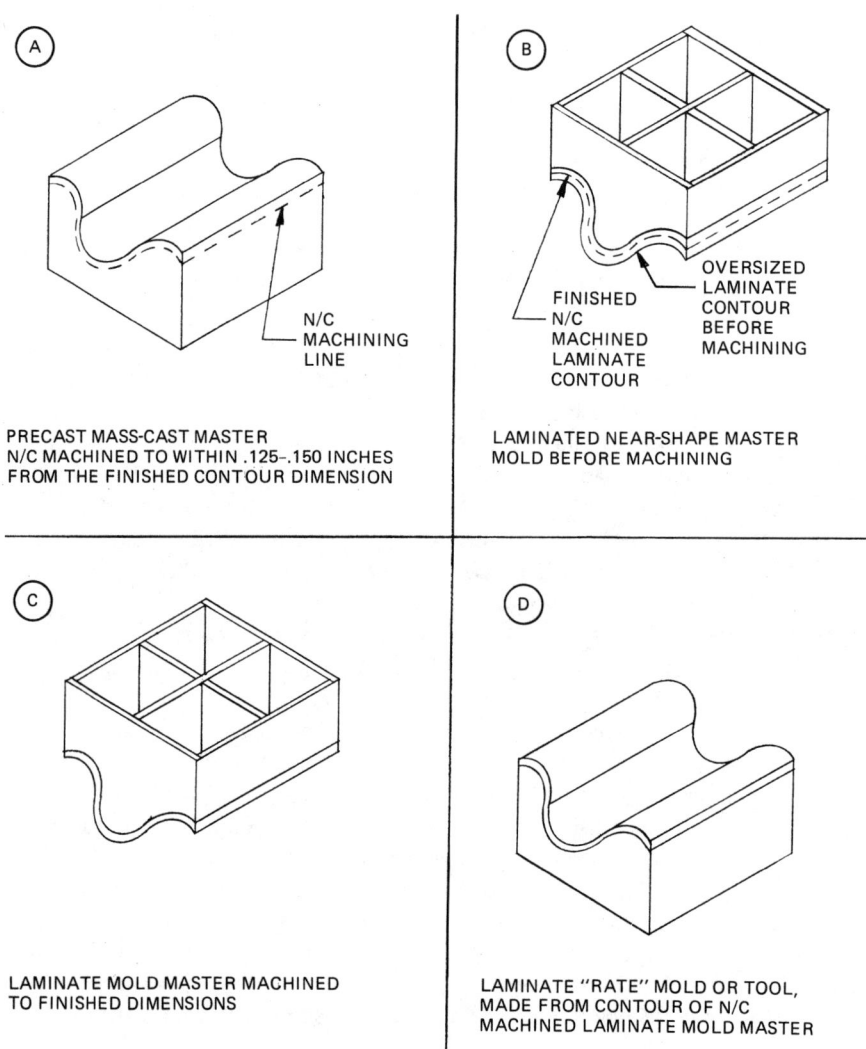

Fig. 6.1. A, B, C, and D. Machined Laminate from N/C Machined Mass-Cast Model.

could be prohibitive. Aluminum, on the other hand, although light and easy to machine, expands nearly 31 times more than cast ceramic. Also, the cost of aluminum and steel range from 3 to 5 times that of ceramic.

Economics thus leads us to choose only a select few of the many materials which can be used to form composite parts. The currently most economically feasible materials for forming composite parts are:

FABRICATING MASTER MOLDS AND NONMETALLIC MOLDS AND TOOLS 157

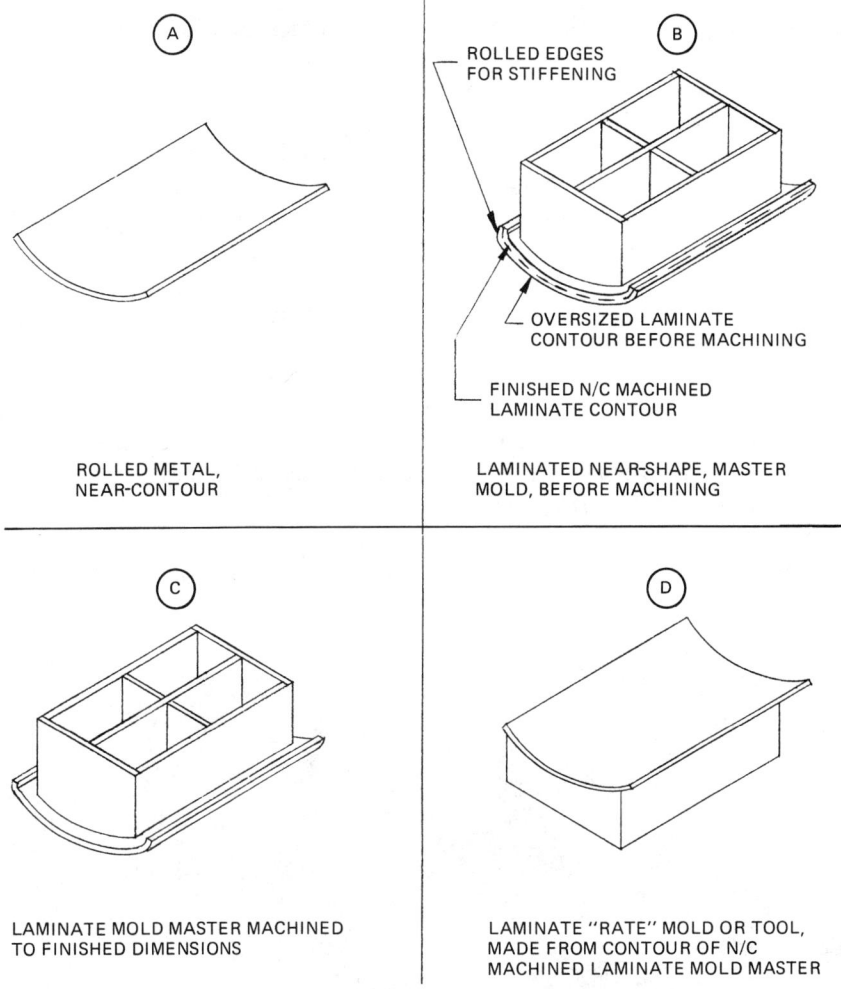

Fig. 6.2. A, B, C, and D. Formed Laminate from Near-Shape Rolled Metal.

- Room temperature and 200°F cure mold and tool prepregs, which, in some cases, are fabricated directly from the master model.
- Nickelplated, electroformed, or deposited over laminate mold forms. (The most economical form of this is done over N/C machined model stock.)
- Machined or cast ceramics and cementitious compounds.
- Carbon/carbon (monolithic bulk graphite).

158 ADVANCED COMPOSITE MOLD MAKING

Wet lay-up epoxy systems reinforced with fiberglass and graphite fabric, metals, and other master mold materials are not shown because economic factors dictate that these materials are feasible only in select cases and should be selected for reasons other than economic ones. Figure 6.3 illustrates a typical high-temperature fiberglass epoxy laminate wet lay-up mold.

Inasmuch as the above materials can be processed fairly inexpensively, they lend themselves to both short and long runs, as well as prototype applications. In fact, a number of molds can be produced for the same cost as one mold made from more expensive materials.

Other factors in addition to basic material cost figures enter into economic considerations.

Regarding the higher temperature, low CTE materials, monolithic graphite block can be machined to contour, and a ceramic material can be cast directly on the machined surface. This allows for a matched mold directly from a machined mold surface, thereby eliminating the need for a master model, and even for master mold fabrication.

The ceramic and machined graphite molds can be strengthened by mounting them on base plates of graphite-reinforced epoxy laminate, or by forming a

Fig. 6.3. Wet Lay-up Mold Made from High-Temperature Fiberglass Epoxy Laminate Material. *(Courtesy of Grumman Aircraft Systems)*

FABRICATING MASTER MOLDS AND NONMETALLIC MOLDS AND TOOLS   159

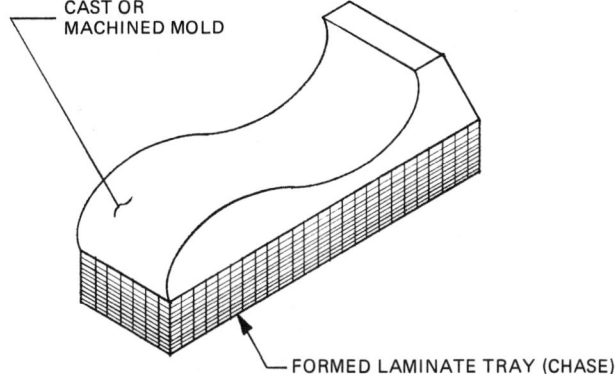

Fig. 6.4. Formed Laminate Tray to Hold Machined or Cast Mold.

tray around the outside, using a tooling prepreg on a laminate (around ceramic castings) formed from ceramic material with ceramic fabric reinforcement (Fig. 6.4).

The reason that the electrodeposited nickel material approach is economical is that the new master model materials available allow for the formation of plated contours directly from the machined master. Mass-cast masters also yield near-shape master molds and the machined master model or form can be used to form a mold or tool using a room temperature or 200°F cure prepreg. The same master can then be used to form the nickel electroform tool.

Using fiberglass and, when required, graphite fabric prepregs is economically advantageous because the 200°F cure systems can be formed over PFP surfaces and directly over the machined master model when the master material can withstand the temperature and the CTE of the model material is acceptable.

The least expensive approach to forming the master model and/or mold is to form a laminate using a room-temperature curing, free-standing post-cure prepreg. This type of material (the least expensive laminate material) does not require autoclave pressures to process, and cures at room temperature.

Master models can actually be formed from any material, even wood. The CTE is not important, since the master model is never exposed to elevated temperatures. An intermediate master mold is not required and the mold laminate can be formed directly over the master model.

Backup structures, which are two-thirds the cost of high-temperature laminate molds and tools, can be eliminated by molding stiffeners onto the back of the mold laminate. These stiffeners are molded during the last stages of the formation of the laminate face. This type of reinforcement, called integral stiffening, can and will replace the costly honeycomb and solid laminate substructures and reduce both the labor and material costs when fabricating laminate molds and tools.

160 ADVANCED COMPOSITE MOLD MAKING

Finally, economic factors are only important if the forming accuracy can be maintained, tolerances met and held, and the quality of the finished mold or tool assured. As mentioned with respect to mold sealing and release systems, the cost investment into the production of the finished part should be judged by what is required to produce only the proper quantity of quality parts, and no more than this.

6.1.2. Selection of Materials

Once the economic factors have been considered, it is time to move on to an investigation of the selection of materials, the fabrication techniques, and the tolerances that can be used to form the various master or part forming molds.

The two methods used to form the PFP are the formation of a solid plaster cast, or utilizing the splash cast technique. In addition, a coating of epoxy is usually applied by a double-coating method, and the epoxy is cured using a hardener that cures in the presence of moisture (hydrophobic). One coat of epoxy is applied to the mold-released master pattern and allowed to gel. The second epoxy coat is applied while the first is still wet, and the neat plaster is either solid cast or splash cast over the surface.

In fact, most contoured or large surface molds or tools are splash cast. The splash casting is formed by pouring about 1/4 to 1/2 in. of neat plaster over the uncured epoxy-coated surface and then allowing the plaster to gel or approach the plastic stage, at which time another mix — usually formed by soaking and impregnating fiber bats 1/2 in. thick and 6 to 8 in. in diameter with plaster, and placed over the entire PFP surface — is applied. Then, the PFP is usually attached to a prefabricated metal or wood frame.

For many years, wood has been used for frames, but metal is by far the better choice. However, if wood must be used, kiln-dried wood is the only kind to use, and the wood frame must never be imbedded in the pattern face, since the moisture from the plaster will swell the wood. Two to three coats of lacquer should be applied over the wood. Then, it must be tied to the PFP face, using fiber hemp bats impregnated with plaster or gypsum cement.

Metal does not have to be prepared. 1 to 1 1/2 in. diameter black pipe is recommended for the above process. It should be tied to the PFP face in the same way as wood (Fig. 6.5).

Sometimes, a welded frame forms the frame or backup structure for the PFP. In this case, the frame must be normalized prior to attachment to the PFP face.

The hydrophobic hardened epoxy surface coat (plastic) is sometimes reinforced with up to four plies of a ten-ounce plain weave fiberglass and, in some instances, graphite fabric impregnated with hydrophobic hardener/resin system. Then, neat plaster is applied as in the reinforced version.

The cast can then be plain plaster and sealed, unreinforced plastic-faced plaster, or reinforced plastic-faced plaster.

FABRICATING MASTER MOLDS AND NONMETALLIC MOLDS AND TOOLS   161

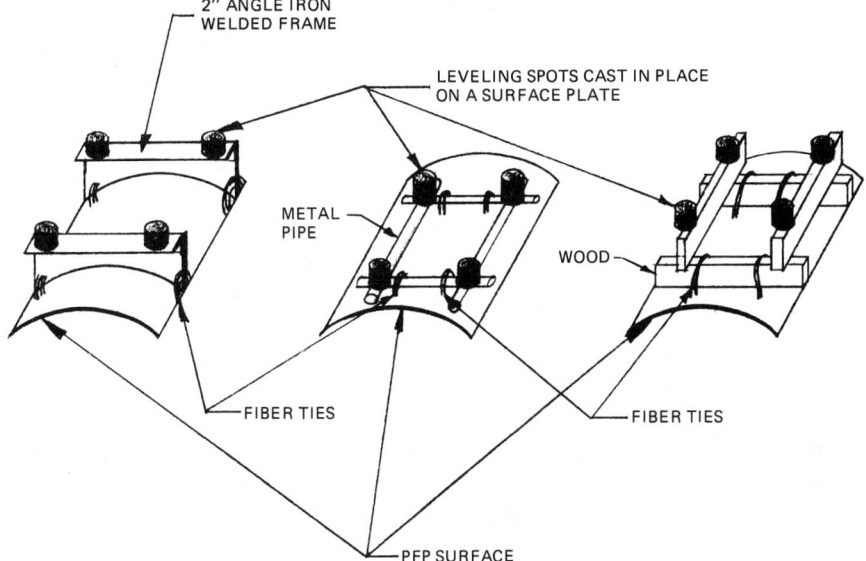

Fig. 6.5. Various Styles of Metallic or Nonmetallic PFP Backup Frames.

An additional construction technique involves the mixing of the hydrophobic hardener/epoxy system with plaster and the plaster splash or cast is then plastic-reinforced. In other words, a plastic and plaster (P and P). Vacuum integrity in the master mold, when the master mold is formed from plaster, however, can be assured to a greater degree if the plaster master mold is a PFP, and not just sealed plaster.

Yet another inexpensive method of forming a master mold is by backing up a room-temperature laminate with "pour-in-place," quick-setting polyurethane or a ceramic syntactic foam. After the two- to four-ply laminate over the surface coat is set, a two-part castable chemical or syntactic foam is poured or a froth foam system is sprayed over the rear of the laminate to stiffen the master mold surface. Sometimes the foam is covered with a few more plies of fiberglass-reinforced lay-up material to form a sandwich foam/laminate structure. This not only reduces the cost to produce the master mold, it also reduces the weight of the mold when very large molds are required.

Another form of inexpensive master mold can be built by producing a composite of a paint-on surface coat backed up with a premixed, frozen syntactic foam material that is supplied in sheet form, about 1/2 to 3/4 in. thick. The frozen sheet of foam coating is laid directly against the surface coat and either vacuum bagged in place or set in place by contact molding methods. This prefrozen foam then cures at room temperature.

This type of syntactic foam is also available in prepackaged, preweighed plastic pouches, which can be easily and quickly mixed together by kneading the resin system together inside the package. The package is then cut open and the nonflowing paste-type material is spread over the paint-on surface coat.

Syntactic foams are also used as casting materials to form master molds. Hollow microspheres, microballoons, or bubbles are mixed into mass-casting resin systems and are poured over master models to form a master mold, when required. The casting material can be adjusted to approach the density of wood, and large, lightweight castings can therefore be formed.

Still another form of master mold that can be used instead of a PFP is a high-temperature laminate. In this case, laminate copies of the master models are formed using room-temperature cured, high-temperature use, epoxy resin systems.

This sort of master mold is valuable for many reasons. Usually, prepreg mold materials that cure in an autoclave at 350°F require this type of master mold. In addition, a very large master mold can be fabricated in this way using the room-temperature/high-temperature cure, reinforced resin system (Room temp/ high temp). This type of laminate tool is usually expended like the PFP. The tool is discarded after the production mold copy is formed from the intermediate tool surface. Tooling prepregs that precure at 200°F are usually made from master molds fabricated as a PFP. Furthermore, master molds as intermediate mold forms can be easily transported long distances. This is especially helpful when the pattern maker is off-site and the master pattern is too large to be duplicated as a PFP and shipped elsewhere.

Recently, prepreg systems have been formulated to cure at room temperature and to be used at high temperatures. This allows for an intermediate mold to be made as a laminate directly from the master model at room temperature. Some fabricators, in fact, have used this room temp/high temp prepreg not only to form intermediate or expendable tools, but prototype and low-rate production tools as well.

This material, like the wet lay-up resin system, provides for the elimination of the PFP or intermediate expendable tool, as the laminate is formed directly from the master model at room temperature without the use of a PFP.

This system is acceptable because the resin system is dispensed into the fabric by machine instead of by hand. It can only be processed on-site if an impregnating machine that can dispense 30%–50% of the resin system into the reinforcement fabric is available. The resin content is then lowered to the required level by controlled bleeding of the laminate. The material and process will yield the most accurate dimensions in the master mold as the master model is being copied directly at room temperature. (This material and process will be covered in more detail in the next section.)

More accurate tolerances can be maintained in a nonmetallic mold if the systems illustrated in Fig. 6.6 are considered when fabricating a master mold.

FABRICATING MASTER MOLDS AND NONMETALLIC MOLDS AND TOOLS 163

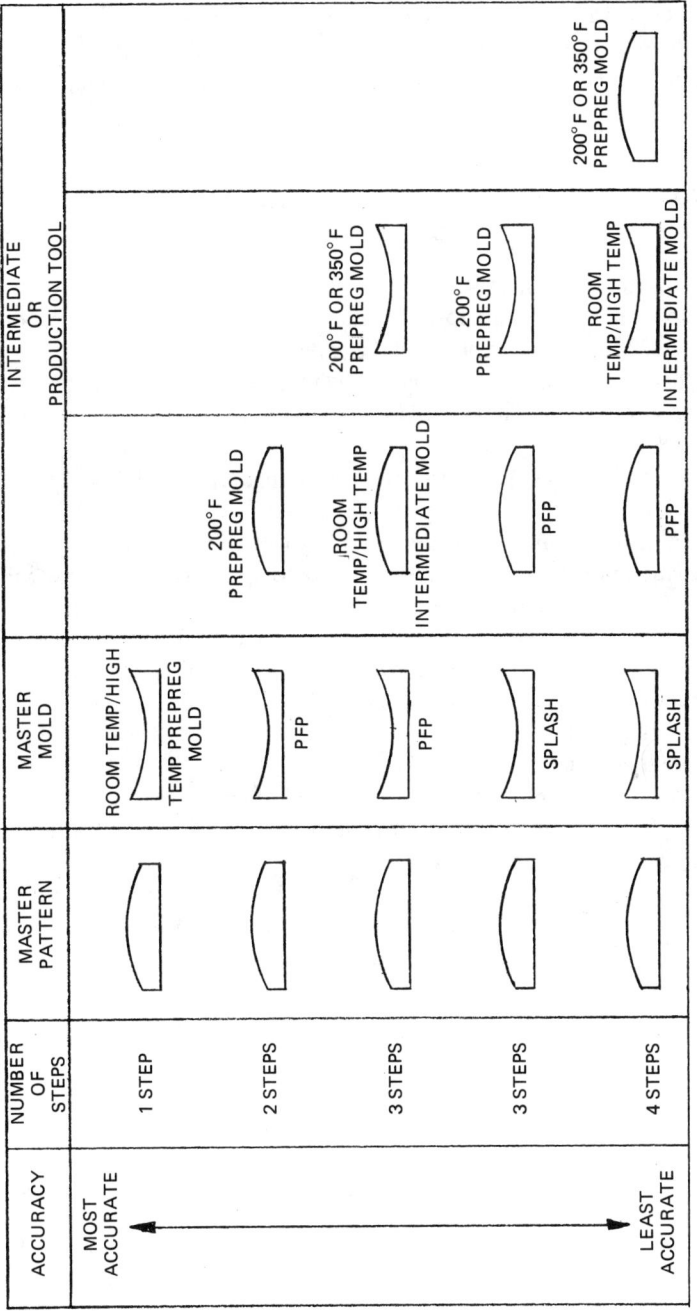

Fig. 6.6. Suggested Dimensional Tolerances Using Various Molding Methods.

164 ADVANCED COMPOSITE MOLD MAKING

The sequences in Fig. 6.6 are shown merely to suggest some of the available choices when fabricating master molds. Other materials such as mass-cast and advanced resin systems are available, but those shown represent the present state of the art.

6.2. NONMETALLIC MOLDS AS LAMINATING MOLDS AND BONDING FIXTURES

In the following sections, some of the more important and most accepted methods used to form nonmetallic low-rate and high-rate, prototype and production molds, tools, and bonding fixtures will be explored.

First, there will be a presentation of the premier materials, with emphasis on their major advantages. These materials, as mass-casting materials, are usable to temperatures of 3000°F.

Second, other high-temperature, machinable materials with low CTEs will be covered.

Third, moderate-temperature use materials will be presented. These precast and mass-cast semirigid compounds are used to form lay-up molds, tools, and bonding fixtures. Both epoxy and polyurethane compounds fall into this class.

Fourth, the lay-up molds, formed by the wet lay-up or prepreg and machine impregnation methods will be demonstrated. The elimination of surface coats, and improved lay-up techniques, will likewise be demonstrated. The advantages and disadvantages of room temperature and 200°F cured systems will be compared to the higher temperature 250° to 350°F systems.

Fifth, backup structures, using conventional eggcrate designs will be supplemented by showing more economical techniques, such as integral stiffening and molds that can be dimensionally adjusted.

The chapter will close with the metal-faced laminated tool fabrication technique on nonmetallic molds, leading us to metallic mold technology, considered presently to be part of the state of the art.

6.2.1. Cast, Rigid, and Flexible Molds

Ceramic. Probably the most important mass-casting material available today is castable ceramic. This type of material possesses a CTE of about $.4 \times 10^{-6}$.

Castable ceramic has a multitude of advantages over other mass casting materials. It is prepared simply by mixing the material with water, it cures at room temperature, and when cast, it has insignificant shrinkage. This versatile material can be cast around heaters, and because the material can be thermally shocked in excess of 2000°F without damage, the heaters and/or cooling coils can be located near the molding surface of the tool (Fig. 6.7).

FABRICATING MASTER MOLDS AND NONMETALLIC MOLDS AND TOOLS   165

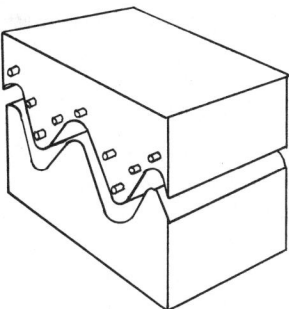

Fig. 6.7. Ceramic Tool with Heaters Cast in Place.

Inexpensive mixing equipment can be purchased for under $1,000, and will handle up to drum quantities of material (Fig. 6.8). The cost per cubic foot is considerably less than some of the mass-casting epoxy systems being produced today.

The weight per cubic foot ranges from approximately 46 to 100 lbs/ft^3. The lightweight versions are processed like wood or precast model stock. This version of the material can be N/C machined, cut, ground, drilled, or shaped, using conventional steel cutters and tools.

Ceramic cements, coatings, paints, and thinners are available to produce various finishes, if desired. Since some of the newer sealers and mold releases are used to form the thermoplastic composites at temperatures to 900°F, the usability of ceramic material at temperatures up to 3000°F puts this well within the range for this use.

Ceramic material is still being evaluated by researchers in order to define all of the available methods of use to produce molds and tools. Thus, variations of the ceramic composite formulation have been developed to allow the ceramic mold to be used in a microwave heating cavity. Microwave energy will heat a large mold to 350°–400°F in 3–4 min. Some composite materials can then be shaped, cured, or formed by heating the part material within the mold first.

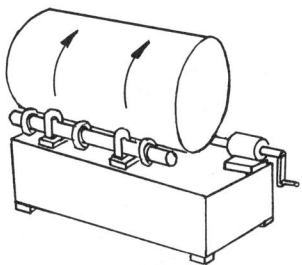

Fig. 6.8. Roller Mixer.

The significant advantages of the microwave sensitivity of ceramics is the choice it affords: The mold or part materials can either be heated first, or the mold and part material can be heated simultaneously.

Usually, the conventionally heated mold, made from conventional materials, lags well behind the part material because of the material heat-up rates. Cast ceramics eliminate this, as the ceramic material is an insulator, not a conductor.

Another advantageous feature of these high-temperature materials is the possibility of combining the pourable ceramic with a ceramic woven fabric to form a laminate (ceramic/ceramic). In this combination, the laminate as well as the pourable form of ceramic can be formed against any and every surface, including wood.

Machinable ceramic block material, on the other hand, can be face-cast or coated using the other available forms of ceramic materials. This provides a surface that is smooth, void-free, and possessing good vacuum integrity.

Each of these qualities make it crucial to consider ceramic molds and tools when distortion-free parts are desired and when the dimensional accuracy of large composite parts is a must. Ceramic molds should also always be considered when the contours are drastic and differences in CTE will lock a formed part in a mold.

There have been attempts, through formulation variations, to alter the form of ultra-high temperature mold and tool compounds. Some of the formulation modifications have resulted in compounds that combine various organic binders, organic and metallic fillers, and other compound additives. It is not recommended that these types of formulations be altered, as the modifiers can only reduce some of the physical properties so important to the final mold material form. Metallic fillers will increase the CTE and organic additives will lower the final temperature of use.

Finally, the low CTE of ceramic mold materials makes them ideal for producing laminate compaction molds, since the laminate part size hardly varies at mold temperatures ranging from the lay-up at ambient room temperature to the midrange temperature compaction or debulk, all the way to high-temperature cures of $450°$ to $750°F$ and higher.

Cementitious Compounds. There are applications for compounds that combine organic binders, organic and metallic fillers, and other compound additives. Thus, the development of various cementitious compounds has evolved. One such compound, unique to this class of formulations, uses a graded aggregate of stainless steel particles embedded in a silica-modified portland cement matrix.

The CTE of materials in this classification approach that of steel or nickel electroform tools. Male and female molds, usually in the form of laminate lay-ups, are used to cast the cementitious compound.

FABRICATING MASTER MOLDS AND NONMETALLIC MOLDS AND TOOLS

Cementitious materials, as well as ceramic compounds, can be cast into molds, tools, and fixtures from the same mold master model as many times as is required, making this a very inexpensive materials and process approach. Therefore, prototype and production tooling can be produced using both high-temperature, low-CTE ceramics and cementitious, mass-casting compounds.

Solid Graphite or Carbon. Graphite, carbon/carbon, monolithic graphite, and other similar descriptive names classify this mold and tool material as graphitized carbon (carbon/graphite). In this process, carbon material is placed in an electric furnace with granular coke tamped around it. Through a series of subsequent steps, the baked carbon material is converted to graphite at temperatures of 2600° to 3000°F. The CTE range of this material is from 3 to 4 $\times$ 10^{-6} per Fahrenheit degree.

Differences in the process will produce various forms of the finished base product that will vary in density, pore size, grain size, and other varying internal and surface qualities. It is therefore suggested that when choosing this type of material as a mold-making substance, careful consideration should be made of all of the base characteristics of the material of choice.

Unlike high-temperature ceramics, high-temperature ungraphitized and graphitized carbon cannot be cast into complicated shapes, only simple blocks. It can, however, be machined. However, ungraphitized carbon is so hard that diamond wheels or equivalent cutters must be used to process the material. Still, formulations have been developed that allow machining to be used to form molds and tools to close tolerances. Sharp male corners and edges will have to be chamfered, since the carbon porosity will cause chips and breaks larger than the pores present in the material.

This material can be bonded, threaded, doweled, and grooved to be attached to larger shapes so that various size mold forms can be produced (Fig. 6.9).

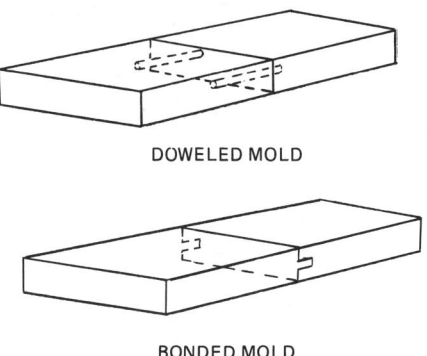

Fig. 6.9. Dowelled and Bonded Carbon Mold.

Impregnated material before use or impregnation after rough machining is a must, since the material porosity will not allow good vacuum integrity or void-free surfaces unless this treatment takes place. The sealer will obviously have to be one that can handle temperatures required to form the new, advanced thermoset and thermoplastic composites. These temperatures sometimes reach 700° to 800°F, so the impregnating material must be capable of withstanding these temperatures as well.

Machining is usually performed by highly qualified carbon machine shops, skilled in the art of machining carbon. The ability to handle these materials differs from that necessary to machine conventional plastics, metals, and the newer ceramic materials.

The uniformity and quality of square or rectangular block molded grades of graphite or carbon are usually better than the extruded versions. Also, large blocks of material can be formed by molding, so that large molds can be formed. Molded, rather than extruded, grades can also be considered nearly isotropic.

In high-temperature environments of 800°F and greater, graphite can be used in a nonoxiding atmosphere such as nitrogen, hydrogen, a vacuum, etc.

Two considerations to evaluate when producing large molds are 1) poor wear characteristics, and 2) the high cost of carbon material. Of course, cost per cubic foot, as well as the total mold weight, will play a part in choosing carbon for a mold or tool form.

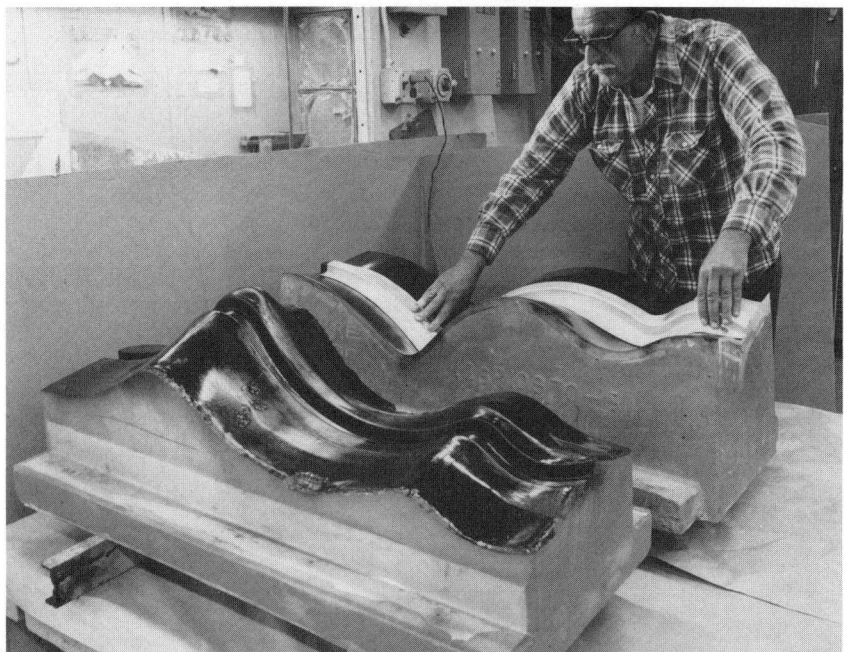

Fig. 6.10. Reinforced Mass-Cast Tool. *(Courtesy of Grumman Aircraft Systems)*

Precast and Mass-cast Semirigid Compounds. These materials are used at temperatures from room temperature ambient to moderate levels of up to 250°F.

The compounds are formulated using polyurethane- or epoxy-modified formulations (Fig. 6.10). Those companies which develop the materials for forming precast and mass-cast molds and tools must be totally sensitive to the needs of the user. So many times, the formulations are based on cost, without a total understanding of the user's needs. The user, then, is forced to choose only the available compounds, and thus may not find a material or compound that is tailored exactly to that user's need.

Low shrinkage when mass-casting leads the list of requirements; thus, new formulations addressing this are constantly becoming available. Blends of epoxies and urethanes, epoxy systems by themselves, and urethanes alone are the base materials usually chosen for producing the mold forms of choice. Prototype and low-production molds are formed directly from the master pattern (Fig. 6.11). This master pattern can be the precast or mass-cast material

Fig. 6.11. Mass-Cast Metal Forming Tool. *(Courtesy of Grumman Aircraft Systems)*

170 ADVANCED COMPOSITE MOLD MAKING

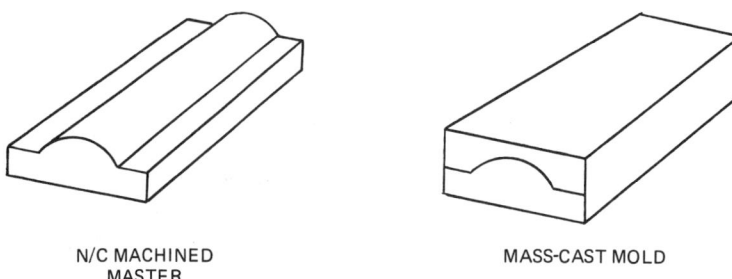

N/C MACHINED MASTER

MASS-CAST MOLD

Fig. 6.12. N/C Machined Master and Mass-Cast Mold.

itself. The precast or mass-cast material can even be machined directly into the mold contour (contour to form a finished part), as shown in Fig. 6.12.

Reaction injection molds (RIM), resin transfer molds (RTM), compression molding forms, autoclave mold forms and bonding fixtures, thermoforming molds, and plating mandrels are just some of the mold forms that can be produced with this material and process technique (Fig. 6.13A, B, and C).

Precast slabs and mass-cast materials, as described in Chapter 5, are used to form master models and molds. Precast materials can also be used directly to form parts as well, by surface casting over the near-shape machined and bonded

A

Fig. 6.13. A, B, and C. Low Cost, Mass-Cast Mold Materials Made from Sheet Metal, Tape, Wood, and Sealant. *(Courtesy of Grumman Aircraft Systems)*

B

C

Fig. 6.13. Continued

172 ADVANCED COMPOSITE MOLD MAKING

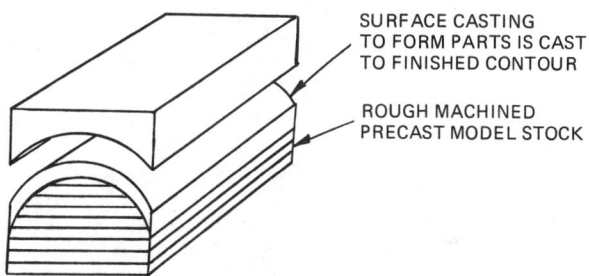

Fig. 6.14. Near-Shape Casting, Machined and Bonded.

surface, when a contour such as a PFP or an existing mold is available for contour duplication, as illustrated by Fig. 6.14.

Substantial savings in mold and tool weight can be experienced with mass-cast molds and tools. A lighter weight also means lower cost, since the fillers reduce the amount of polymer required to form the mold.

The thermal coefficient of expansion is another important characteristic to consider when mass-casting the mold form. This characteristic, however, is not important if the mold is used at room temperature, as when producing a laminate from the surface, using a room-temperature-setting laminating resin system, or the formation of an RTM, RIM, or foam composite part.

Mass-casting compounds are easy to mix, and are usually formulated as one-to-one ratio compounds (resin to curing agent). Castings can be poured to any size, thickness, durometer (hardness), density, or impact resistance.

When cast metal molds or tools are replaced by plastic ones, the mass-cast plastic tool obviously eliminates the need for the foundry, and if the tool is made from kirksite alloy, the plastic casting will be less than half the cost (Figs. 6.15 and 6.16).

Fig. 6.15. Mass-Cast Plastic Tool Made from the Surface of a Foundry Casting. *(Courtesy of Grumman Aircraft Systems)*

FABRICATING MASTER MOLDS AND NONMETALLIC MOLDS AND TOOLS 173

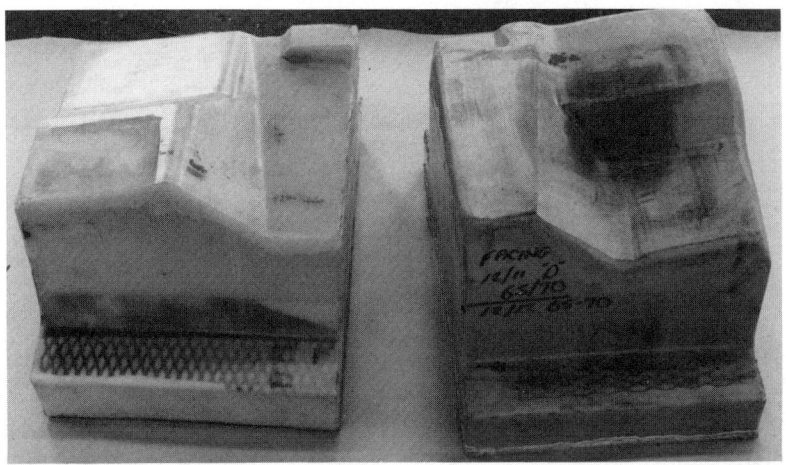

Fig. 6.16. Reinforced Mass-Cast Mold Set. *(Courtesy of Grumman Aircraft Systems)*

Another advantage of the mass-cast mold is that alterations and modifications are easily accomplished by sandblasting the original surface and pouring a near-shape cast surface over the original one. Chipped, worn, or broken molds and tools can be altered, repaired, or otherwise modified by merely casting over the existing surface (Figs. 6.17 and 6.18).

Fig. 6.17. Mass-Cast Alteration of a Large Tool Surface Prior to N/C Machining. *(Courtesy of Grumman Aircraft Systems)*

174 ADVANCED COMPOSITE MOLD MAKING

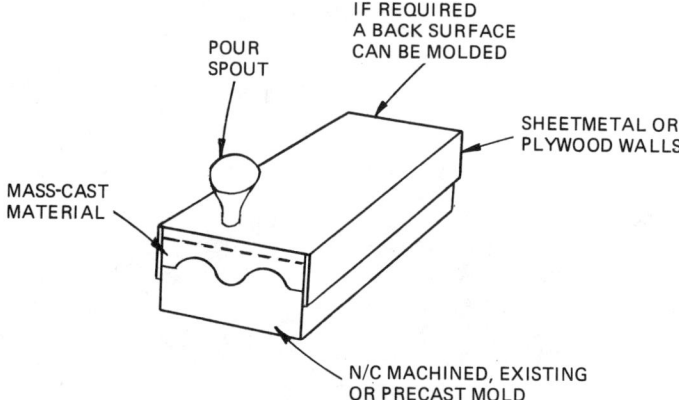

Fig. 6.18. Mass-Casting from N/C Machined Master or Existing Mold.

The advantages continue with the capability for short lead times, because mass-cast molds are produced in one step.

Expanded sheet metal, reinforcing rods, wire mesh, and other forms of reinforcement are sometimes required to secure the cast mold internally and maintain contours during shipment, use, and storage (Fig. 6.19).

Mass-cast mold forms for use in producing 200°F cure and room-temperature cure/high-temperature use prepreg laminated molds, as well as nickel electroformed mandrels, have become very popular. The use of cast or cast and machined plastics is, in fact, a technology that is spreading worldwide. Automobile, aircraft, aerospace, appliance, and marine industries are areas in which the materials and processes are being used.

Fig. 6.19. Mass-Cast Tool Showing Reinforcement Cast in Place. *(Courtesy of Grumman Aircraft Systems)*

FABRICATING MASTER MOLDS AND NONMETALLIC MOLDS AND TOOLS 175

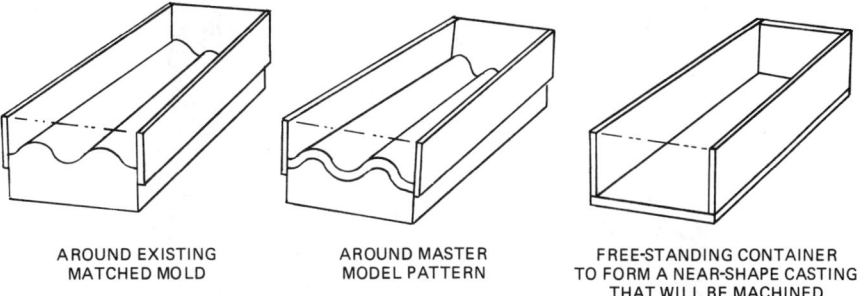

AROUND EXISTING MATCHED MOLD

AROUND MASTER MODEL PATTERN

FREE-STANDING CONTAINER TO FORM A NEAR-SHAPE CASTING THAT WILL BE MACHINED

Fig. 6.20. Mass-Casting Against Master Pattern or Mold Form Surface.

Mass-cast moderate-temperature materials are prepared and poured in the same manner as mass-cast ceramics. The mixing, pouring, and curing methods are all the same, done by hand and at room temperature. No elaborate machinery is involved or required, hand or basic pedestal mixers, with or without vacuum, are part of the apparatus. The mold forms are simple cavities formed around existing molds, master patterns, or free-standing containers (Fig. 6.20).

Castings can be formed exactly to an existing contour (net) and then used for forming autoclave or oven laminate molds and tools, plating mandrels, etc., can be double-cast if shrinkage becomes a problem. The casting surface is simply waxed and the first casting is poured. Then, the casting and wax are removed, the surface is sandblasted, and a 1/2 to 3/4 in. layer is surface-cast on the surface of the mass-casting. This is shown in Figs. 6.21A, B, and C and 6.22.

A

Fig. 6.21. A, B, and C. Foundry Tool Set Up for Face Casting and the Face-Cast Foundry Tool. *(Courtesy of Grumman Aircraft System)*

176 ADVANCED COMPOSITE MOLD MAKING

B

C

Fig. 6.21. Continued

FABRICATING MASTER MOLDS AND NONMETALLIC MOLDS AND TOOLS   177

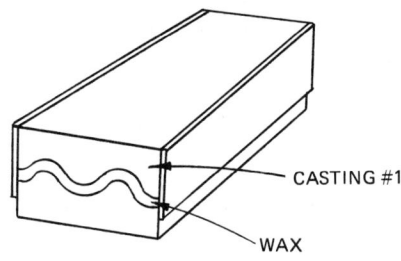

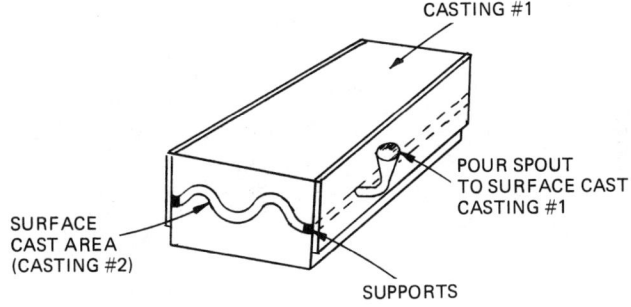

Fig. 6.22. Surface Casting.

In some instances, wax is not necessary. If wax is not used, simply space the first casting, after sandblasting the surface, 1/2 to 1 in. away from the original contour and surface cast a net face on the original casting.

Eyes or lifting attachments can be cast into the mold. Two suggested forms are illustrated in Fig. 6.23.

Mass-cast and machined surfaces sometimes require that the scallops or machining ridges be removed. This is easily accomplished by spraying the surface with a colored lacquer, sanding sealer, or suitable coating that will reveal the machined ridge shapes and amplify the area being scraped flat. After scraping, the cutter ridges can be sanded further with a 120 grit paper and completed with a 400 grit paper.

If bubble-free castings are required, moderately priced mixing equipment that has a vacuum degassing capability to deaerate the mixed compound before casting can be attempted. Other methods of eliminating bubbles also exist, such as slight vibration of the mold, painting some of the mass-cast material on the contour surface before casting, surface-tension-reducing additives, and even a process of completely pressurizing the mold and casting in a pressure vessel at low pressure prior to the gelling of the compound.

It should be pointed out at this point that recent advances in mass-casting technology have expanded the use of resin systems to include the forming of

178 ADVANCED COMPOSITE MOLD MAKING

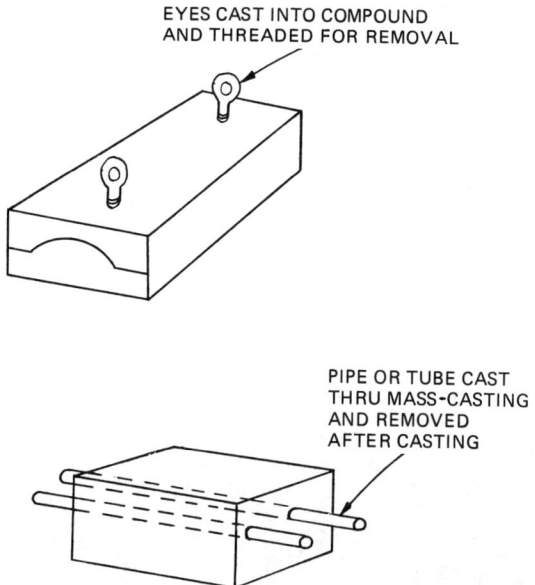

Fig. 6.23. Lifting Attachments Cast into Mold Form.

metal sheet materials in the same manner as materials produced by the drop-hammer process, the stretch and stretch-draw form processes, and metal stamping and forming processes (Fig. 6.24). Plastic tools operate more quietly than metal ones, eliminate the presence of unsafe metals such as lead, eliminate overhead operations such as the foundry, and can be simply processed by the operators within the individual part manufacturing departments. Molds and tools can even be N/C machined to the finished contour and duplicated by casting the mating mold half over the machined half surface, using inexpensive hand mixers. The application variations are limitless.

Flexible Mass-Casting Compounds. Flexible materials are usually processed in the same fashion as ceramic and moderate temperature semirigid mass-casting compounds.

These materials allow the polymer to be poured and cured at room temperature to form a flexible mold. These flexible molds are used to form composites which are free-standing, or to form a component part of a composite mold such as a caul pad, bag side tool, or a pressure pad or intensifier (which is used to intensely define a radius or curve in a molded laminate part).

Both nonsilicone and silicone elastomers, reinforced and nonreinforced, fall into this category. Mass-cast flexible molds and mold components allow undercuts to be formed in molds or tools that would otherwise have to be formed

FABRICATING MASTER MOLDS AND NONMETALLIC MOLDS AND TOOLS 179

Fig. 6.24. Mass-Cast Tool in Hammer Press, Used to Form Sheet Metal Parts. *(Courtesy of Grumman Aircraft Systems)*

using "knockouts" or "breakaways." The breakaways form a contour, but must be removed from the part after molding.

Molds can range from simple to complicated, with compounds cast and used at room or elevated temperatures, or cast at room temperature and then postcured at elevated temperatures for elevated temperature use.

Some common flexible mold materials used for mass-casting are urethanes, polysulfides, and silicones. These materials can be mass-cast over PFP's, machined and sealed wood, machined metal, or plastics.

Flexible mass-cast materials possess a high CTE (greater than 50×10^{-6}), especially the silicones. For this reason, the material can be used to take advantage of its expansion upon the application of heat. This feature will allow the flexible elastomer to swell during use, and thus apply increased pressure in designed areas. Usually, mass-cast flexible molds are used within rigid molds that retain their growth during cure cycles.

Another feature of the flexible mass-casting compound is its ability to form hollow composite structural members such as T's, Z's, C's, beams, etc (Fig. 6.25).

180 ADVANCED COMPOSITE MOLD MAKING

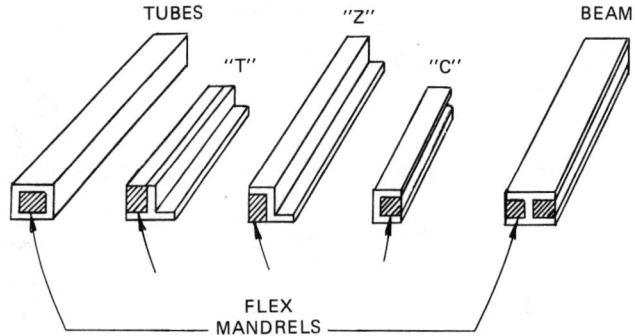

Fig. 6.25. Casting to Form Hollow Composite Structural Members.

By varying the mass-cast formulation, thickness and elastomer form, it is possible to control the elastomer growth and thereby the internal pressure. This feature also allows the mass-cast material to be used as a localized bonding fixture as well as a mold form (Fig. 6.26).

The rubber pressure pad technique used under vacuum bags will be addressed later in the text.

Integral stiffening can be effected by using precast mass-cast flexible forms (Fig. 6.27).

Molds and tools mass-cast using flexible compounds are processed in the same way that semirigid and rigid compounds are. Care should always be taken to assure that the mass-casting material is poured over a sealed, dry, and properly released surface. Any moisture present can affect the cure of any mass-casting compound.

6.2.2. Laminated Molds

As demonstrated in the preceding section, mass-cast ceramic molds and tools produce the most accurate large or small parts because of their CTE of $.4 \times 10^{-6}$.

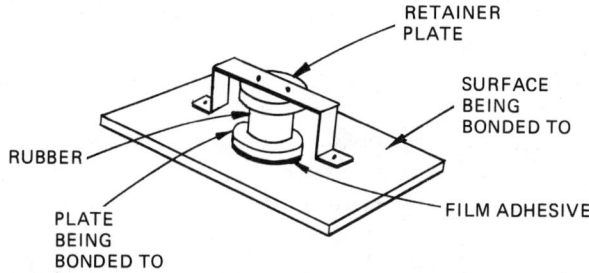

Fig. 6.26. Localized Bonding Fixture Using Expandable Rubber Casting.

FABRICATING MASTER MOLDS AND NONMETALLIC MOLDS AND TOOLS   181

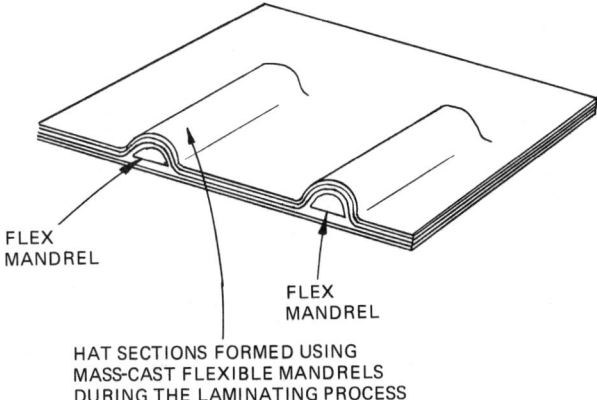

HAT SECTIONS FORMED USING
MASS-CAST FLEXIBLE MANDRELS
DURING THE LAMINATING PROCESS

Fig. 6.27. Mass-Cast Hat Sections.

Carbon block is dimensionally close to graphite fabric reinforced epoxy prepregs with CTEs in the range of 2 to 4×10^{-6}.

Large net lay-up parts and parts with inclusions such as inserts, integral stiffeners, hardware, tooling pins, toolholes, and tight trim lines cannot be molded in or on anything but low CTE mold and tool materials.

Metal mold and tool materials, as well as fiberglass reinforced epoxy prepregs, will create problems when used to mold or bond such accurate parts, since their CTEs fall into the 7 to 12×10^{-6} CTE range.

Graphite fabrics are unique in that, when heated, they shrink slightly, so that if they are impregnated with epoxy resins, which are in the CTE range of $50-75 \times 10^{-6}$, the resulting composite possesses a CTE of approximately 3×10^{-6}. Therefore, a major advantage of the fiberglass or graphite/epoxy prepreg mold or tool is that the part made from its surface will be exactly the dimensional mate of the master pattern from which the mold laminate was made. So, to make a mirror copy, one simply makes a graphite/epoxy part from a graphite/epoxy mold and a fiberglass/epoxy part from a fiberglass/epoxy mold.

Another major advantage of the graphite fabric/epoxy mold or tool is its surface and internal hardness coupled with its strength and weight advantage over metal molds. Therefore, mold forms 30 to 40 ft in length could be easily fabricated with a laminate thickness of only .250 to .400 in.

The cost of graphite/epoxy molds can be as little as half that of nickel electroform tools, when the support or backup structure is carefully designed. A fiberglass/epoxy prepreg mold is much less expensive, since the material cost of fiberglass fabric is much lower than that of graphite fabric. Wet lay-up or machine impregnated molds compare similarly, since the material and labor costs are about the same.

Thus, the prepreg materials provide the mold and tool designer and fabricator less expensive alternative materials and methods for making molds to metal or wet lay-up techniques, and inexpensive, mass-cast, precast slab, plaster, or even wood models which can provide the contour forming surfaces.

It then becomes accurate to state that the state of the art in laminating molds and tools are the preimpregnated fiberglass or graphite-reinforced epoxy materials and processes.

There are exceptions. Wet lay-up materials are only recommended where it is necessary, or the mold or tool maker is forced by some requirement to use this material and process technique. Room-temperature precure, high-temperature post-cure prepregs exist that replace the labor intensive, inconsistent, uncontrollable, hand wet lay-up mold-making process. Intermediate molds and tools, which produce the final production prepreg laminate mold or tool, could be fabricated by the wet lay-up process if the mold form is made from a master model that cannot be subjected to heat, or the intermediate mold is to be expended after completion of the final production mold or tool.

It is obvious, though, that room-temperature-cure prepregs, supplied and stored, assure simplicity and quality.

Some methods of approach to the fabrication of impregnation equipment and its use will be described in this section, as well as a description of the suggested methods for processing room-temperature-cure prepregs as prototype, short run production, and intermediate expendable molds and tools.

Filament-wound materials have also been used in the laminated mold making process. The filament-wound material performs as a prepreg mold material in most cases and, depending upon the resin system/process parameters, is cured to whatever temperature the chosen resin system dictates. This method will be described, showing the approach and explaining the preparation details.

Room-temperature-cure, vacuum-bagged, prepreg laminates, as well as 200°F, free-standing post-cure, and even 350°F vacuum-bagged/autoclave-compacted debulked and cured prepreg laminate molds will be covered.

Finally, the use of film surface or gel coats, processed, applied, and cured like a prepreg, will be discussed, as well as such unique process methods as picture framing and peripheral bagging.

Laminated Wet Lay-up or Prepreg Molds and Tools. This section will cover the wet lay-up process, prepreg mold and tool making, and filament wound and "impreg" materials and process technology. Included will be room-temperature, 200°F, and 350°F cure systems combined with unique manufacturing materials and process approaches such as frozen film surface coats and "picture framing."

The prepreg process. It must be emphasized that the wet lay-up process is a process of last or special resort. The most accepted method of producing a laminated mold or tool is the prepreg process.

FABRICATING MASTER MOLDS AND NONMETALLIC MOLDS AND TOOLS   183

The selection of the right tooling material is crucial to the success of any mold and the resulting parts. However, until recently, wet lay-up composite molds and tools have been limited in use by a number of factors:

- High resin contents
- Improper mixing of resin systems
- Short service life
- Inconsistent and nonuniform fabrication results
- Unacceptable service and cure temperatures
- High labor costs
- Toxicity problems with fabricators

The prepreg of both fiberglass and graphite fabrics offer choices of resin formulations to suit all manufacturing methods. These formulations allow the use of wood, foam, plaster, machinable plastic, and metal as the master model material (Figs. 6.28A, B, and C).

When choosing the prepreg system, look for high hot flexural strength and thermal stability, the longest work life, but the shortest cure time at the lowest temperature, the best coefficient of thermal expansion (CTE), the highest use temperature, good drapability, and handling characteristics allowing for low vacuum bag pressures to be used to cure the prepreg.

A

Fig. 6.28. A, B, and C. Large Prepreg Laminate Tool. *(Courtesy of the Toolrite Div. of Fiberite Corp.)*

184 ADVANCED COMPOSITE MOLD MAKING

B

C

Fig. 6.28. Continued

FABRICATING MASTER MOLDS AND NONMETALLIC MOLDS AND TOOLS 185

Good handling characteristics should be complemented by the fact that the plies must be as thick as possible, to reduce labor time, while simultaneously not sacrificing drapability and handling. In addition, these features should reduce the need for compaction and debulking cycles.

New 200°F cure prepregs and newer room temperature systems allow the user to lay-up on foam, wood, plastic, plaster, and metal master models. Automotive and marine applications, because of mold labor and material cost, as well as curing equipment availability, require the use of these newer room temperature systems.

N/C machined plastic and plastic-faced plaster materials, however, still continue to be the least expensive master model materials. Mass-cast machined plastic, the advanced state of the art, leads the group for quality, low cost, materials, and processes.

When a high-temperature cure prepreg must be used to form a mold laminate, ceramic mold materials, mass-cast at room temperature, lead the group.

Otherwise, a metal master model is required, or a PFP that is expended during the laminate cure because of the high cure temperature.

Toxicity in the handling of wet lay-up tooling materials has plagued all industries in the manufacture of molds and tools for many years.

The "white room"-type environment surrounding the use of prepreg systems reduces or, in most cases, eliminates the hazards of direct operator contact with unmixed and reacted resins and curing agents. The fabrication of molds and tools in the same manner as parts also allows the part fabricators to laminate all types of tools. Thus, combined skills allow for a more efficient operation (Fig. 6.29A, B, and C).

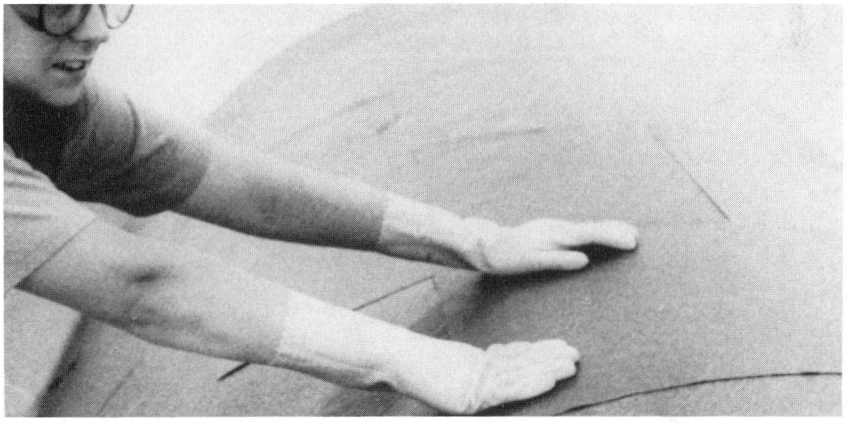

A

Fig. 6.29. A, B, and C. Prepreg Being Applied to Form Laminate Mold Form. *(Courtesy of Ren Plastics Div. of Ciba Geigy Corp.)*

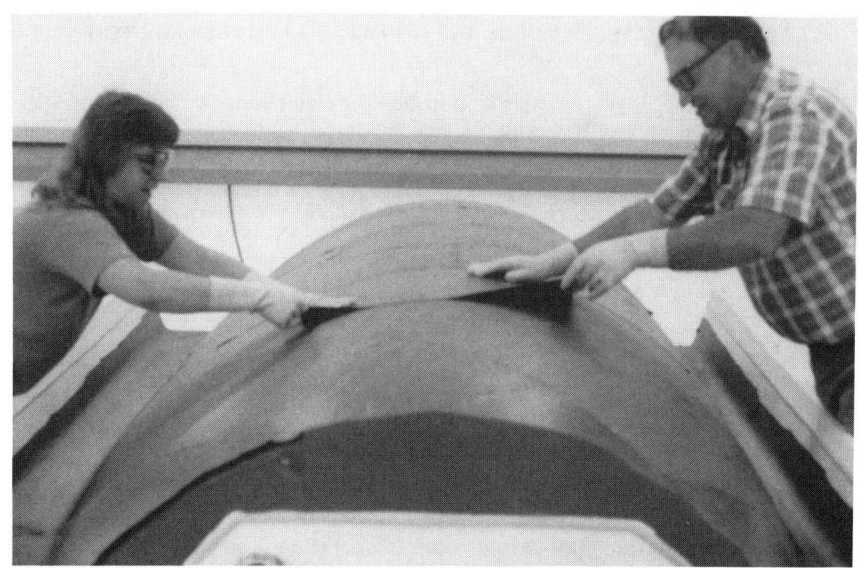

B

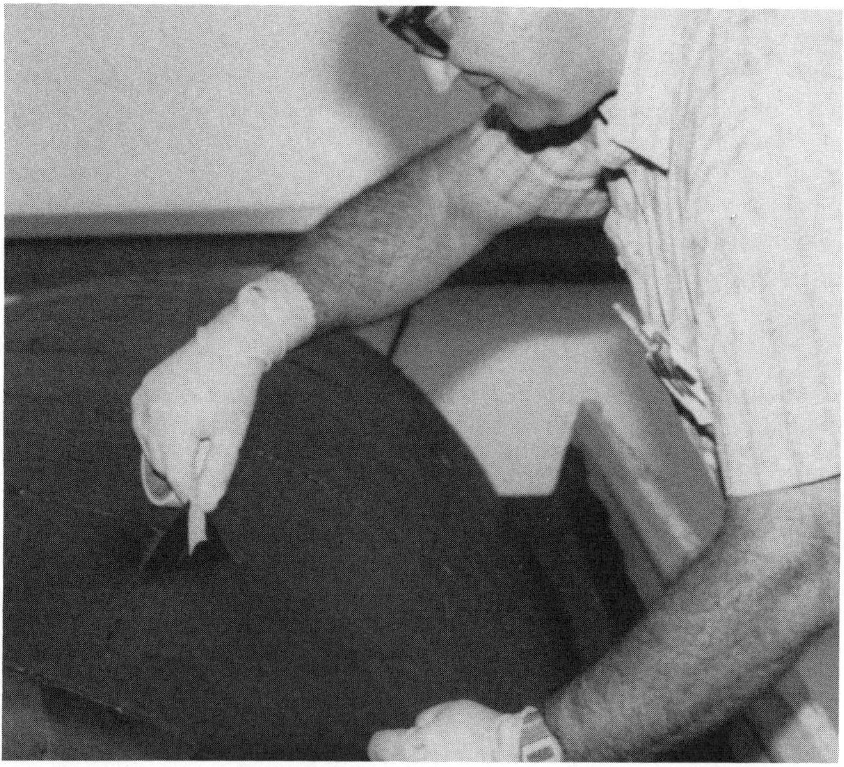

C
Fig. 6.29. Continued

FABRICATING MASTER MOLDS AND NONMETALLIC MOLDS AND TOOLS 187

Besides allowing for a clean and safe working environment, the molds and tools produced by the use of prepregs are more consistent, longer lasting, and of better quality. This is partially a result of the fact that if a parts fabricator were to make a mold, the problems encountered when using a mold or tool in production would be the basis for the inputting of information to improve future tools.

Consistency and quality also come from the fact that the reinforcing fabric is impregnated by machine. This allows for tighter control of the resin system to fabric content, as well as more uniform dispersion within the material. The uniformity of material then transfers to the finished tool.

Another positive advantage of this method is that the tool is fabricated in substantially less time than with a wet lay-up, since thicker materials can be used to build up a laminate faster than thin ones (Fig. 6.30A and B).

This cost reduction is of particular importance in large tools. Because of the extended working time that prepregs allow, very large molds and tools can be fabricated. Recently formulated prepregs also allow cost reductions in operating expenses, since the cure cycles are reduced by the use of room temperature and 200°F precure prepregs. The cost reductions continue during the use of the larger tools, especially with graphite epoxy prepreg molds or tools. These heat

A

Fig. 6.30. A and B. Prepreg Tool. *(Courtesy of Hexcel Corp.)*

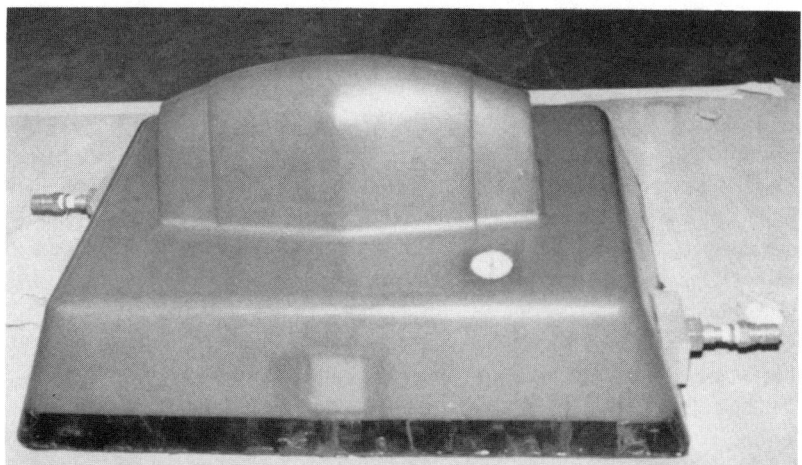

B

Fig. 6.30. Continued

up and cool down faster than metal tools, are lighter and less costly to move, thus requiring less sophisticated transportation equipment.

350°F cured prepreg materials. The original materials used to fabricate molds and tools, these prepreg materials were soon replaced by 250°F systems. The 250°F systems were then replaced by 200°F and room-temperature-cure prepregs, since a temperature above 212°F causes a PFP to decompose through calcination of the plaster.

Two important advantages are: 1) room temperature master materials like wood, foam, and plastic sheeting can be used, and 2) very accurate mold master pattern copies can be made.

Room temperature prepregs. Room temperature prepregs, new to the area of mold and tool making, offer many advantages over temperature set and cure systems.

These room temperature prepreg systems eliminate the problems of thermal growth when duplicating master models. This is very advantageous when large molds or tools must be made. For instance, molds or tools as large as 30 ft long have been constructed and the dimensional accuracy of the mold copy was well within .150 in. over the length of the mold.

Application is not easy, but can be aided with room-temperature-cure surface coats, some precatalyzed and frozen in their shipping containers.

Room temperature prepreg materials arrive in frozen form, and usually only have a 16- to 24-hour working life. The material therefore requires careful handling, transportation, and storage, since a few hours of thawing will ruin a batch of material or can be the cause for improper lamination of a mold or tool. Thus, it is wise to check each batch of material used in order to assure that the material has the shelf life and working time stated by the manufacturer.

It is advisable to have the prepreg packaged in precut rectangular sheets instead of rolls, since the material can then be used immediately upon thawing and will not require cutting, which uses up the work life of the material. Furthermore, room-temperature-cure prepregs are very soft, flexible, and "wet," so handling is more difficult than the 200°F and 300°-350°F. systems.

Room temperature prepregs are customarily laid up directly over a properly prepared master of wood, plaster, plastic, or metal, vacuum bagged and left at room temperature under vacuum for 2 to 3 days. Dry plies of veil mat and fabric are usually applied in back of the surface or gel coat and laminate stack to assure void-free surfaces and laminates, and no-bleed resin lay-ups are recommended, as the resin content of the prepreg is usually lower than 200°F cure systems.

200°F cure prepregs. The ideal material for a high-temperature-use prepreg mold or tool is a 200°F cure prepreg. It offers a longer working time (up to 30 days at room temperature), higher thermal and physical properties, and much easier handling or transportation characteristics than room temperature prepregs. Furthermore, this type of prepreg can be applied as a surface or gel coat.

In fact, the 200°F cure system will usually provide the best combination of properties when compared to the room temperature and 300°-350°F cured systems.

Surface coats or gel coats. These are sometimes required to provide an extremely smooth surface to the front of the tool laminate. Because these coatings, in some instances, are painted on, it is necessary to use certain material choice and application precautions:

- Choose a flexible, not rigid, paint-on surface coat
- Do not apply too thick a coating (.012-.015 in. is ideal)
- Choose a room-temperature-set and heat post-cure system
- Do not puddle resin in female corners

Fiberglass epoxy prepregs, cured into a laminate, have approximately a 10 to 14×10^{-6} CTE. Graphite epoxy prepregs, on the other hand, are about 3 to 4×10^{-6} CTE.

It is always a good idea to use a liquid gel or surface coat when "vacuum-bag-only" forming a mold or tool. The low pressures will not compress the laminate sufficiently to effect vacuum integrity, so a coated mold or tool surface is required.

Fig. 6.31. A and B. Application of (A) Film Surface Coat, and (B) Prepreg Laminate Tool Material Over Film Surface Coat. *(Courtesy of CTL Aerospace Corp.)*

FABRICATING MASTER MOLDS AND NONMETALLIC MOLDS AND TOOLS 191

Paint-on surface coats are also used to form spray-metal-faced RIM, RTM, and SMC tools by painting on a gel coat layer in back of the sprayed metal, thereby bonding the metal to the laminate during the co-cure of gel coat and laminate resin system (see Section 6.2.3).

Film gel or surface coats are available in various forms: Frozen films of epoxy are reinforced with lightweight fabrics (2 to 4 oz/yd), veil mats, or unreinforced but supported nylon or dacron scrims (Figs. 6.31A and 6.31B). It is also not unusual to apply a fiberglass epoxy prepreg (.002–.003 in. thick) as the surface coat of a fiberglass or even graphite prepreg laminate mold.

The film gel coats usually eliminate mark-through of the laminate prepreg (Fig. 6.32A and B).

Paint-on surface coats. These range in CTE from 40 to 75 $\times$ 10^{-6}, since various fillers and milled reinforcements alter the cured coating's CTE. Some liquid surface coats are formulated as pastes, and must therefore be wiped or screeded onto the master model surface. Usually, these coatings cure at room temperature or low temperature, and as a reflow system, and finally co-cure with the laminate.

Some mold and tool makers use a fine swirl or veil mat in back of the screeded coat. This dry, fine, and light weight layer will absorb resin from the laminate and the surface coat, thereby becoming the transition layer between the surface coat and the laminate.

A

Fig. 6.32. A and B. Graphite Epoxy Prepreg Tool Showing Pattern of Frozen Film Gel Coat. *(Courtesy of Grumman Aircraft Systems)*

192 ADVANCED COMPOSITE MOLD MAKING

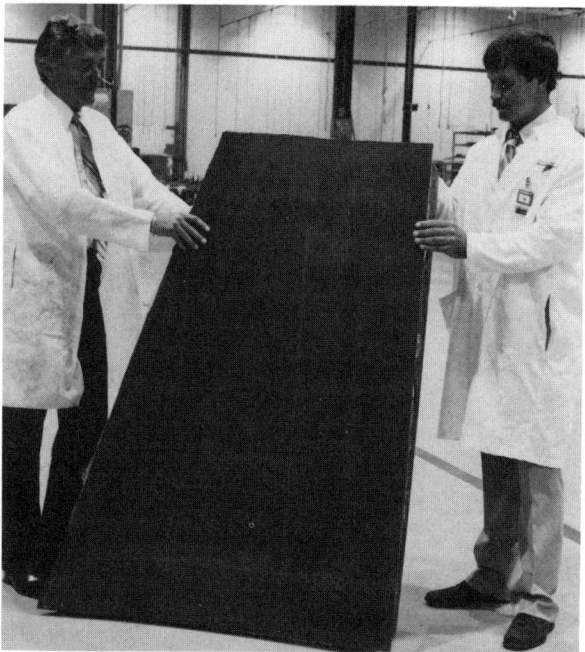

B
Fig. 6.32. Continued

Since the surface coat is usually softer than the laminate resin system, a transition from filled resin system gel coat to reinforced fabric lamination is suggested.

Some mold and tool makers prefer a paint-on surface coat, in which the first resin coat is allowed to gel and "tack off." The first resin system gel coat, about .010 in. thick, is followed by a .004 to .006 in. thick coat and the prepreg layers are applied to the wet resin system, providing a chemical and mechanical transition bond layer.

The veil mat is also sometimes laid down on a painted-on surface coat, following its application to the master model surface (Fig. 6.33).

Face bagging a tool during the post-cure is very desirable. Whether a paint-on surface coat or film coat is used, the laid-up laminate must be post-cured. Before post-curing the laminated mold or tool, vacuum bag the front surface of the mold, over the cured gel coat. The bagged front of the tool will allow pressure to be applied to the tool face during the post-cure. Any imperfections, voids, etc., will easily show up after the post-cure and will therefore be readily repairable. This is an advised practice, whether a paint-on or film coat is used. The vacuum bag may be sealed to the tool periphery using disposable sealant tape, after the tool surface is covered with a perforated release film and one ply of 10 oz/yd^2 breather.

FABRICATING MASTER MOLDS AND NONMETALLIC MOLDS AND TOOLS   193

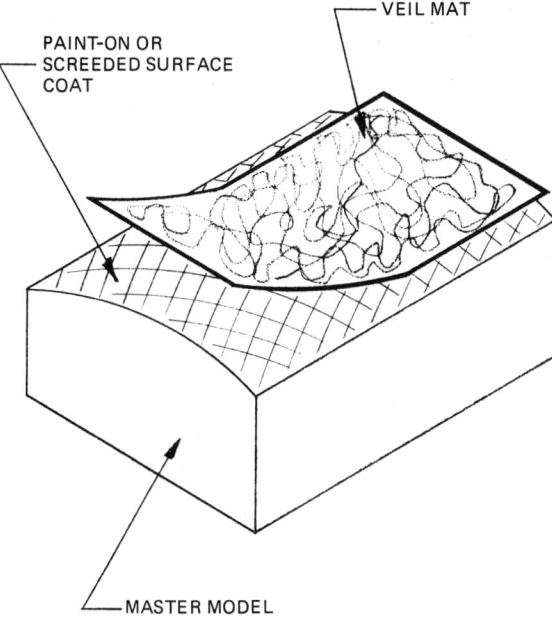

Fig. 6.33. Application of Veil Mat.

Special cases. There have been cases in which both paint-on and film surface coats have been eliminated and the face of the tool is formed by subjecting the laminate to 100 psi in an autoclave during the cure. There are companies which produce molds and tools using this approach. The higher the pressure applied to the laminate during cure, the better the quality of the mold or tool.

Overlapping the rectangular pieces of film surface coat and vacuum bagging the film to compact it to the master model surface prior to prepreg application is a good procedure to follow. In fact, the laminate itself is usually compacted under atmospheric pressure, at room temperature, every few plies, by applying a vacuum bag and atmospheric pressure at room temperature for approximately 15 minutes. Thus, if the lay-up of a laminate is interrupted by the end of a workshift, the 200°F prepregs allow the tool maker to leave a bag over them, under compaction, until the start of the next day's workshift.

Heat debulking, especially with paint-on surface coats, is a good process to practice. Simply apply a peel ply, perforated release film, breather stack, and vacuum bag and, at a low temperature, subject the uncured laminate to temperature and pressure (debulk) to allow the paint-on or surface film and laminate to soften and intimately contact the contour surface, densifying and pressing the laminations together.

194 ADVANCED COMPOSITE MOLD MAKING

Some mold and tool makers have partially cured or "B" staged the surface film or gel coat prior to application so that the long cure cycle that is usually required for these materials is reduced. Care, however, must be taken when using this approach, since inconsistent results could occur if even, controlled curing is not achieved.

Picture framing. This approach to applying the prepreg laminations is employed to further eliminate the possibility of leaks in the laminate tool face. If a cut, nick, or gouge is present in a mold or tool face or edge, the leak path may be interrupted by using this laminating approach.

The first two plies of prepreg are applied to the gel or surface coat or film, in the usual rectangular shape. Then, "picture frame" or tape strips are applied around the mold periphery. These 3 to 4 in. wide strips are staggered, butt seamed, or overlapped for 2 to 3 plies (Fig. 6.34).

In some instances, this will also further stiffen the mold or tool edge.

Impregnation machines. These are used to produce "prepreg" type tooling materials in-house, although one could say that the material processed through this equipment is considered "impreg" rather than "prepreg."

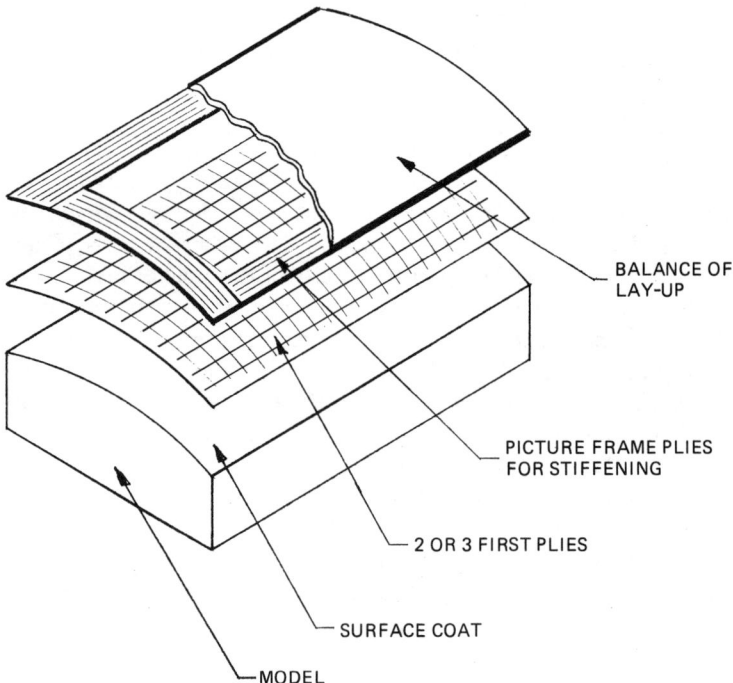

Fig. 6.34. Application of Picture Frame Plies.

FABRICATING MASTER MOLDS AND NONMETALLIC MOLDS AND TOOLS   195

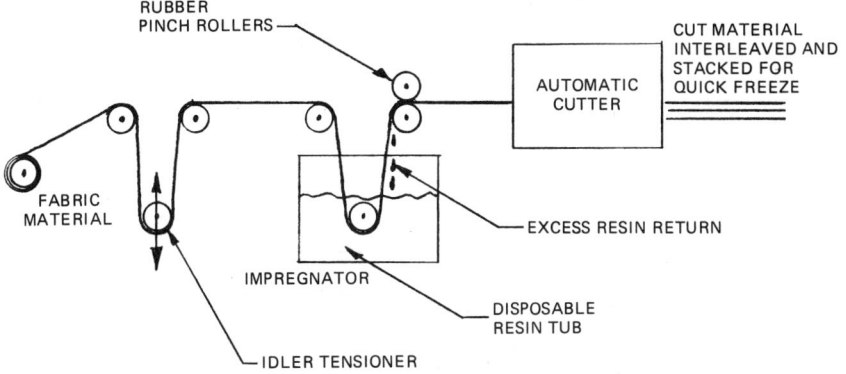

Fig. 6.35. Simple Impregnation Machine.

The machinery can be something as simple as that used to impregnate reinforcement material for filament wound structures. A bath or resin system container, reinforcement feed and tension system, squeeze rollers, and a take-up is all that is really necessary (Fig. 6.35).

On the take-up side is a cutting system which automatically cuts the material into sheets.

The resin system is similar to that used in the room-temperature-setting prepreg system, but purchasing and processing such a resin system in-house, with this equipment, provides a much less expensive approach to an impregnated reinforcement than buying a prepregged material.

This type of equipment is portable, and can be located near the mold or tool work stations or adjacent to the freezer area, so that impregnated material can even be quick-frozen after impregnation, for future use.

There have been instances in which precut fabric patterns are fed into the equipment, impregnated, kitted, and frozen for future use.

The wet lay-up process. Some of the unique features shown in the preceding section on the prepreg process, such as picture framing, special post-cure techniques, reinforced wet and prepreg surface coats, veil mat reinforcements, and others, should be considered before proceeding with the wet lay-up process.

The wet lay-up process, which is not considered the present state of the art in mold making, will not be emphasized in this text, because inconsistent and uncontrollable results occur when this process is used to prepare mold forms. The only recommended use of wet lay-up is to form contact hand lay-up or vacuum-bagged laminates for short run production, or producing prototype parts and intermediate expendable molds and tools.

If, however, it is absolutely necessry to use this process because of cost and the fact that the master pattern cannot see any temperature exposure, then the

196 ADVANCED COMPOSITE MOLD MAKING

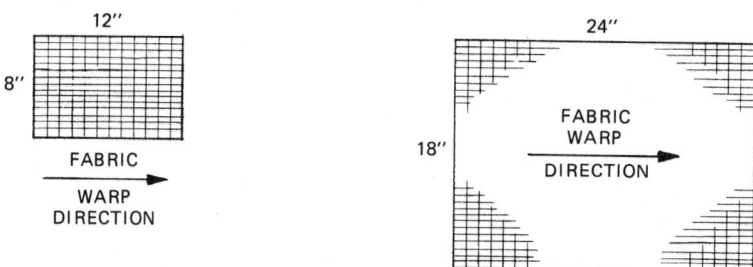

Fig. 6.36. Rectangular Lamination Patterns.

resin system must be preweighed to assure that a 50% maximum resin to fabric ratio is maintained. It is sometimes also necessary to preheat the resin system from 100°–110°F to lower the viscosity and allow for easier wetting of the reinforcement.

The order of the remainder of the process is given here.

Always impregnate the fabric reinforcement before laying it up on the mold. The fabric used in the wet lay-up process, as well as in all other lamination patterns, should be cut into rectangles, with the largest leg of the rectangle never more than 24 in. long (Fig. 6.36).

Prewetting the fabric reinforcement will reduce vertical drainage of the impregnating resin system into corners or pockets in the mold cavities. The resin system, used to laminate male molds, should drain into a peripheral bleeder. If a recess exists that cannot be bled, then a puddle may form in a female radius. Even with prewetting the fabric, a puddle may form, so care should be taken with the lay-up and, if necessary, an intensifier or bag side tool aid should be used to maintain the radius and thickness of the laminate in the radius (Fig. 6.37).

Paper or plastic patterns can be used as locating templates for spotting, bushing, insert, or other hardware locations. The bushing or insert hole should be punched or cut in the exact location in the dry fabric prior to impregnation.

Usually, a plain or square weave 10 ounce per square yard fiberglass fabric is used for the wet lay-up process with 2 plies of 4–5 ounce fabric in back of the

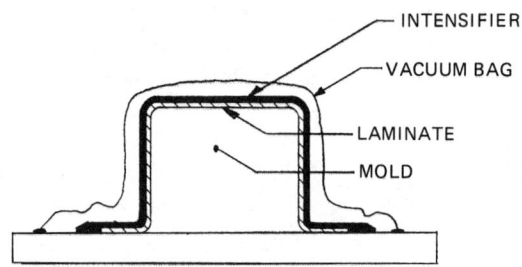

Fig. 6.37. Use of an Intensifier or Bag-Side Tool.

FABRICATING MASTER MOLDS AND NONMETALLIC MOLDS AND TOOLS 197

gel or surface coat and 2 plies on top of and at the end of the tool laminate sequence. As a guide, about 15 plies of 10 ounce fabric reinforcement will produce a 1/4 in. thick laminate weighing about 2 lbs/ft^2.

Assuming that the mold master pattern has been prepared with the proper sealer, paste release, and mold release systems, the gel or surface coat is now ready to be applied.

To prepare this, mix the resin and curing agent thoroughly in their original containers, and weigh out and mix the proper resin system amounts together. With a stiff bristle brush, usually 2-3 in. wide and trimmed on a slant, as shown in Fig. 6.38, apply a layer of gel or surface coat on the mold to a thickness of approximately .012-.015 in. (The thickness of the surface coat may vary from manufacturer to manufacturer and user to user. Some users, in fact, apply two layers, allowing the first (.008 in. thick) application to gel, and sometimes, to set.)

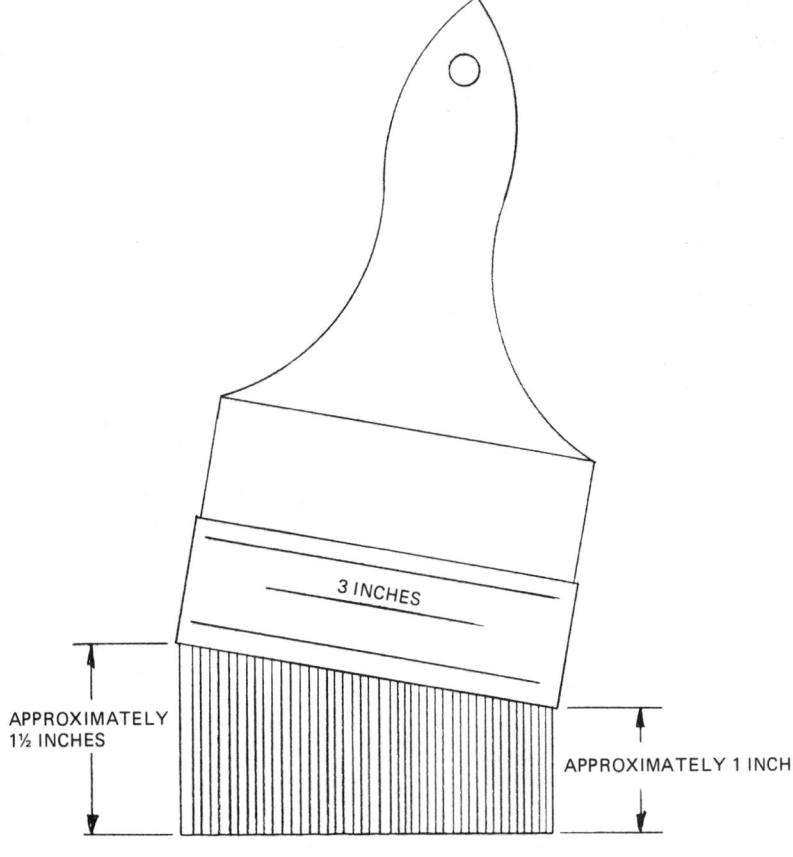

Fig. 6.38. Trimmed Lay-up Brush.

If necessary, apply the laminating resin system to the preset gel coat, since the laminating resin system will cure harder than the surface coat. It is important to remember that the more resin on the surface of the mold, the more problems may arise, i.e., mark-through of the fabric reinforcement, cracking, crazing, lifting, etc.

If a surface or gel coat has cured hard before the laminating process has started, the exposed surface must be re-activated for adhesion by abrasion or solvent cleaning. This must take place before continuing the laminating process, and the surface coat must be in total contact with the contour it is applied against before starting the lay-up.

Once the surface coat has been applied, the lamination process continues by the application of pieces of precut reinforcement fabric in a pattern that allows the pieces to butt seam together. Fiberglass and graphite fabric, as well as other forms of reinforcement (mat, rovings, scrims, etc.) are sometimes used to produce a laminate mold face by the wet lay-up process. Plain weave .010 to .050 in. thick woven fabrics are recommended over mat roving or other thick reinforcements.

A vacuum-bagged laminate should be prepared as per the detailed information contained in Chapter 8. Refer to this chapter also for instruction in the installation of dams and mold walls, used to produce a net molded edge that will eliminate post-laminating trimming around the periphery of the laminate.

When contact molding using the wet lay-up process, a mixture of milled fibers and resin system is sometimes used to round off, fill in, or reinforce a female radius before the reinforcement laminations are applied. This step should reduce and even, at times, eliminate the possibility of voids occurring between the laminated reinforcement and the surface coat. Any resin puddling in a corner would also be reinforced with the fibers.

The next step in the process is the coating of the gel-coated surface with the thoroughly mixed laminating resin system, and the careful laying down of the first prewet ply of reinforcement material. Once this has been accomplished, butt seam the rectangular pieces together, simultaneously, with whatever tooling aids suit the contour, working out all of the excess or puddled resin and entrapped air. Choose the tool carefully. Metal or plastic rollers, plastic wedge wipers, and stipple trimmed brushes are just a few of the many tools available. A short knapp paint roller works very well.

The butt seamed plies should be applied in layers, with the seamed laminations staggered and the warp direction of the woven reinforcement rotated at each layer. If a selvage edge (i.e., one that will not unravel) fabric has been used to form the rectangular pieces, rotation of the pieces is a must. Try to apply the fabric without twisting the material, since a twisted fabric ply could be the cause of warpage of the tool laminate during the part cure cycles.

Room-temperature-cure, free-standing post-cure, laminating resin systems are recommended, but a total room-temperature-cure system that does not require a post-cure is also a common material for laminating.

For a mold with a nearly flat contour, rotate the plies $0°$ and $90°$. For molds with more complex or compound contours, use $0°$, $+45°$, $90°$, $-45°$, and so on.

If a thickness greater than 1/4 in. is anticipated, the laminate should be allowed to gel at room temperature.

The lamination process is continued after the heat of reaction (exotherm) is over and the gelled laminate cools. Whether or not vacuum pressure is used to set the laminate, it is recommended that the laminate be covered with a layer of peel ply or release fabric while the resin is wet. This will provide a rough molded surface when the release fabric — sometimes called a "peel ply" — is removed. This rough or bondable surface is also the surface that an eggcrate or backup structure would be bonded to, prior to removing the laminate from the contour surface that is being duplicated. Besides the release or peel ply, coarse metallic and nonmetallic granules, sprinkled on the wet laminate surface prior to cure, have also been used to provide a rough bondable surface.

If a vacuum-bagged laminate is planned, the release fabric or peel ply is covered with release film and the appropriate bleeder and/or breather stack and required vacuum bag.

If no vacuum is planned, and contact molding is decided upon, the release material is merely covered with an inexpensive bagging material and the excess resin is wiped to the outer laminate edges using a tapered plastic wiper, as previously described. The bleeder and/or dam previously installed will define the edge and the physical contact pressure, and careful wiping action will assure the elimination of puddling and excess resin. A laminate thickness of approximately 1/4 in. is recommended.

If an eggcrate or backup frame is to be attached to the laminate, this frame should be prepared, thermally stabilized (post-cured) and attached to the back of the rough surfaced laminate, as per the subsection on backup structures. (Rolled tool edges, incorporated into the laminate design for added stiffness, are also demonstrated in this section.)

Hollow Composite Parts. These are made from a number of materials and process approaches. One of the more reliable methods has been through the use of eutectic salts, melted at temperatures of $400°$ to $500°F$ and poured and cooled into ceramic or laminated epoxy-reinforced molds. These split molds are then removed and part laminate material is applied to the outside surface of the "salt" mandrel, cured at the required temperature and pressure, and cooled. The salt, soluble in water, is then washed out of the laminated part. The split mold is reassembled and the process is repeated. The salt can be also "slush" cast, coat-

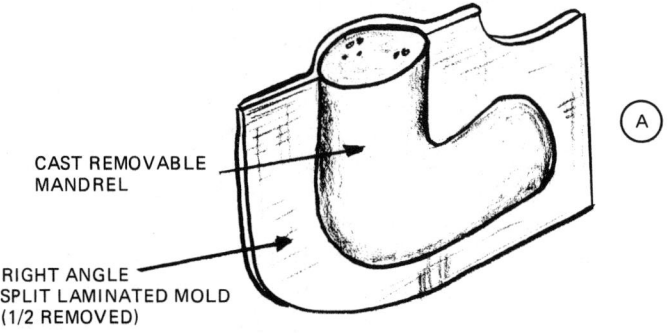

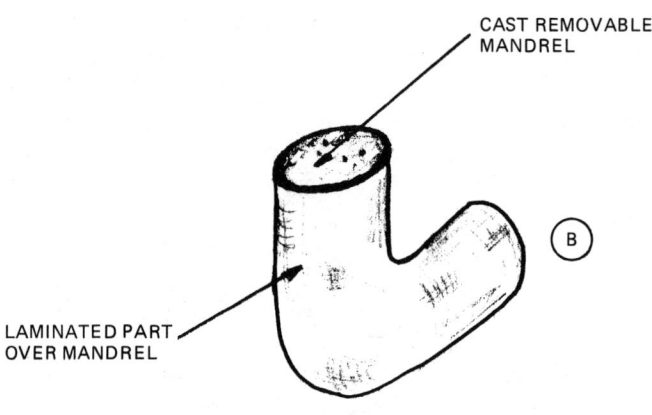

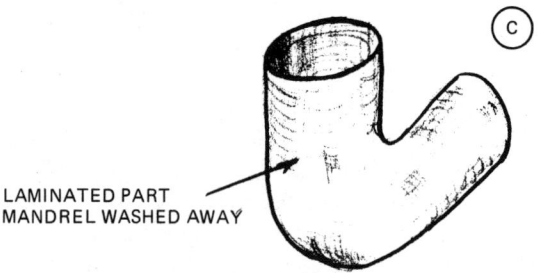

Fig. 6.39. Typical Part Fabrication Process of a Laminated Part Over a Removable Mandrel.

ing the inside walls of the split mold (desired method) or solid cast by pouring and filling the split mold. This process is illustrated in Fig. 6.39.

Some salts or removable mandrel materials cause laminate casting molds to delaminate, separate, and thermally decompose after repeated castings. Recently developed cast ceramic mold materials will outlast the laminate casting mold if large quantities of mandrels are required from the same mold form (Fig. 6.40).

Filament Wound Parts. These constitute another process used to form hollow laminates. Removable mandrels can be either broken out — if thin walled — or washed out, if water-soluble material is used.

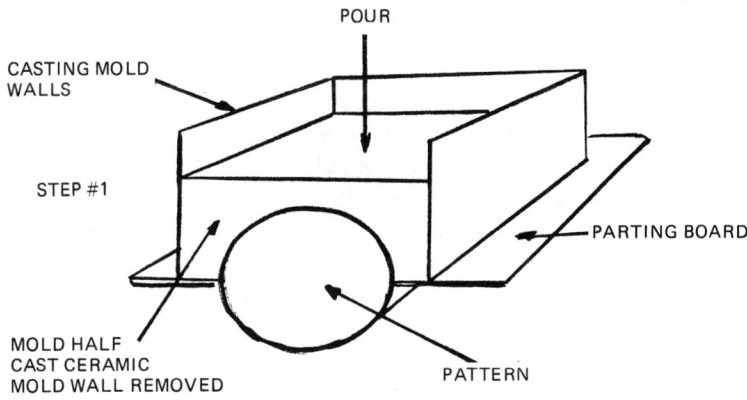

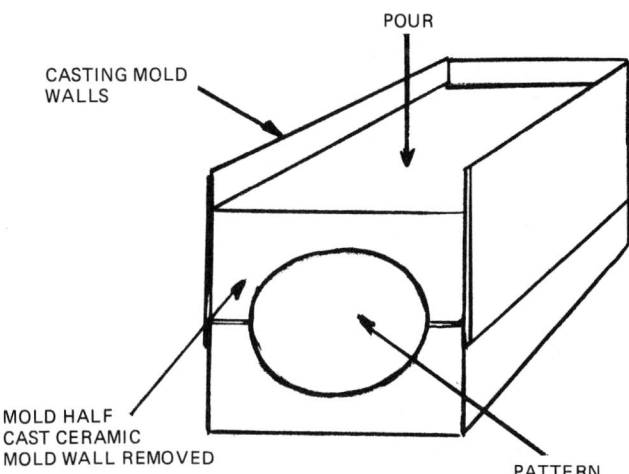

Fig. 6.40. Ceramic Salt Mandrel Mold Casting Sequence.

Table 6.1. Various Types of Filament Winding Mandrel Materials.

NONMETALLIC	METALLIC
Laminated plastic	Steel, collapsible
Plaster, water-soluble, and nonsoluble cast plastics (polyurethanes/epoxies)	Aluminum, collapsible
Wood	Low-temperature-melting alloys
Ceramic, thin-walled*	
Salts, water-soluble*	
Fiber- and paperboard	
Thermoplastic rotational or blow-molded materials	
Solid or hollow flexible bladders, reinforced and/or inflatable	

*Best for large diameter, stable part forming.

Some manufacturers of composite parts have wound laminate material on top of mandrels with an automatically (computer-controlled) applied, preimpregnated (prepreg) or manually wet-out reinforced fabric laminate material. The wet laminate is then compacted, slit from the mandrel, transferred to a master model or existing intermediate mold, bagged and cured into a laminate mold.

Mandrels for both helical or polar wound parts have been formed from metallic or nonmetallic mold materials, some of which are listed in Table 6.1.

In most cases, internal heating can be used to aid in mandrel heat transfer and, in the case of ceramic mandrels, laminate parts have been cured using radiofrequency (R.F.) and microwave energy, since the ceramic mandrel can be subjected to this environment without the energy being absorbed or reflected by the ceramic.

Backup Structures. Unlike cast or ceramic reinforced ceramic, laminated molds and tools require reinforcement or strengthening of the laminate face from the rear of the mold or tool. A framework of some type is either formed during the face laminate fabrication (integral stiffening) or is added to the rear of the tool after the laminate is formed. Most mold and tool designers choose the same frame materials as the front mold or tool laminate, since distortion may occur if dissimilar materials are arbitrarily used (Fig. 6.41).

When other frame materials are utilized, attachment precautions must be considered. These will be discussed later, but an examination of the various materials is appropriate at this point.

Some of these materials include steel and aluminum sheet, plastic or metal angles, or tubes, or a combination of these. Others include solid fiberglass or graphite fabric-reinforced epoxy sheet materials (usually 1/4 in. thick), square and round tube formed from braided material, and pultruded or woven fabric-reinforced tubes (Fig. 6.42).

FABRICATING MASTER MOLDS AND NONMETALLIC MOLDS AND TOOLS 203

Fig. 6.41. Composite Tool Showing Solid Board Eggcrate Substructure. *(Courtesy of Grumman Aircraft Systems)*

Additionally, honeycomb panels, paper-based phenolic, and even wood have been used. These sheet materials or panels are fabricated by using prepreg materials or the wet lay-up process and forming the laminate panels in a press or autoclave. Finally, the precut and assembled sheets are connected, thermally stabilized, and attached to the laminate face using mechanical fasteners, laminate

Fig. 6.42. Large Composite Tool Showing Round Tube Substructure. *(Courtesy of Ren Plastics Div. of Ciba Geigy Corp.)*

tie-ins, threaded fastenings, or some or all of these. In all cases, the laminate is "bagged" to the master model before the frame is attached, as will be explained later, in the subsection on peripheral bagging.

If a support structure is used, the structure should always float from the back of the laminate. Approximately a .100 in. gap should be used to space the eggcrate or laminate box frame away from the laminate rear surface. Removable shims or removable tabs shaped into the edge of the contour board can be used to set the gap.

Above all, the frame must not be in direct contact with the mold or tool laminate during use, since it will distort the laminate by expansion and contraction. Also, the framework which supports the laminate, and provides a sturdy base upon which it rests, must not be constructed so solidly that it restrains the mold or tool's normal expansion and movement during cure cycles, or restricts the heated air flow under the front laminate surface.

The framework should be attached to the laminate mold face before the laminate is removed from the reference contour.

The contour board material or panels that will be cut into the shaped boards should have patterns traced upon their surfaces using thin plywood, polyester film, or cardboard templates, (Fig. 6.43). If plywood is used, the eggcrate support backup can be preassembled and set on the back of the laminate to check

Fig. 6.43. Solid Eggcrate Laminate Material Being Laid-Up to Net or Finished Size to Eliminate Trimming. *(Courtesy of Ren Plastics Div. of Ciba Geigy Corp.)*

FABRICATING MASTER MOLDS AND NONMETALLIC MOLDS AND TOOLS 205

Fig. 6.44. Thin Plywood Cut to Contour and Used as a Template to Shape Composite Eggcrate Material. *(Courtesy of Ren Plastics Div. of Ciba Geigy Corp.)*

the fit before cutting the actual boards (Fig. 6.44). The outside edges of the tool laminate face should either be flush with the outside edge of the backup structure or recessed (Figs. 6.45 and 6.46).

This will protect the face laminate edges of the mold during movement, use, and storage.

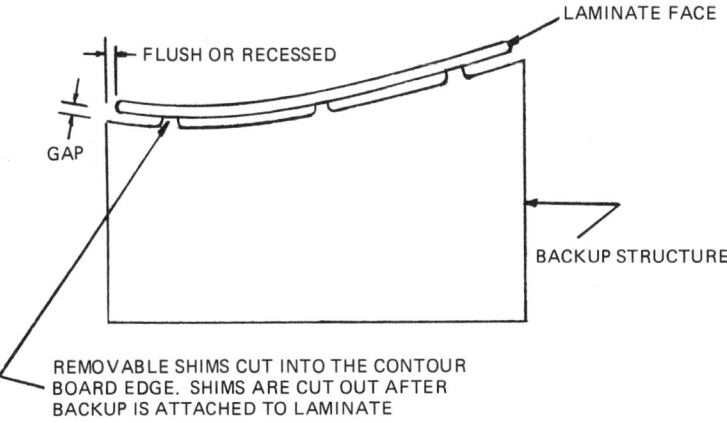

Fig. 6.45. Recessed Laminate Tool Face.

206 ADVANCED COMPOSITE MOLD MAKING

Fig. 6.46. Flush/Recessed Composite Tool Eggcrate. *(Courtesy of Grumman Aircraft Systems)*

Eggcrate backup structure. As mentioned earlier, the eggcrate is formed by various materials, even wood. Usually, the wood is laminated over with a few plies of reinforced polyester or an epoxy resin system to seal the surface and form a strong "sandwich" panel.

Whatever the material used to form the backup structure, contour boards must be cut, slit, and preassembled. The contour boards or headers may be N/C machined or cut manually. When manually cut, polyester film tracings, cardboard, or even sheet metal templates may be used (Fig. 6.47).

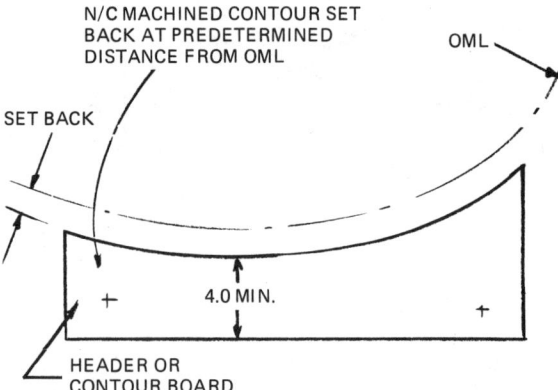

Fig. 6.47. Contour or Header Board.

FABRICATING MASTER MOLDS AND NONMETALLIC MOLDS AND TOOLS 207

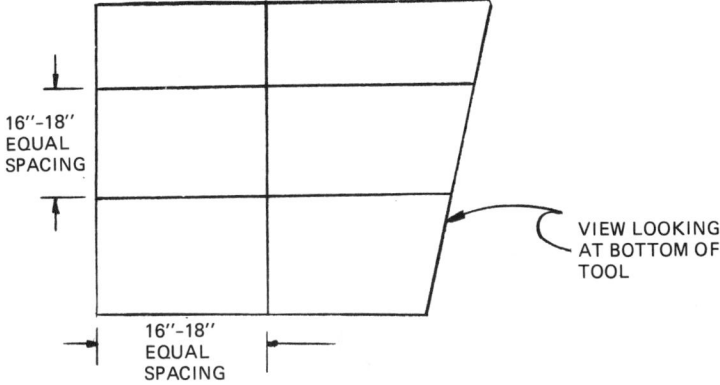

Fig. 6.48. Backup Board Spacing.

A good procedure to follow in establishing the eggcrate opening size is an equal spacing of 16–18 in. (Fig. 6.48).

The eggcrate is laminated together, using corner "tie-in" coverage, 100% internally and 50% edge contact, when attached to the face laminate (Figs. 6.49 and 6.50).

The "tie-ins" are 6 in.-wide strips of reinforcement fabric folded in half, impregnated, laminated to the eggcrate surface, cured, and post-cured prior to attachment to the face laminate.

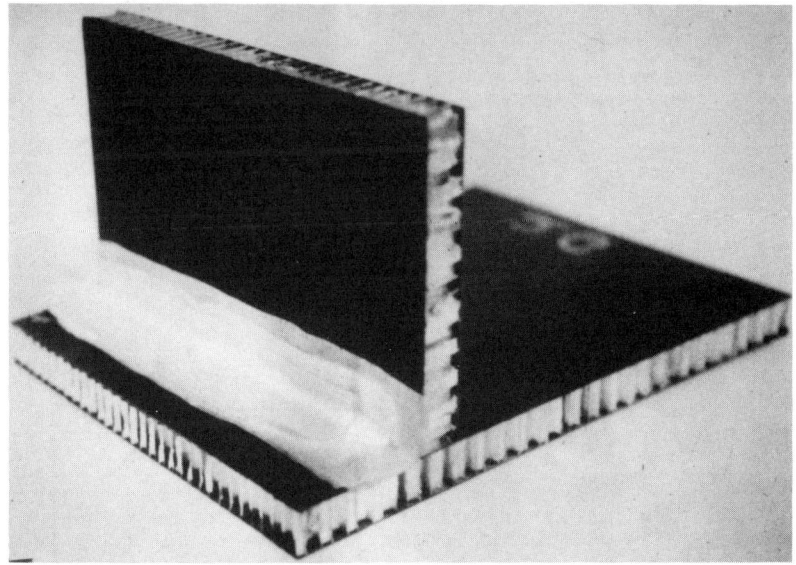

Fig. 6.49. Eggcrate Laminated Tie-In. *(Courtesy of Hexcel Corp.)*

208 ADVANCED COMPOSITE MOLD MAKING

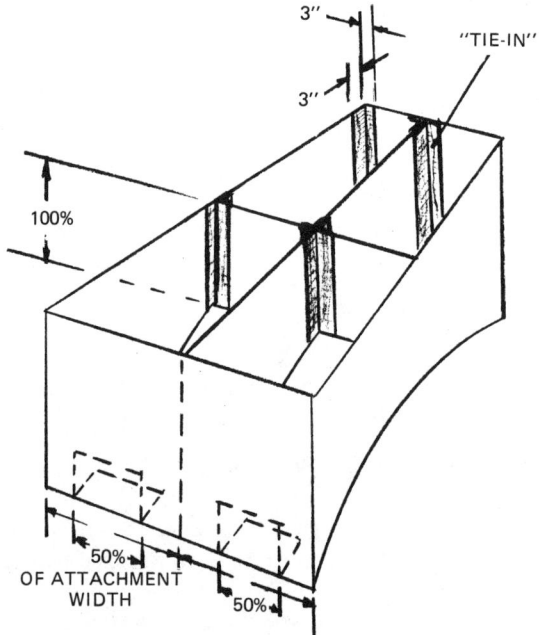

Fig. 6.50. Configuration of Backup Board "Tie-Ins."

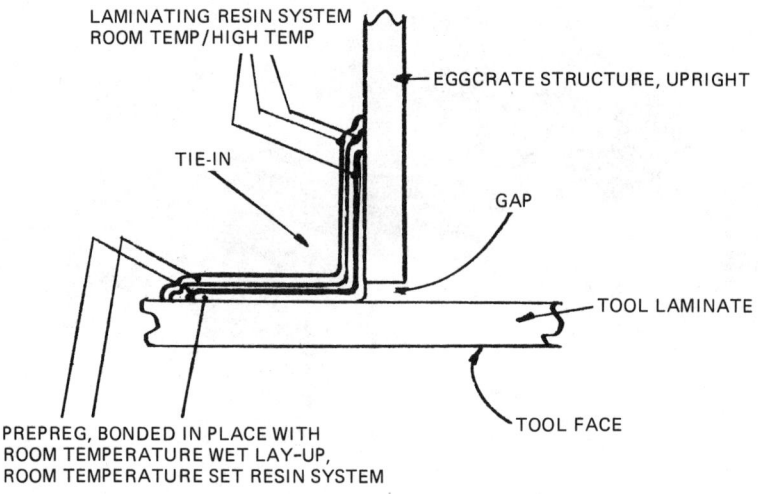

Fig. 6.51. A Tie-in.

FABRICATING MASTER MOLDS AND NONMETALLIC MOLDS AND TOOLS 209

The use of a 200°F cure prepreg is also suggested to form the tie-in. This is accomplished by precutting the 6 in.-wide tie-in strips and laminating them in place by using a room-temperature-cure/high-temperature-use lamination resin system. The preimpregnated material is then held in place and encapsulated with a room-temperature-set laminating system. Whether wet lay-up or prepreg, 4 plies of reinforcement (10 oz or about .010 in. thick/ply) are recommended (Fig. 6.51).

If honeycomb material is used to form the front laminate sandwich (Hexcel Flexcore*), or for the backup (rigid) structure, a number of considerations should be taken into account and some precautions should be taken. One is the design of the honeycomb material (Figs. 6.52A, B, and C and 6.53). For molds and tools, the hexagonal core is recommended.

The following configurations can be considered for attaching honeycomb mold and tool corners together (Fig. 6.54). The honeycomb backup thickness is usually 1/2, 1, or 2 in. thick, depending upon tool size.

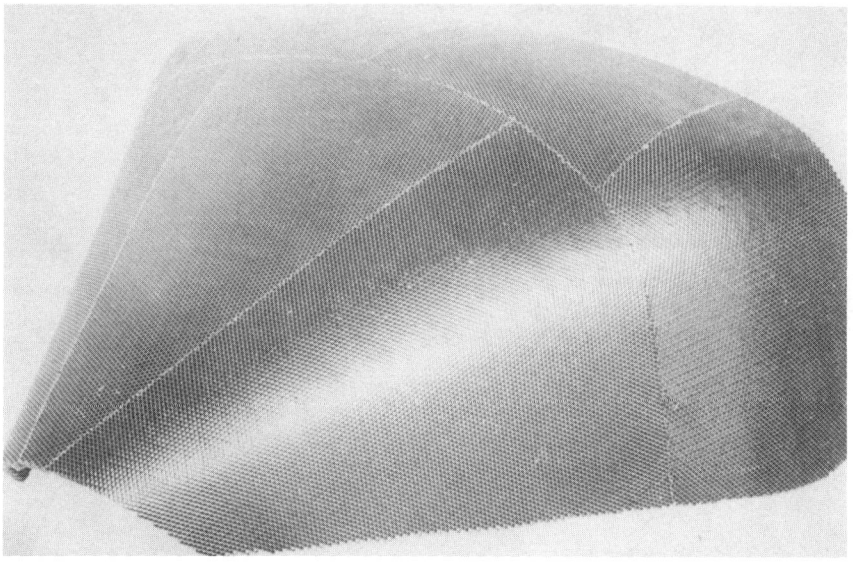

A

Fig. 6.52. A, B, and C. Flexcore Honeycomb Reinforcement. *(Courtesy of Hexcel Corp.)*

---

*Flexcore is the reg. trade mark of the Hexcel Corp.

210 ADVANCED COMPOSITE MOLD MAKING

B

C

Fig. 6.52. Continued

FABRICATING MASTER MOLDS AND NONMETALLIC MOLDS AND TOOLS

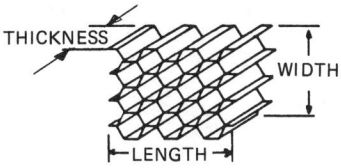

HONEYCOMB HEXAGONAL CORE

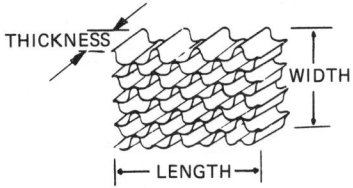

HONEYCOMB FLEXCORE

Fig. 6.53. Honeycomb Flexcore and Hexagonal Core. *(Courtesy of Hexcel Corp.)*

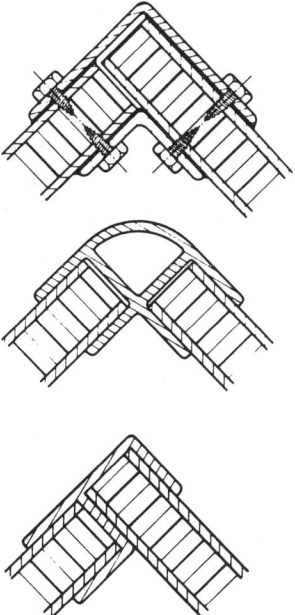

Fig. 6.54. Prefabricated Honeycomb Mold and Tool Corners.

212 ADVANCED COMPOSITE MOLD MAKING

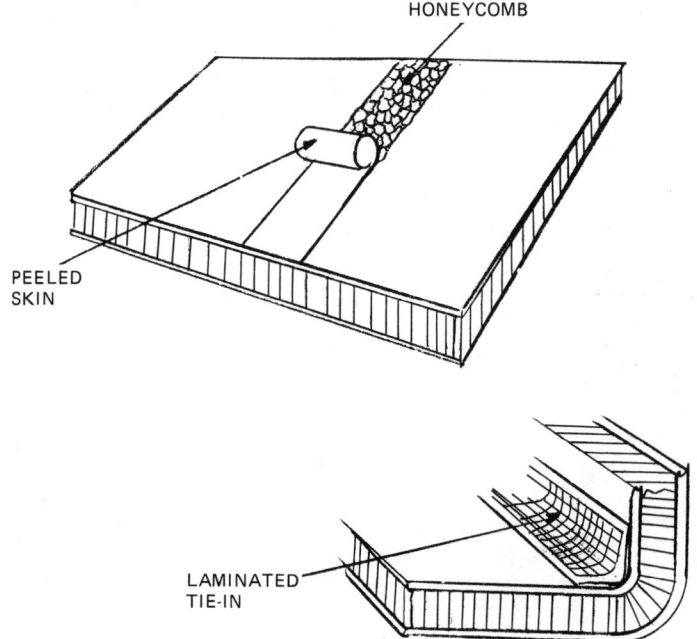

Fig. 6.55. Honeycomb Corner, Bent and Reinforced.

When laminating the corners of a honeycomb sandwich backup structure, the skin may be scored and peeled, and the honeycomb bent to form a corner, as illustrated in Fig. 6.55.

Another and suggested laminated corner is formed by cutting, assembling and laminating the corner together (Figs. 6.56 and 6.57).

If inserts of various kinds are to be bonded into the honeycomb board, the following factors must be considered:

- Corrosion (steel inserts into aluminum honeycomb must be precoated with epoxy or polysulfide adhesive)
- Environmental exposures (tool use and storage)
- Panel thickness of backup
- Specific strength requirements

Three types of structural fasteners are defined, as shown in Fig. 6.58.

When attaching an eggcrate substructure to the rear of a laminate mold or tool, place the laminate in intimate contact with the master model or pattern. When practical, this can be accomplished by a bagging method referred to as "peripheral bagging."

Fig. 6.56. Laminated Honeycomb Eggcrate Tie-ins. *(Courtesy of Hexcel Corp.)*

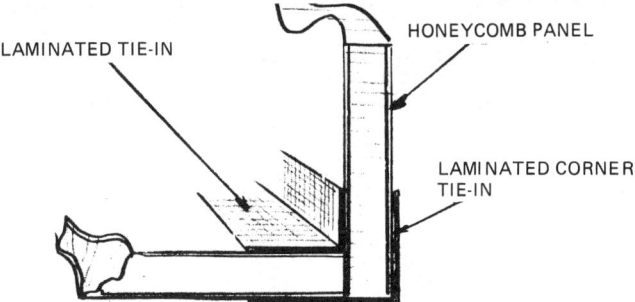

Fig. 6.57. Cut and Laminated Corner, Honeycomb Material.

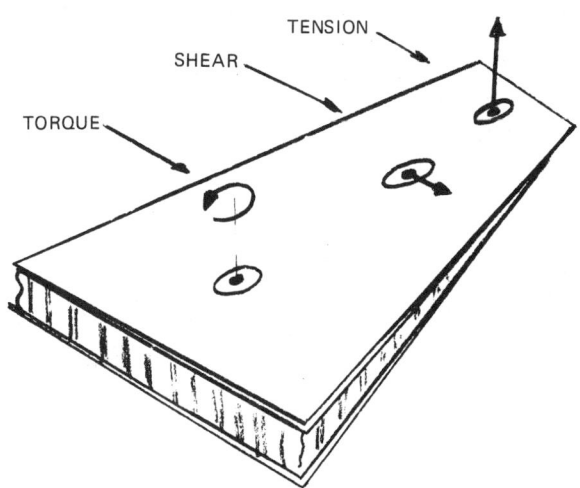

Fig. 6.58. Three Types of Honeycomb Fasteners.

214 ADVANCED COMPOSITE MOLD MAKING

Peripheral bagging. This process is accomplished by simply applying a vacuum bag around the periphery of the back of the face laminate and pulling a vacuum. Atmospheric pressure will then press the laminate against the master pattern or model.

Then, the preassembled, post-cured, and thermally stabilized eggcrate is attached to the back of the face laminate, as illustrated in Fig. 6.59.

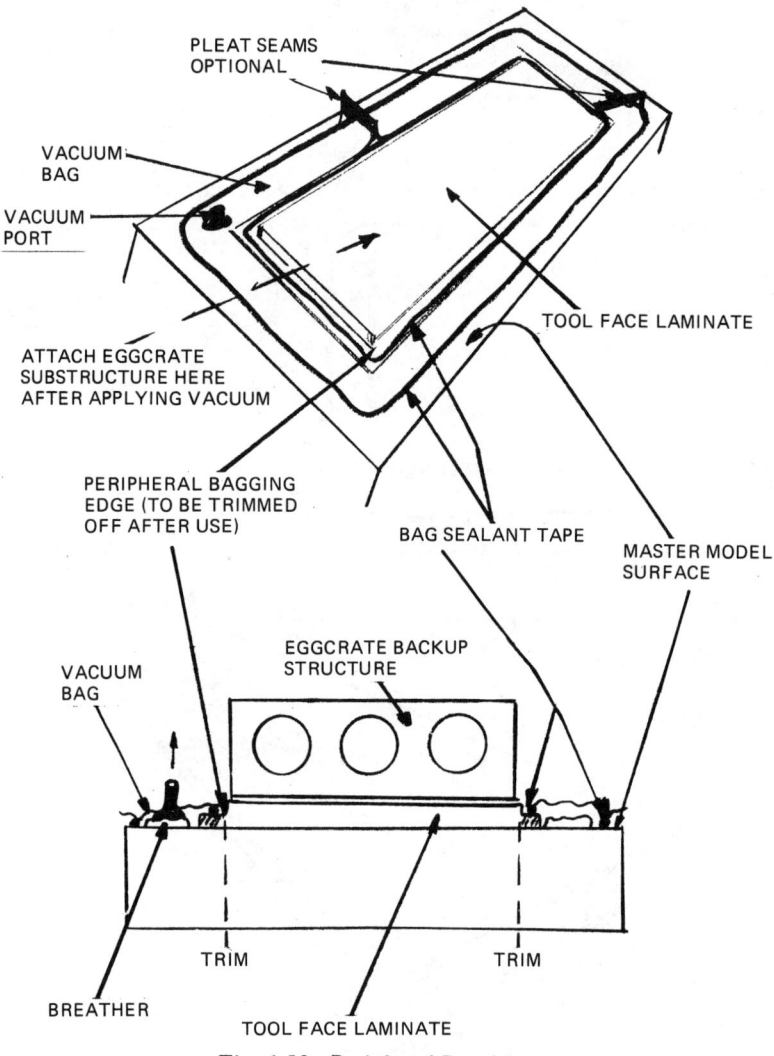

Fig. 6.59. Peripheral Bagging.

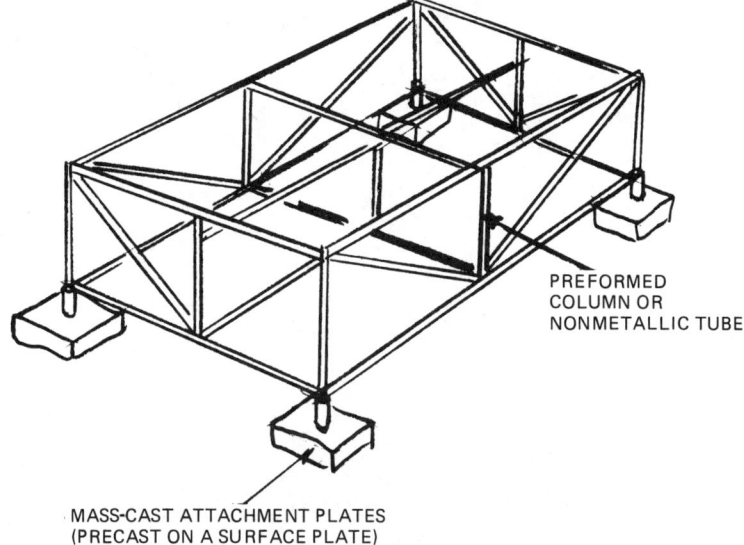

Fig. 6.60. Tubular Mold or Tool Backup Support Frame.

Nonmetallic tubular frames and backup structure. An eggcrate substructure is only one form of support available for a laminate mold or tool. A lightweight (and preferred) method is the use of tubular material or preformed columns. In this process, the tubes are cut, attached to each other, and formed into a frame. This frame or structure can be used to form an assembly fixture, size gauge, or — attached to the rear of a laminate mold — as a support frame (Fig. 6.60).

The tube is available in round, square, rectangular, or other forms, but the square variety is recommended. For plug gauges, robotic assembly fixtures, checking, test, and inspection fixtures, 1 and 2 in. square tubes are usually used.

The tube is formed from a wrap of resin system impregnated tape, strands, braided fabrics, woven fabrics, hybrids, etc. Precoated or impregnated toe materials have also been used to form the woven braid or tube, and then, after the tube shape has been woven, the precoated fabric is cured.

Probably the least expensive, strongest, and lightest form of tube is that made by the pultrusion process. Pultruded tube possesses the capability of having the fibers aligned so that the tensile and flexural strength of the tube is greater than any of the other forms.

A number of methods may be used to connect nonmetallic tubular frame members together, as in Fig. 6.61.

Metal eggcrate, angle, and tubular frames have been used to form inexpensive support structures. When the CTE is critical and thermal expansion or contrac-

216 ADVANCED COMPOSITE MOLD MAKING

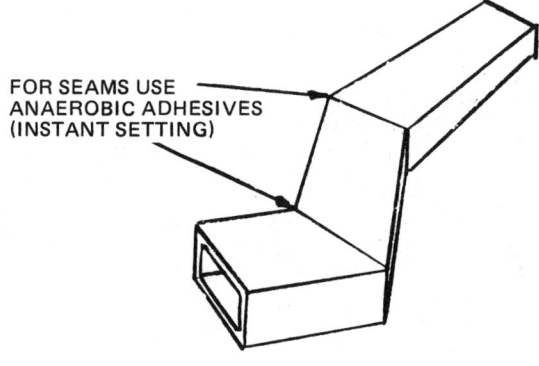

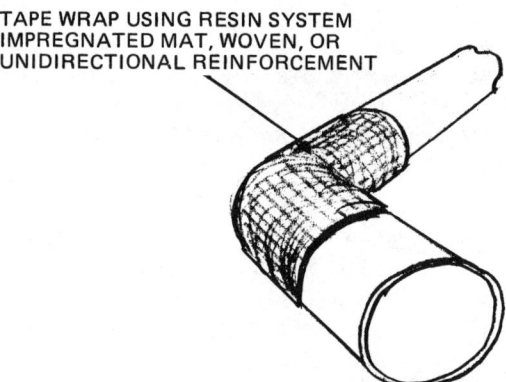

Fig. 6.61. Nonmetallic Tube Attachment Suggestions.

tion must be considered, a slightly different attachment approach must be used (Figs. 6.62 and 6.63).

The expandable pad attachment approach shown above can also be used in conjunction with an attachment element that contains a series of expansion slots machined so that the tool laminate can float on pins that move in the slots (Fig. 6.64).

Still another form of backup structure is one that is used for nickel electroform molds and tools. This method is also used with laminate molds when the mold surface or contour must be dimensionally adjusted.

Adjustable molds and backup structures. In this form, the metal or nonmetallic shell is attached to a thread pad that is plated or laminated into the mold face (Fig. 6.65).

FABRICATING MASTER MOLDS AND NONMETALLIC MOLDS AND TOOLS 217

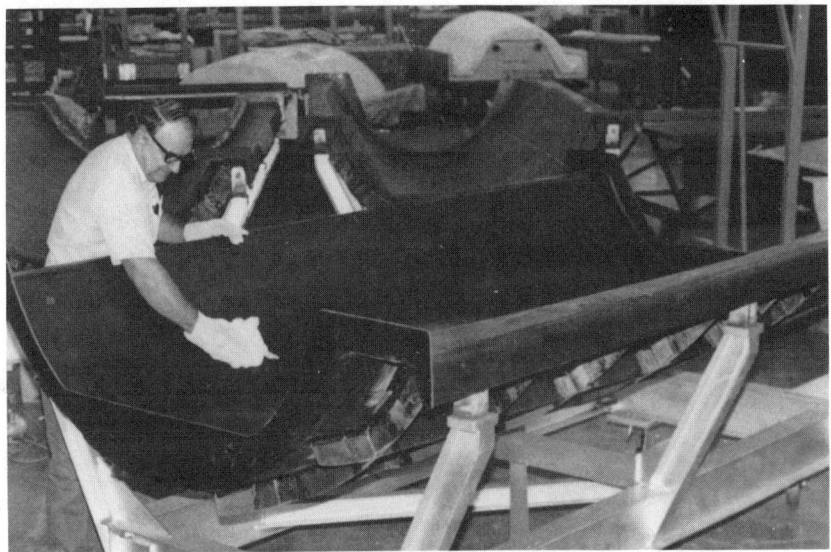

Fig. 6.62. Composite Tool with Aluminum Frame Attached to Elastomeric Pads Applied to the Rear of the Tool Laminate. *(Courtesy of Ren Plastics Div. of Ciba Geigy Corp.)*

There are many design reasons besides cost and weight that dictate the use of an adjustable frame. The most important of these is termed *springback*.

Springback of the mold from the master and/or the part from the mold can be compensated for by adjustable molds and tools. Bonding fixtures and forming molds can be fabricated to be adjustable. It is *imperative* to design a mold,

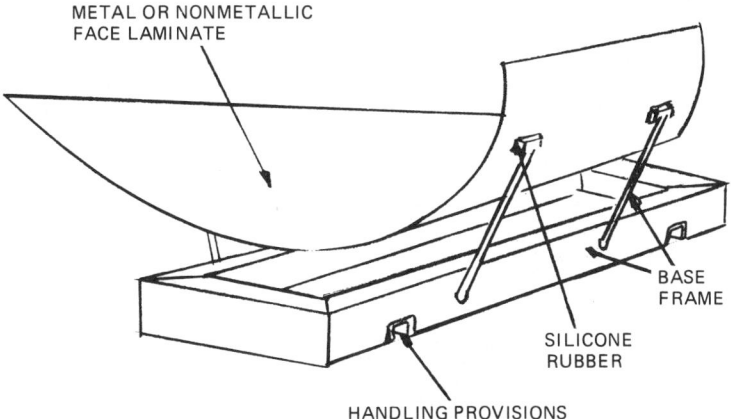

Fig. 6.63. Backup Structure with Expandable or Thermal Adjustable Attachment.

218 ADVANCED COMPOSITE MOLD MAKING

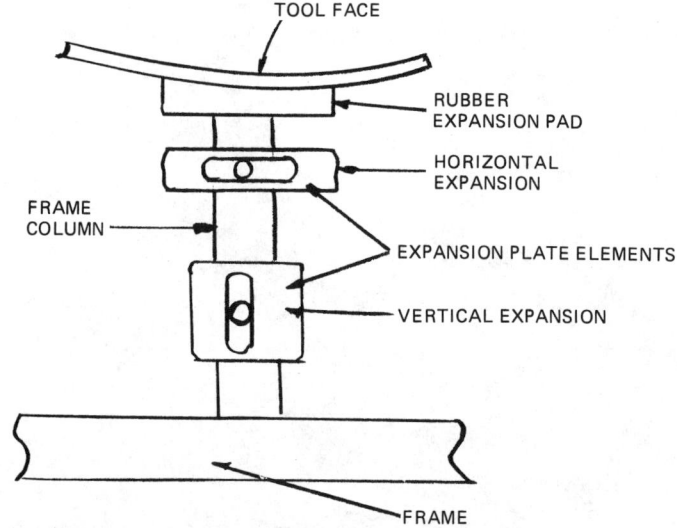

Fig. 6.64. Expansion Slot Attachment Element.

tool, or fixture to adjust out the springback by putting the mold or tool face under compression, not tension (Fig. 6.66).

Of course, designing out as much springback as possible during the structural design of the part will eliminate adjustment requirements.

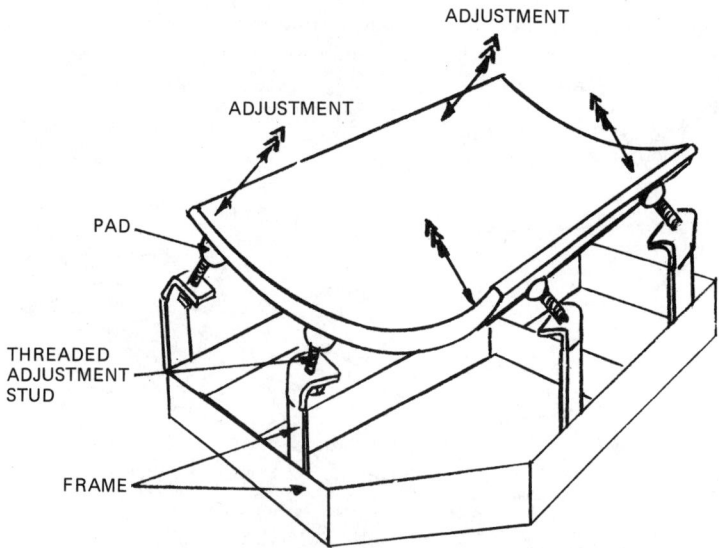

Fig. 6.65. Adjustable Molds and Tools.

FABRICATING MASTER MOLDS AND NONMETALLIC MOLDS AND TOOLS 219

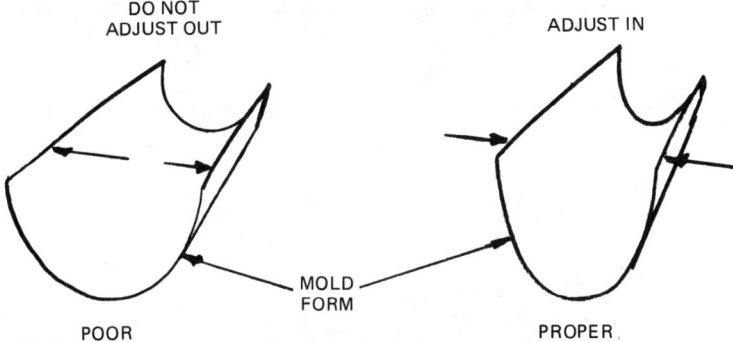

Fig. 6.66. Proper Adjustment Direction for the Removal of Springback.

Integral stiffening. This is still another form of mold and tool reinforcement. This is the least expensive form of reinforcement for laminate molds and tools. The reason for this is that the substructure is formed at the same time as the face laminate. This substructure elimination can save as much as two-thirds of the cost/ft of a mold or tool.

Figures 6.67 and 6.68A and B shows molded-in stiffeners using a simple technique whereby high-temperature foam is used to support the laminating material (wet lay-up or prepreg) during cure.

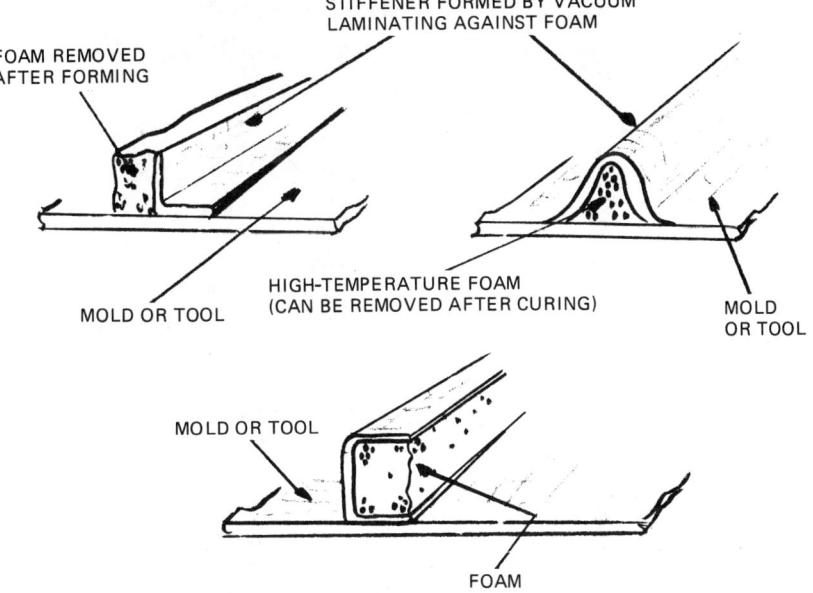

Fig. 6.67. Integral Stiffener.

A

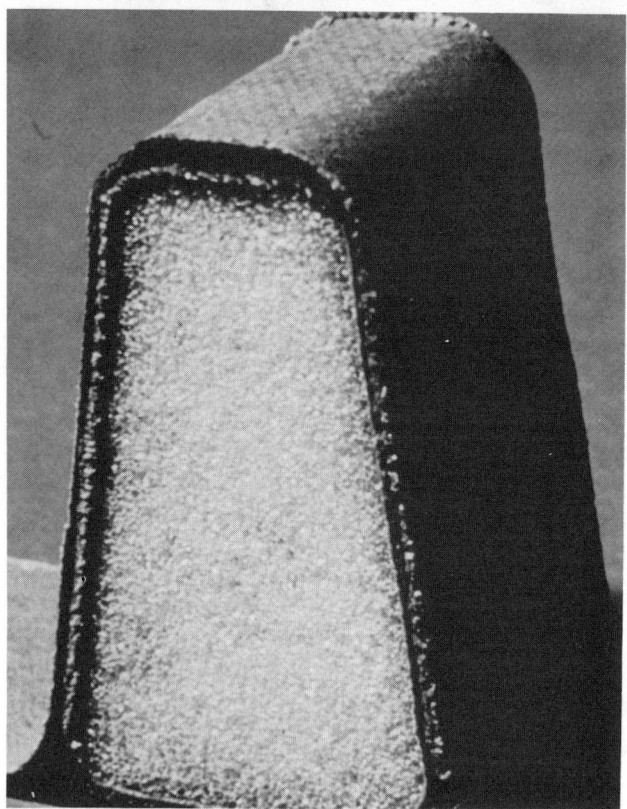

B

Fig. 6.68. A and B. Close Views of Foam Stiffener. *(Courtesy of Rohacell Corp.)*

FABRICATING MASTER MOLDS AND NONMETALLIC MOLDS AND TOOLS 221

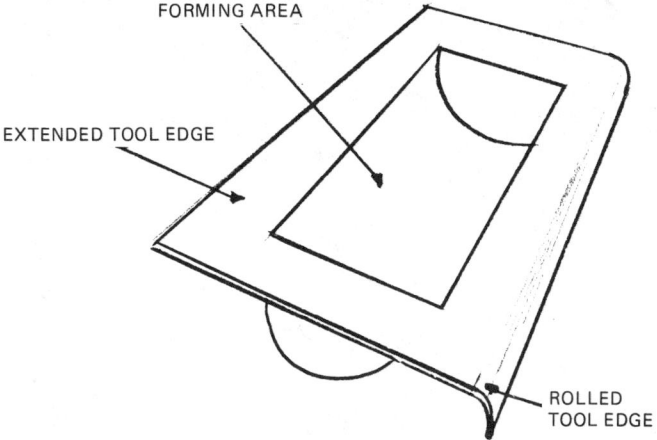

Fig. 6.69. Integrally Stiffened Tool.

Integral stiffening has been used in structural part manufacture for many years. Its use in mold and tool manufacture can create substantial cost savings (Figs. 6.69 and 6.70A and B).

A final form of integral stiffening is a technique whereby stiffeners are formed by designing the mold or tool to have a built-in brace. A *rolled tool edge* is a

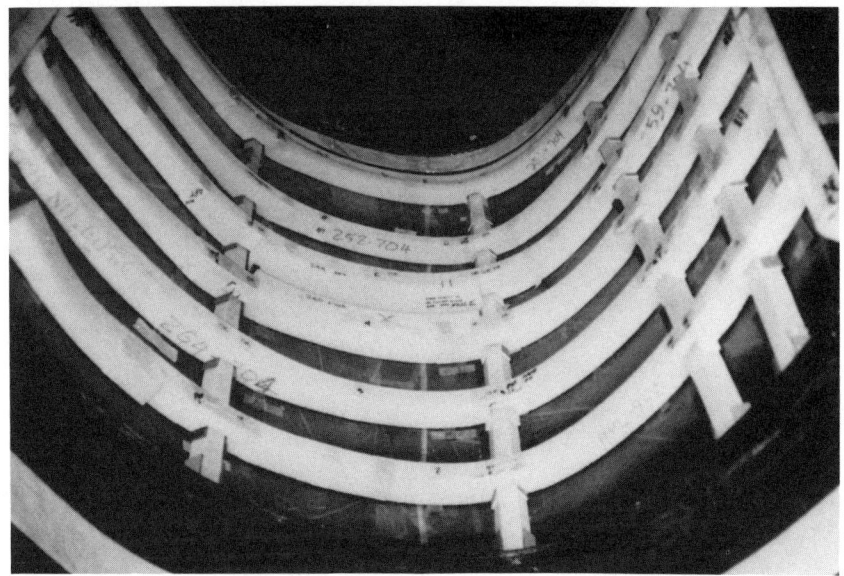

A

Fig. 6.70. A and B. Composite Part Showing Integral Foam Stiffeners. *(Courtesy of Grumman Aircraft Systems)*

222 ADVANCED COMPOSITE MOLD MAKING

B
Fig. 6.70. Continued

good example. This rolled edge or, as it is sometimes called, extended edge, also allows for a part bagging surface when a vacuum-bagged laminate part is required.

The rolled edge will also provide a base upon which the mold or tool can sit or rest (Fig. 6.71).

6.2.3. Metal-Faced, Laminated, Nonmetallic Molds

With the innovation of metallic faces applied and co-cured to the face of the laminate, fiberglass- and graphite-reinforced prepreg molds, which are state-of-the-art in the aircraft and aerospace industry, have also gained application in the automotive, marine, and appliance sectors.

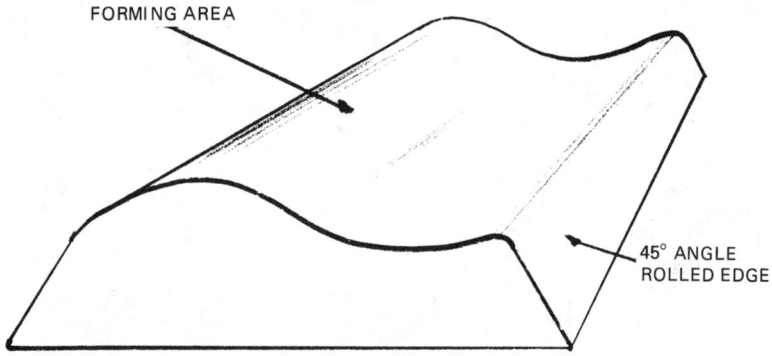

Fig. 6.71. Rolled Tool.

Fig. 6.72. Sprayed Metal Face of RIM Tool Being Applied to Wooden Pattern. *(Courtesy of Jotko Enterprises Corp.)*

In this process, a spray metal layer (like a gel coat) is the face of a laminate mold or tool. The spray metal is co-cured to the reinforced epoxy laminate and becomes the surface upon which part material is formed. Imbedded heating and cooling coils maintain the temperature of the molding face so that parts made using the vacuum bag, autoclave, RIM (reaction injection molding), RTM (resin transfer molding), rotational molding, blowmolding, and other plastic-forming processes can be produced (Fig. 6.72).

The process has evolved over the past ten years and has a solid six to eight years of proven application background in the mold- and tool-making area. Many thousands of tools presently in use produce a variety of commercial and military products. Metal faces eliminate many problems encountered when using wet and film epoxies as the surface coat on lamination or bonding tools. Before the evolution of this process, the flame spray process was considered. This method produced excessive heat that built up in the master model and which generated excessive heat in the wire and over the wire being sprayed. A flame, blowing onto the pattern surface, damaged both it and the coating being sprayed.

The arc spray method, on the other hand, heats only the wire metal being sprayed, and this molten metal, atomized, is what coats the master model. The maximum temperature of the metal material coating the surface does not cause the surface of the master pattern to rise above 150°F.

This metal application process is manual and the coating applied is dense.

Master and Model Preparation. The master model or pattern to form the mold can be made from wood, plaster, plastic, or metal. Even rubber materials and wax have been used to form a master model, since the cool spray does not deform a pattern made from any of these materials. In fact, the spray copies every detail with no distortion and a minimum of shrinkage.

Master patterns must have a sealed surface, which, since moisture will cause bubbles in the metal coating, must be absolutely moisture-free.

When plaster models are used, the use of PFPs is highly recommended. Urethane or epoxy coatings over wood patterns will provide a smooth surface upon which the parting agent can be applied. This parting agent will allow the molten metal to adhere to the master model and at the same time provide a release layer for removing the reinforced metal-faced shell. Parting agents must also protect the master pattern from the impinging effects of the molten metal and the stream of propelling air. Parting agents formulated around a PVA (polyvinyl alcohol) formulation work extremely well.

Surfaces to be duplicated also take the form of finished parts and, sometimes, existing molds. In these cases, the surfaces must be smooth, contain no "lock-on contours," and, as with all master patterns, molds, etc., the surfaces should be solvent-cleaned of oil, waxes, fingerprints, and various residue.

The reason for this care is simple: The metal spray process copies intimate detail, in the same manner as the electroform plating process.

Finally, the master pattern, existing mold, part, or object being copied must be mounted on a base. Sometimes, the base may have sides that form returns on the sprayed mold (Fig. 6.73).

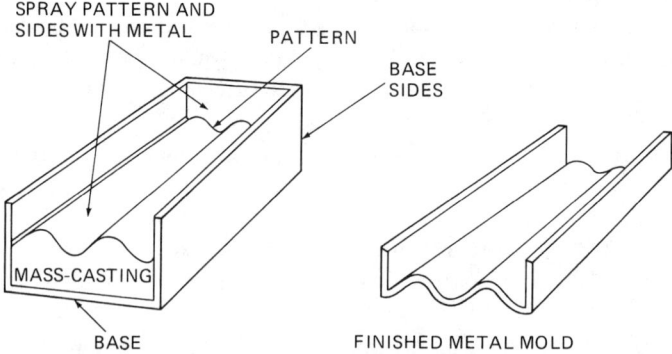

Fig. 6.73. Mass-Casting Used to Create a Sprayed Metal Mold.

FABRICATING MASTER MOLDS AND NONMETALLIC MOLDS AND TOOLS 225

The sides or base may be metallic and could become part of the mold. In addition to bases and sides, hardware, threaded inserts, attachments and other inclusions may be sprayed into the metal tool surface.

Applying the Metal Spray Face. Visualize a paint spray gun. That is the general appearance of an arc spray gun.

Instead of paint, wire of various metal alloys is automatically fed into the gun. The arc transforms the metal into a molten state and automatically controlled compressed air propels the molten metal, thus producing a consistent 3 in. diameter spray coverage on the mold surface when the gun is held at a distance of 8 in. from this surface.

To continue the comparison: The spray motion of a skilled painter, wide and long sweeping motions with coats that overlap, is suggested. The coats should be applied thinly and in opposing directions (Fig. 6.74).

The use of a kirksite-type wire alloy is recommended. Metals such as nickel, aluminum, and various alloys of steel require such a high temperature that the master model or release system would be damaged, or (at the very least) have its contour distorted.

Simple shapes may be duplicated using metals other than kirksite and tin/zinc alloys, but kirksite-type alloys appear to be the most popular choice, providing finishes that are surprisingly sharp. Consider that, at the required and suggested thickness of .090 in. areas of about 5 ft^2/hr can be coated.

The spray metal technique produces tools with significant cost and time savings. Spray metal application also allows molds and tools to be made in-house with quick turn-around times. Mold and tool repairs are also an excellent application area for the spray metal process.

Other applications will suit individual needs. Practice and experimentation, with trial runs, always provides information concerning good techniques to produce a quality finished mold or tool.

Laminated Support and Other Backups. Face laminates must be supported by a backup structure so that the tool does not deform when in use. Because the

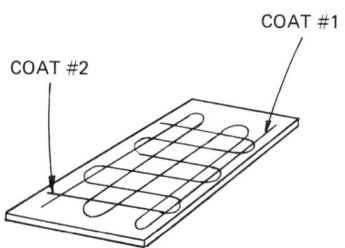

Fig. 6.74. Applying Metal for a Sprayed Metal Face.

laminate faces of a mold or tool, when used as a composite lay-up mold or bonding fixture, will usually reach temperatures up to 350°F, it is suggested that the backup structure chosen to support the laminated face be one that is thermally stabilized and post-cured to a temperature above the use temperature of the mold. After post-cure, the backup eggcrate or cast support is attached.

Fiberglass- and graphite-reinforced epoxy laminates have been used successfully for supporting spray metal faces. Both room-temperature and low-temperature curing laminates bond intimetely to the sprayed metal. These laminates allow for the fabrication of large tools, since a mass-cast backup is sometimes too heavy and massive for use as a support.

Master models, as described earlier, are usually made from room- or low-temperature materials like wood, plastic-faced plaster (PFP), machined plastic, etc.

The preimpregnated face laminating materials that cure at room temperature or 200°F, and are free-stand post-cured, are the best suited to be used as backup materials. The basic reason for this is that the resin to reinforcement ratio is controlled accurately by machine, and the CTE (as well as shrinkage) is dependent upon the amount of resin applied to the reinforcement.

Hand-applied wet lay-ups can be and are used for laminated supports for the spray metal, but prepregs are preferred. Frozen 200°F cure prepreg materials will allow the mold maker many days to fabricate a laminate, since the prepreg is initially activated for precure by temperatures of 200°F, and will not cure at room temperature. Room-temperature-cure prepregs, on the other hand, will give the mold maker about 12 hours of working time.

Wood, honeycomb laminates of various kinds, such as paper, plastic, and metal faces and cores, solid epoxy reinforced laminates, and other laminated configurations are used to support the laminate face (Fig. 6.75).

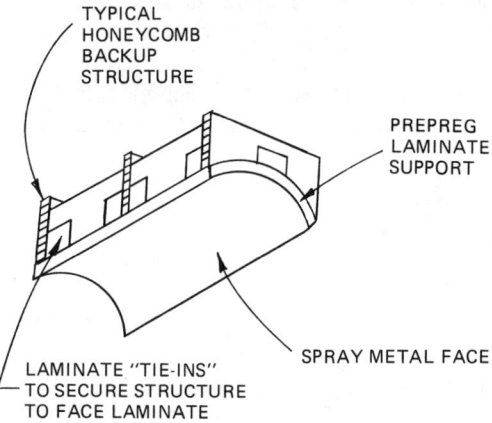

Fig. 6.75. Typical Backup.

FABRICATING MASTER MOLDS AND NONMETALLIC MOLDS AND TOOLS

Room-temperature-cure and 200°F prepregs are used as backup laminates because plaster (PFPs) should not be subjected to temperatures of more than 212°F.

Machined master models, as well as existing parts and tools, sometimes cannot be exposed to temperature, as they will distort, expand, and cause the finished tool to be dimensionally inaccurate.

After the initial cure (room temperature or 200°F), the thermally prestabilized or post-cured backup is attached to the spray metal face laminate and then the entire structure is removed from the contour surface to be post-cured, "free-standing." This step post-cures the face laminate, and now the mold or tool is ready to have the front face sealed.

In addition to the above, there are other backup materials that have been used to support the spray metal face and form the backup structure, a few of which are listed below.

Adhesive material forms like pastes are formulated by dispersing milled or chopped fibers into an epoxy resin. Then, the paste is applied over the rear of the spray metal face. Both the laminated backup and paste form of reinforcement are usually applied over a paint-on epoxy surface coat that has been allowed to gel after being applied .025 to .030 in. thick to the spray metal back surface.

Dry fabric reinforcement, impregnated with epoxy or epoxy-fiberglass prepreg layers, are usually applied over the paste layer and the paste layer/fabric is repeated until a final thickness of 1/4 to 3/8 in. is reached.

Vacuum bagging the prepreg laminates to the metal face and paste layer, and the use of a low-temperature heat gun to aid in contouring the backup material, is recommended. The use of peel ply material over the last ply of laminate material provides a rough surface to which the backup structure will bond.

The support on the back of the laminate can be a mass-cast material, cast to and in back of the rough-surfaced laminate. In fact, ceramic syntactic foam mass-cast materials are available where the entire backup is cast over the spray metal directly, thereby eliminating the laminate (Fig. 6.76).

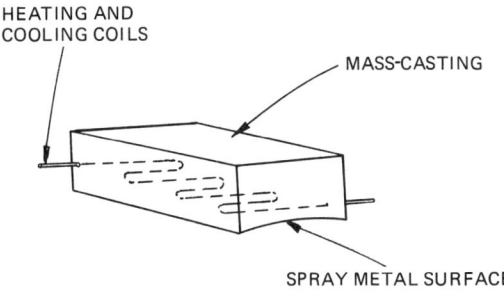

Fig. 6.76. Mass-Casting with Embedded Heating Coils and Metal Sprayed Surface.

228 ADVANCED COMPOSITE MOLD MAKING

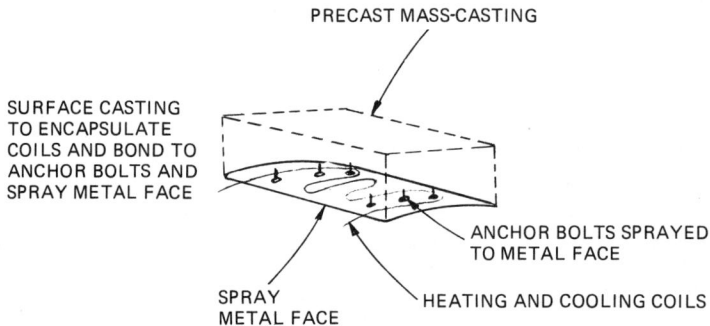

Fig. 6.77. Heating Coils Included in Sprayed Metal Face.

Cooling and heating coils are laminated, paste bonded in place, or mass-cast near the surface, when necessary, to provide mold or tool temperature control. Mass-casting materials such as metal-filled epoxies, ceramic syntactic foam and filled urethanes have been used for this.

Anchor bolts are another support approach. Bolt assemblies have been sprayed into the spray metal face and then attached to a support laminate by encapsulation, or mass-casting in place (Fig. 6.77).

Sheet metal eggcrate backups, tubular frames, honeycomb substructures, solid laminates, and a host of other methods (including plywood), can be used to support the tool face.

Spray metal, laminated, or paste layer/fabric tools have s significant weight range, as shown in Table 6.2.

Surface Filling. Various types of materials and processes are used to finish the molding or tool surface of a spray metal tool. Electroplating, brush plating of nickel, and organic and inorganic sealers are just some of these materials.

Table 6.2. Average Weights of Spray Metal Tools.

	APPROXIMATE WEIGHT
Kirksite spray-type surface, about .070 in. thick:	2 lb/ft
Laminated plastic, spray metal support back, about 3/8 to 1/2 in. thick:	4 lb/ft
Mass-cast backup (metal-filled and ceramic heavier) about 3 in. thick:	10 lb/ft
Wood, laminated honeycomb, or sheet metal eggcrate backup, about 6 in. deep:	2 to 20 lb/ft

FABRICATING MASTER MOLDS AND NONMETALLIC MOLDS AND TOOLS 229

Silane coatings and organic surface impregnants, applied to the molding surface, assure that the resins used to form laminated or cast parts will not penetrate the mold surface and cause porosity of the molded part, rough part surface, or part "lock-on" condition. Frekote Corporation manufactures these types of coatings.

Some of the newer release coatings and sealers are usable to temperatures of 900°F.

Any plating equipment or materials supplier will provide the necessary materials and processes required to electroplate the tool surface.

The finished spray-metal-faced composite tool is a superior tool in terms of utility, and produces parts equal to those produced by other methods (Figs 6.78 and 6.79).

Fig. 6.78. Spray-Metal-Faced Composite Tool Set Mounted in a RIM Press. *(Courtesy of Jotko Enterprises Corp.)*

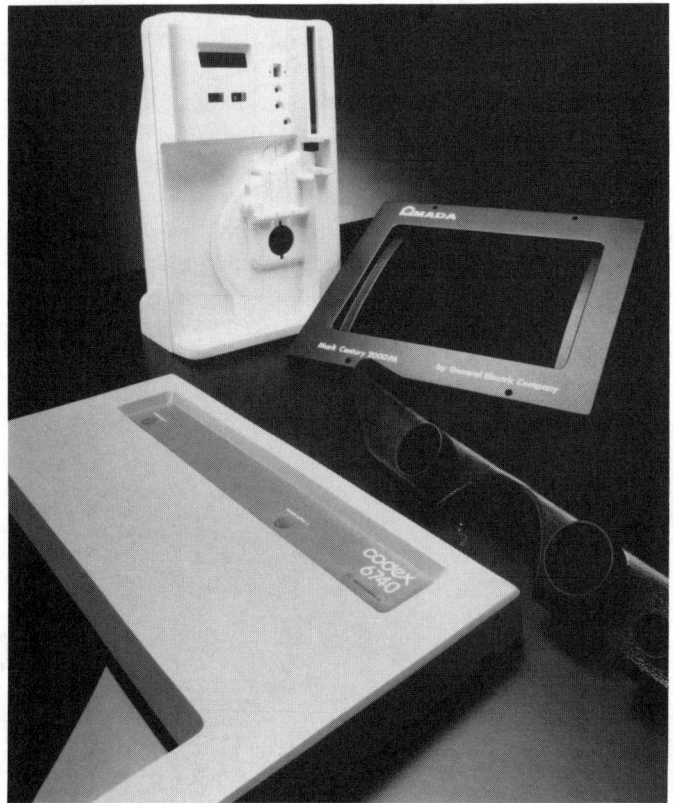

Fig. 6.79. RIM Parts Molded from Spray-Metal-Faced Composite Tools. *(Courtesy of Jotko Enterprises Corp.)*

7
Fabricating Metallic Molds

There are many methods of fabricating metallic molds for use as tools in manufacturing advanced thermoplastic and thermoset composite materials. These metallic molds are used as autoclave molds, bond fixtures and tools, and as compression, RIM, and RTM process aids.

The choice of a material and fabrication process is dependent upon the size of the tool required, complexity of contours, tool reliability and the number of finished parts in the production run.

Another important factor which determines mold fabrication choice is cost feasibility. A reasonable mold fabrication decision can only be made after careful consideration is given to all of the factors stated above.

This chapter presents current and state-of-the-art techniques for metallic mold fabrication. Three major materials and process topics will be discussed: Formed metal molds, machined metal molds, and electroformed nickel molds.

The first two major topics contain subtopics: roll forming, brake forming, stretch forming, stretch drawing, and formed and bonded thin sheet will be presented.

Machined molds and tools can be created by either N/C machining or through the use of Keller models, pantographs, or hydrotel machined. Each of the previously mentioned metallic mold processes have certain advantages and disadvantages, which should be considered when choosing a mold fabrication technique.

As we proceed through the different methods, it will be possible to categorize each process. It will soon be apparent that each has its own special purpose, but that electroformed nickel is by far the most universally chosen method for detailed and complicated molds.

7.1. FORMED ALUMINUM AND STEEL MOLDS AND TOOLS

Although the fabrication of formed aluminum or steel molds and tools is identical, the characteristics of the tool material determine the properties of the finished mold or tool. For this reason, it is important to comprehend the properties of the material from which the mold or tool will be created. The advantages and disadvantages of steel and aluminum as mold materials are as follows:

Steel
  Advantages

- Dimensionally stable with temperature
- Reasonable CTE
- Readily adaptable to combination-type fixtures
- Can be brake- or roll-formed to shape

  Disadvantages

- Slow machining
- High tool weight and mass
- Slow heat-up rate
- Oxidation prone if left untreated — should be nickel plated

Aluminum
  Advantages

- Dimensionally stable with temperature
- Readily adaptable to combination-type fixtures
- Can be brake- or roll-formed to shape
- Easily machined
- Low tool weight and mass
- Good heat-up rate
- Less costly per pound in comparison to steel

  Disadvantages

- High CTE
- Less scratch- and dent-resistant than steel
- Less durable than steel

From the above comparison, it is possible to choose a material that will reasonably match specific tool design requirements. Since tool material properties and reliability affect the finished part, the importance of materials selection in tool design cannot be overemphasized.

The fabrication of the individual constituents of metallic formed tools can be achieved in any of a number of methods, but the final construction is identical.

FABRICATING METALLIC MOLDS 233

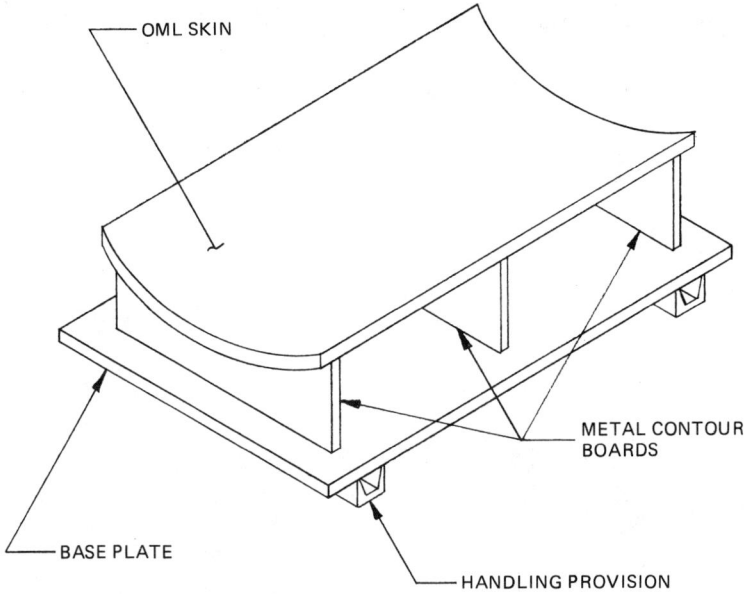

Fig. 7.1. Metallic Formed Tool.

Formed metallic tools have three major components: formed outside mold line (OML), skin, contour boards, and a base frame (Fig. 7.1).

The base frame is the main supporting structure for the entire tool. Usually, it is welded together.

Welded to the base frame are the contour boards. Since they are the backup structure for the OML skin, their curvature must match the back of the OML skin at a given station. Curvature can be generated on the contour boards through the use of master contour templates, or by N/C machinery. The data for both of these techniques is obtained from engineering geometry.

The tooling surface or OML skin can be fabricated by many methods and then welded to the contour boards. The techniques, beginning with the simplest and progressing to the more complicated, are discussed in the following material.

7.1.1. Roll Forming

This is a process in which a sheet of steel or aluminum is fed into a series of rollers and is actually rolled into shape. This process is illustrated in Fig. 7.2. The shape can be controlled by the position and pressure of the rollers. It must be emphasized that this process lends itself only to *straight line elements.*

234 ADVANCED COMPOSITE MOLD MAKING

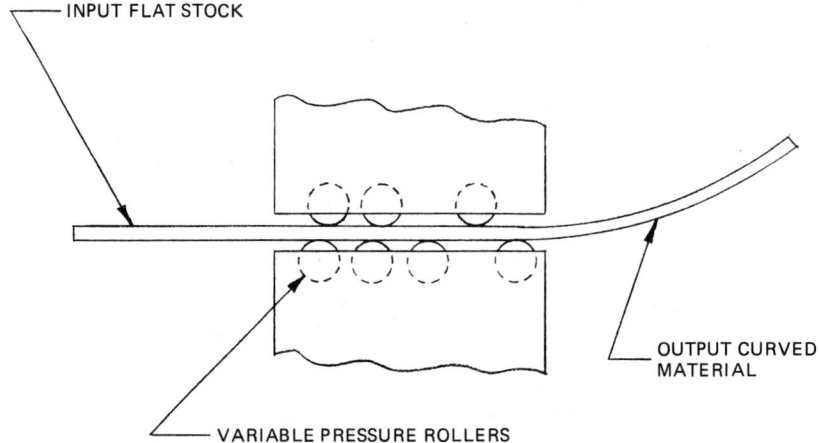

Fig. 7.2. The Roll-Forming Process.

7.1.2. Brake Forming

This is a process by which a series of bends is imposed on steel or aluminum sheet in order to generate a contour. This process is illustrated by Figs. 7.3, 7.4, and 7.5. This type of forming is not recommended for high-tolerance tools, since it can create a series of creases. Along with roll forming, brake forming should only be used for straight line elements.

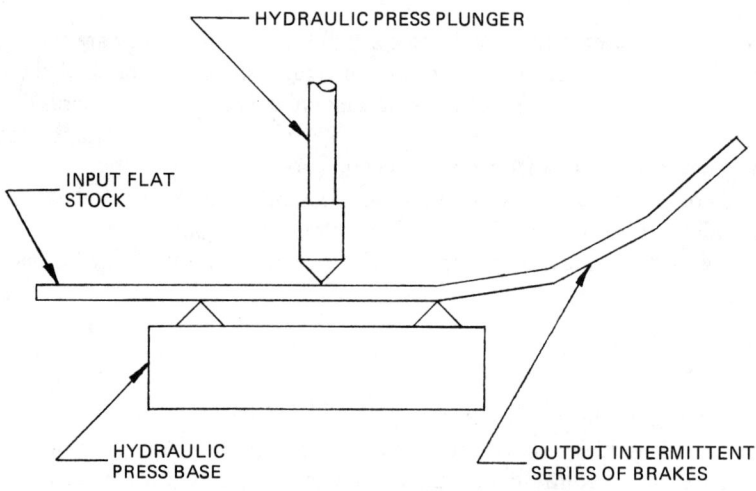

Fig. 7.3. The Brake-Forming Process.

FABRICATING METALLIC MOLDS 235

Fig. 7.4. Brake-Formed and Bent Aluminum Sheet Metal Composite Laminating Tool with a Metallic Substructure. *(Courtesy of Grumman Aircraft Systems)*

Fig. 7.5. Graphite Epoxy Prepreg Part, Integrally Stiffened, Made from Tool Shown in Fig. 7.4. *(Courtesy of Grumman Aircraft Systems)*

236 ADVANCED COMPOSITE MOLD MAKING

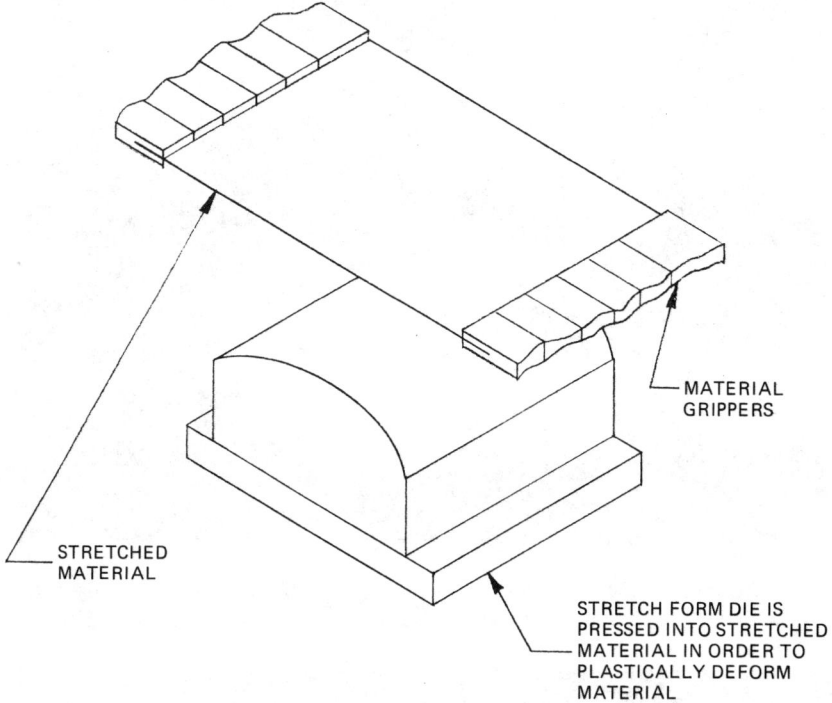

Fig. 7.6. The Stretch-Forming Process.

7.1.3. Stretch Forming

This process is intended for tool surfaces with complex curvatures. Sets of jaws grip the edges of the metallic sheet and a male die is pushed into the sheet to form the contour. This is illustrated in Fig. 7.6. The contour is created by stretching the material past its elastic limit, which deforms the molecular lattice, forming the contour.

7.1.4. Stretch Drawing

This is similar to stretch forming, except for the addition of a matched female die. The sets of jaws grip the metallic sheet and the male die is then pushed into the sheet to obtain the general contour. Continuing in the same direction, the male die with the metal stretched across the surface is pushed into a stationary matching female die. The female die can accommodate undercuts and reverse complex contours. This is illustrated in Fig. 7.7.

FABRICATING METALLIC MOLDS 237

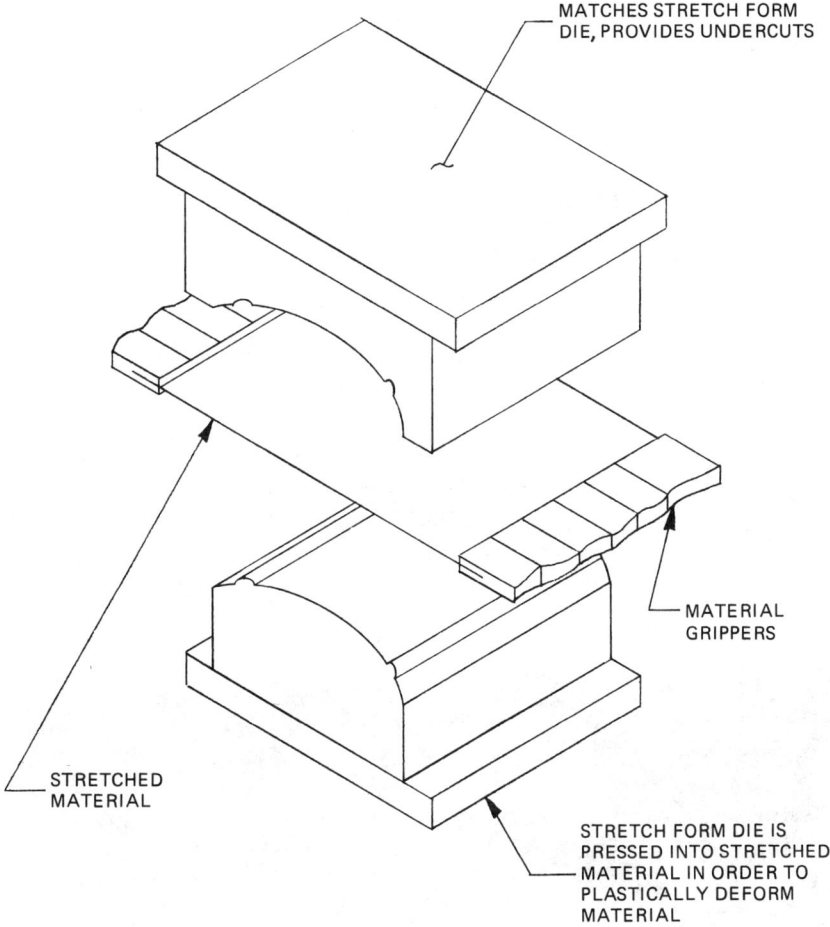

Fig. 7.7. The Stretch-Drawing Process.

7.1.5. Formed Bonded Thin Sheet

This provides for radical changes in contour which would otherwise crack or fracture thicker material in the transition region. Thin sheets of metal are formed in any of the previously described methods. Then, an adhesive film sheet is placed between successive layers of metal until a desired thickness is obtained (Fig. 7.8).

When any of the forming processes previously discussed are completed, the tool skin is now ready to be welded or bonded to the contour boards (Fig. 7.9A and B). In order to accommodate vacuum ports, the back of the tooling surface should be machined to allow for welding or bonding of the vacuum boss (Fig. 7.10).

238 ADVANCED COMPOSITE MOLD MAKING

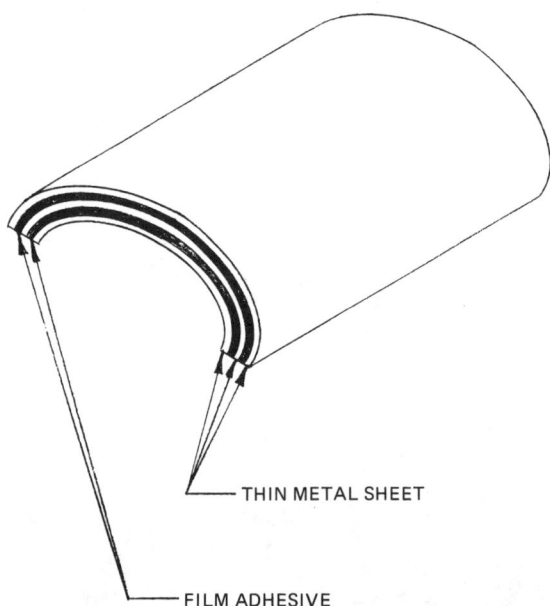

Fig. 7.8. Formed and Bonded Thin Sheet Process.

Fig. 7.9A. Roll-Formed Aluminum Metallic Tool, Ready for Machining. *(Courtesy of Grumman Aircraft Systems)*

FABRICATING METALLIC MOLDS 239

Fig. 7-9B. Roll-Formed Aluminum Metallic Tool Being Machined Using a Keller Pattern or Tracing Laminate Master Model (Hydrotel Machined).

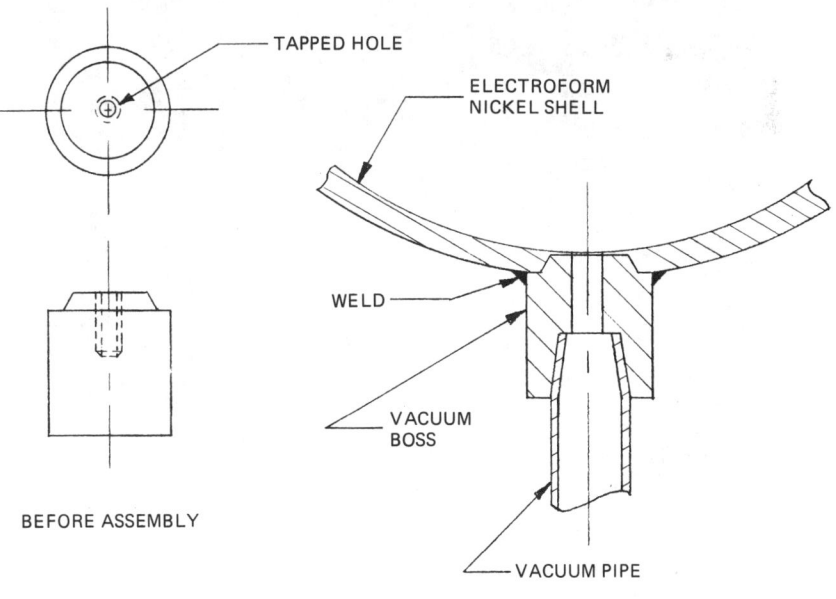

Fig. 7.10. Vacuum Boss.

240 ADVANCED COMPOSITE MOLD MAKING

7.2. MACHINED ALUMINUM AND STEEL MOLDS AND TOOLS

The properties of machined aluminum and steel tools are the same as those mentioned in the previous section.

Basic straight line or simple curved shapes can be machined in the traditional manner with the use of lathes, milling machines, etc. When the complexity increases and the contours become compound, more elaborate machining is inevitable. Through the use of Computer-Aided Design — Computer-Aided Machining (CAD–CAM), complex contours can be achieved by using numerically controlled (N/C) machinery. Here, the N/C data is transformed from engineering geometry to an N/C tape, which becomes the machining data that creates the final mold or tool (Fig. 7.11A, B, and C).

Another method of fabricating a machined tool is through the use of a Keller model. This model can be a duplicate tool, a plaster casting, a laminate, or even a plastic casting of the desired finished tool. The tool model is then used as a three-dimensional tracing template. A hydrotel machine does the tracing while simultaneously machining the final mold or tool.

A

Fig. 7.11. A, B, and C. N/C Machining of Aluminum Metallic Laminate Molds. *(Courtesy of Grumman Aircraft Systems)*

FABRICATING METALLIC MOLDS 241

B

C

Fig. 7.11. Continued

242 ADVANCED COMPOSITE MOLD MAKING

Choosing whether to use N/C machining or a Keller model is dependent upon the existence of engineering geometry in CAD–CAM, the feasibility of creating CAD–CAM data, the existence of a master model, and the feasibility of creating a Keller model (Figs. 12A–E).

7.3. ELECTROFORMING NICKEL

Electroforming is an electroplating process whereby pure hard nickel ions are deposited on either a conductive or conductive-coated master. The electroplating process is actually a nickel ion transfer from a solution of sulfamate nickel electrolytes to deposits of hard nickel particles on the conductive master.

Since the plating is an ion transfer, the process occurs very slowly, at a rate of .0005 to .001 in./hr, until a thickness of .250 in. or more is obtained. Although the plating rate is very slow, the plating process occurs 24 hours a day without interruption. Hence, for a typical plated thickness of .250 in., the in-tank plating time required can range between 4 and 11 days. Therefore, ample lead time should be anticipated and provided for in any electroplating process.

Although the ion transfer during the electroplating process is provided by electrical energy, little or no heat is generated. The lack of heat generation enables the master to be fabricated from materials with low thermal stability, such as mass-cast and machinable plastic, wax, mixed forms of rubber, and various other forms of thermoset and thermoplastic materials.

Fig. 7.12A. Machined and Segmented Metallic Molds. *(Courtesy of Grumman Aircraft Systems)*

Fig. 7.12B. Hand Finishing One Half of a Metallic Mold Set. *(Courtesy of Grumman Aircraft Systems)*

Fig. 7.12C. Matched Metallic Mold Set. *(Courtesy of Grumman Aircraft Systems)*

244 ADVANCED COMPOSITE MOLD MAKING

Fig. 7.12D. Large N/C Machined Metallic Mold Set. *(Courtesy of Grumman Aircraft Systems)*

Fig. 7.12E. Graphite Epoxy Sine Wave Spar Produced from the Metallic Mold in Fig. 7.12D. *(Courtesy of Grumman Aircraft Systems)*

Furthermore, unlike metal castings, electroformed nickel does not shrink, has zero porosity, and creates extremely accurate reproductions. Thus, it may be deduced that the electroplated surface can have a mirror finish if and only if the master surface has a mirror finish.

Exact reproduction of contour and surface finish coupled with master fabrication materials of low thermal stability and cost are the major advantages of the electroforming process.

The outstanding major disadvantages are the lengthy time needed to create a tool surface of substantial thickness, and the fact that if a large mold or tool is required, the master pattern must be transported to the plater. Large plating tanks exist at only a very few locations in various countries.

7.3.1. Master and Plating Mandrel Preparation

As stated previously, the master or plating mandrel must be fabricated from conductive-coated materials. A conductive master such as copper, aluminum, or steel can be heavy and expensive. Therefore, it is usually advantageous to create a master with low weight, high durability, and low cost. This is accomplished by using lightweight, mass-cast plastic materials, which are cast near-shape, or precast and bonded together, N/C machined, and hand finished. The mass-cast method allows for quicker and easier machining than any of the metallic master mandrel materials.

For applications that demand less tolerance, wax, thermoplastics, rubber and other thermosets can be utilized. The thermoplastic materials can be heat-formed to shape, but their durability is less than that of other materials.

Since it would be advantageous to know how the master is created prior to plating, let us examine some of the techniques for creating masters.

Plastic-Faced Plaster (PFP). This is created from a plaster material that has a coating of epoxy on the front surface (see fabrication of a PFP, Chapter 6). To aid in the drying of the PFP before using it as a plating mandrel, an oven (at low temperature) can be utilized. A low oven temperature is required to avoid cracking of the plaster.

Next, a sealant coating or masking material is applied to the bare plaster or any area to which the plating is not supposed to adhere.

Precast and Bonded Materials. Precast and bonded materials such as laminates and precast syntactic foam boards are cut and bonded near-shape. Once the bonding adhesive is cured, the plating surface can then either be N/C machined or trace-machined from an identical tool. Since the cutting leaves ridges or cusps (N/C machining), the master is then scraped and sanded as required. The precast

246 ADVANCED COMPOSITE MOLD MAKING

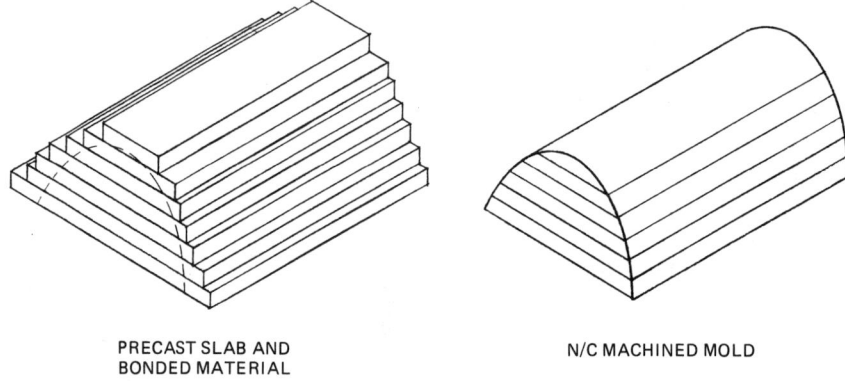

PRECAST SLAB AND   N/C MACHINED MOLD
BONDED MATERIAL

Fig. 7.13. Precast and Bonded Fabrication.

and bonded tool is now ready for plating preparation. Scraped and sanded surfaces may be coated with a fairing compound, if required. Figure 7.13 shows the precast and bonded fabrication technique.

Wet or Prepreg Lay-ups. Wet or prepreg layups using paint on surface coats are usually carried out on a tool or a simulated tool surface. Successive layers of fiberglass cloth along with a resin matrix are applied to the tool surface, using a contact molding technique until the desired thickness is obtained (See Chapter 6).

Preimpregnated fabric reinforcement (Prepreg) may also be applied to produce a laminated plating mandrel. (See Chapter 6).

The backup structure can be bonded to the back of the laminate. Upon curing, the fiberglass-reinforced matrix resin (epoxy or polyester) is ready for plating preparation.

Mass-Casting. This is the newest technique, and has the potential to eliminate many manufacturing steps, which in turn allows for an extreme reduction in cost.

A syntactic foam, in the form of a castable resin system, is formulated, usually with a lightweight filler such as hollow microballoons or spheres.

The mass-cast form is cast near-shape and cured. Upon curing, the syntactic foam reaches nearly the density of wood, for lightweight and easy machining and handling.

The near-shape, cast plating mandrel can be either Hydrotel or N/C machined to shape. Since the mandrel was cast near-shape, little machining is required.

After machining, the syntactic foam mandrel surface can be scraped to remove machining ridges (cusps), or covered with a fairing compound to avoid the finish hand work. The mandrel can now be coated with a conductive film for plating (Fig. 7.14).

FABRICATING METALLIC MOLDS 247

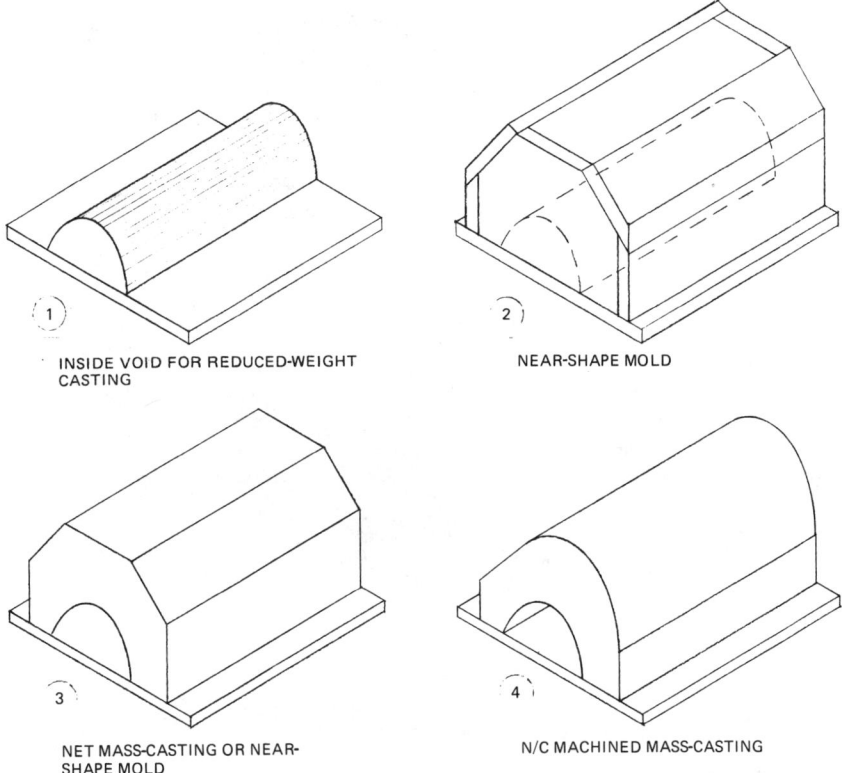

Fig. 7.14. Mass-Casting Mold.

Whatever material is used for the nonconducting master, or the fabrication technique, the preparation prior to electroplating is the same. The mandrel is solvent-cleaned to remove dirt, oil, grease, or other foreign debris. Then, a suspension of silver powders, combined with a variety of organic binder systems, is applied to the mandrel surface by either dip, spray, brush, stylus, syringe, or screen print.

Since it is usually preferred that the substructure of the mandrel not be plated, it can be coated or painted with a nonconductive paintable masking material. The nonconductive coating does not permit ion transfer, preventing electroplating from taking place on the covered areas. See Figs. 7.15A–E for an illustration of the steps of this procedure.

7.3.2. Electroplating

Regardless of the plating mandrel material, the actual electroplating process is basic and similar from material to material.

248 ADVANCED COMPOSITE MOLD MAKING

Fig. 7.15A. Tool Contour Duplicated by Mass-Casting. *(Courtesy of Grumman Aircraft Systems)*

Fig. 7.15B. Metallic Mold Frame Walls to Cast Inside of. *(Courtesy of Grumman Aircraft Systems)*

FABRICATING METALLIC MOLDS 249

Fig. 7.15C and D. Large Mass-Cast Plating Mandrel or Composite Lay-Up Tool. *(Courtesy of Grumman Aircraft Systems)*

250   ADVANCED COMPOSITE MOLD MAKING

Fig. 7.15E.   Electroformed Nickel Tool Showing Adjustable Substructure.
(Courtesy of Grumman Aircraft Systems)

Since electroplating is a transfer of ions, stimulated by electrical energy, it is obvious that positive and negative terminals are required for the process to take place. The negative terminal is usually the conductive or conductive-coated mandrel, and the positive terminal is submerged within the sulfamate nickel electrolyte bath. When the polarity is arranged in this fashion, the nickel ions from the electrolyte bath migrate to the negatively charged mandrel, where the ions form hard deposits of nickel. If the polarity is reversed, so that the mandrel is positively charged, the electroplating process reverses, and the hard nickel deposits on the mandrel revert to nickel ions within the electrolyte bath (reverse plating).

When on-off or forward-reverse plating is carried out in a series with a longer on or forward plating time duration than off, (reverse plating time duration) a more uniform tool thickness is achieved. Although on-off plating yields a better, more consistent tool, it should be noted that about twice the amount of time is required, compared to straightforward plating. For example, straightforward plating occurs at .0005 to .001 in./hr, but for on-off plating, the rate is half that. Hence, a trade-off between tool uniformity and plating time duration must take place.

7.3.3. Backup Structure

In order to insure finish part stability and reliability, some type of backup structure must be attached to the nickel electroplate.

There are two basic types of backup structures for electroformed nickel tools, the first of which is designed for small tools of two square feet or less.

The small tool backup structure consists of a series of metallic angles which are screwed to both the nickel electroplate and the main supporting structure, as shown in Fig. 7.16. This allows for rigid tool support, while not inhibiting the quick heat-up rate during autoclave or oven cycles, which is inherent to nickel tools.

For larger electroformed tools, a different, more flexible backup structure may be utilized (Fig. 7.17).

This type of backup structure has bronze bosses electroformed into the tool and attached to the backup structure via threaded rods. In order to electroform over the bronze bosses, the plating is allowed to continue until half the desired plating thickness is obtained. At this point, the plating is interrupted, and the bronze boss is bonded to the tool with an epoxy adhesive. Then, the plating is

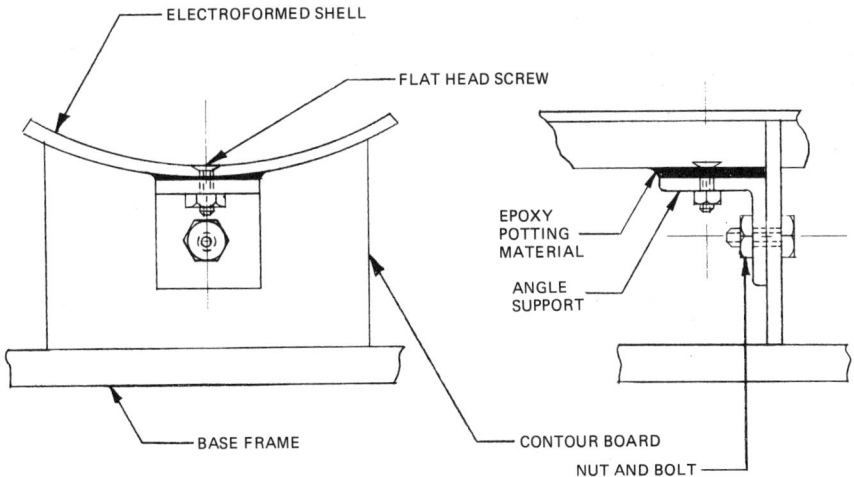

Fig. 7.16. Small Electroformed Backup Structure.

252 ADVANCED COMPOSITE MOLD MAKING

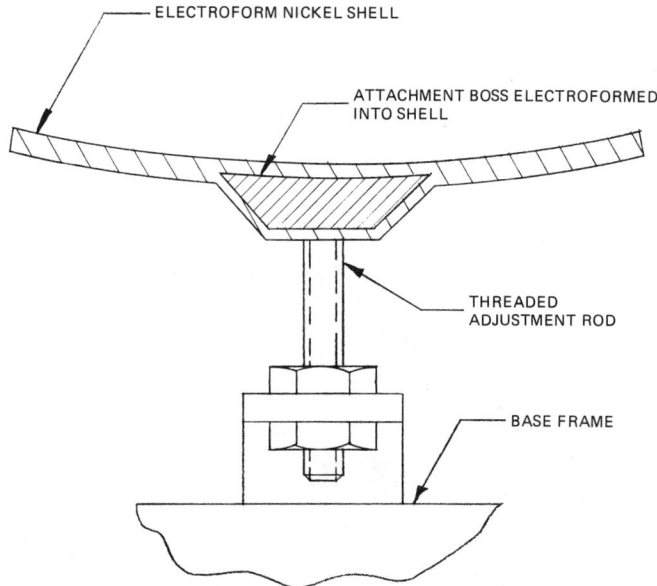

Fig. 7.17. Large Adjustable Backup Structure.

allowed to continue, encapsulating the bronze boss within the nickel tool. The boss area can now be drilled and tapped for attachment to the backup structure or for vacuum port provisions. Using threaded rods for attachment to the backup structure enables the tool maker to adjust the tool against distortion or springback of the finished part.

7.3.4. Testing Porosity and Shape

Testing porosity can be accomplished easily through a vacuum check, in which the vacuum leak rate determines the tool integrity.

A shape test is carried out by laying-up two plies of the appropriate composite material. This shape test or tool prove-out is an extremely important operation, and must be carried out prior to production.

8
Conventional or Permanent Reusable Vacuum Bags

Vacuum bagging is usually a topic covered under the production parts section of a book such as this. But because of its importance to and impact upon advanced composite mold making, the following technical information is presented in detail, for use in preparing a vacuum-bagged laminated mold master, laminate production rate mold or tool, and reusable elastomeric, plain, or custom-shaped vacuum bag systems or accessories.

For those who know about the materials and process for fabricating and applying vacuum bags, this section should serve as a review. For those who are just beginning to acquire such technical information and data, the theory and purpose of vacuum bagging is explained below.

There are a number of ways that a laminated mold or tool (and even a part) can be formed when using a laminate construction approach. Component parts of a laminate mold, tool, or part assembly can also be attached together by using various forms of adhesive, cured under pressure. Several methods are used to apply this pressure. Some of these might include mechanical or hydraulic presses, autoclaves, pneumatic or mechanical clamps, compression chambers or cavities, and various kinds of solids or liquid pressure.

Pressure is continually acting upon us and all that surrounds us, downward (towards the earth's center) at a pressure of 14.7 lbs/in. To take advantage of these forces for forming laminates, laminated parts, or bonded assemblies, one can apply the vacuum bag theory. If pressure must be applied over a large surface and a mechanical or hydraulic press, clamp or other means is not practical, a vacuum bag, installed over the entire surface can provide the necessary pressure.

Here is the method of bag installation:

1. A flexible sealant is installed around the master mold or tool periphery.
2. A material (breather) is provided beneath the bag and over the surface to be pressed, so that air can circulate and travel to the vacuum exit or port.

254 ADVANCED COMPOSITE MOLD MAKING

3. The bag material is installed, by pressing or sealing it to the tape sealant around the mold periphery.
4. The vacuum exit or port is installed and connected, and the air is drawn from beneath the bag. Atmospheric pressure then squeezes whatever has been sealed beneath the bag at whatever pressure has been chosen, up to the capacity of the vacuum pump and/or the quality of the edge seal. The pressures usually vary from approximately 10–13 psi, unless the bagged assembly is inserted into an autoclave. Figure 8.1 illustrates a typical vacuum bag installation.

Vacuum bags also fulfill another function, which is to allow the escape of volatiles, moisture, and other by-products of the reactions of the materials beneath the bag during the cure cycle. The by-products escape and are removed from the laminations or components beneath the bag and through the vacuum system.

Vacuum bags allow for the production of large, high quality, lower cost composite parts that cannot be fabricated by hand pressure alone.

In the following chapter, conventional, disposable vacuum bagging materials and techniques, using bag films, vacuum breathers, resin bleeders, peel plies, and other accessories of metallic or nonmetallic materials will be presented.

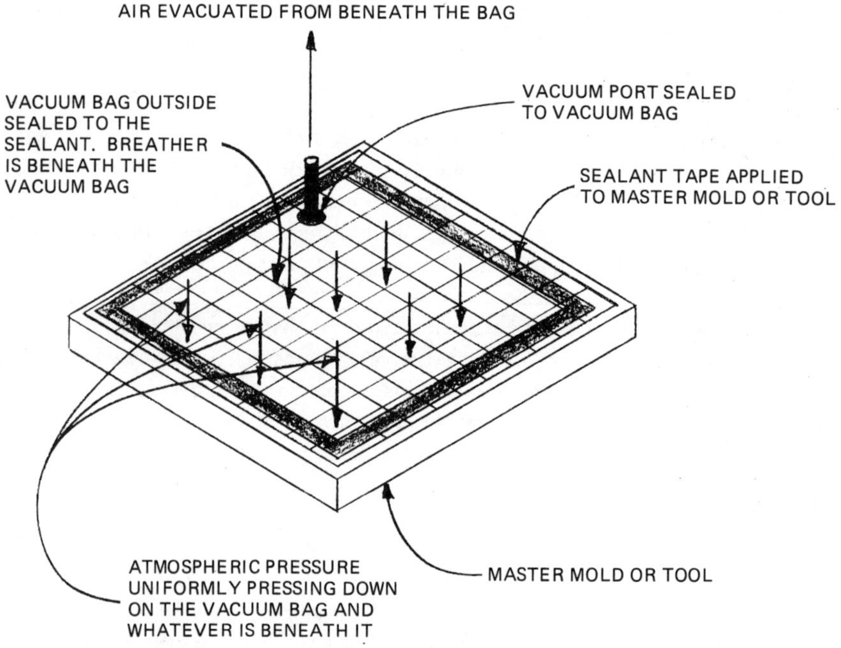

Fig. 8.1. Typical Vacuum Bag Installation.

This will be followed by an examination of semiautomatic, reusable rubber bag materials and processes — very important mold and tool material and process accessories which should be included in the scope of the mold and tool designer's library. Thus, for the first time, this chapter will include detailed fabrication techniques which will provide the designer and fabricator with the basic information to be used to create even the most complicated and complex reusable vacuum bags. In this review of techniques, sprayed, laminated, cured/uncured, reinforced, and unreinforced materials will be examined.

Following this, a presentation will be made of materials and methods that can be used to produce the seals around the periphery of a reusable vacuum bag, using rigid or flexible closures and edge seals, applied either to the tool or bag edge.

This will be followed by an explanation of elastomeric materials for pressure pads used in both autoclave and press molding. Pressure pads help to form radius corners or other part contour surfaces that the vacuum bag or press alone cannot shape.

Elastomeric accessories such as molding radius intensifiers, rigid and flexible, flat, and contoured caul plates and caul pads for conventional and permanent vacuum-bagging processes will be presented. In this section, the caul pad and caul plate materials will be shown prepared with and without internal reinforcements. Molding techniques using unique pressure intensifier methods will also be demonstrated.

The final elastomeric material and process technique to be presented will be elastomeric thermal expansion molding. For convenience, the use of rubber-type and foam-type expandable materials will be used to demonstrate thermal expansion molding, thermal expansion foam molding, and thermal expansion resin transfer molding.

8.1. CONVENTIONAL DISPOSABLE VACUUM BAG AND ACCESSORY MATERIALS AND TECHNIQUES

The trend in production vacuum bagging systems seems to be towards high temperature, reusable vacuum bagging materials and systems. This topic, as well as other advanced state-of-the-art vacuum bagging materials and processes will be covered in more detail later in the chapter, but in order to fully appreciate the reasons for this trend, an examination of conventional materials and techniques is necessary.

The trend in conventional disposable materials, bagging films, release fabrics, peel plies and films, vacuum bag sealant tapes, and other materials in the disposable category is toward the development and production of higher temperature materials (600°F or more). One of the leading factors contributing to this trend is the recent development of castable mold-making materials usable to

256 ADVANCED COMPOSITE MOLD MAKING

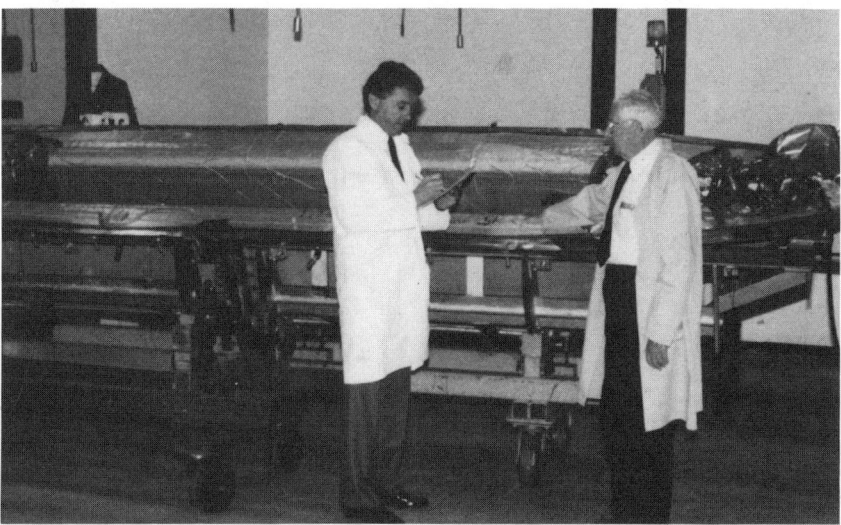

Fig. 8.2. Large High-Temperature Polyimide Part, Vacuum Bagged Using Disposable Film Materials. *(Courtesy of Grumman Aircraft Systems)*

3000°F, and part-making materials requiring process forming temperatures in excess of 700°F (advanced thermoplastic composites) (see Fig. 8.2).

Usually, the vacuum bag is used to apply pressure to surfaces of a composite part laminate or structure, either prior to curing a thermoset laminate lay-up, or for forming and shaping an advanced thermoplastic composite laminate part material. Foils of stainless steel, high-temperature films, and other diaphragm materials are presently used to fabricate vacuum bags for forming these materials.

For forming a composite mold laminate or tool, the conventional disposable vacuum bag materials and system are used to compact the ply pieces of a laminate composite mold during lay-up, when necessary. These bag materials are also used to form the vacuum bag surface pressure required during the initial or final cure of the tool laminate.

Room-temperature-cure mold and tool laminate materials can be vacuum bagged using inexpensive bagging films such as Polyvinyl Alcohol (PVA) and even bags made from disposable thin latex rubber sheet and sealants. Pressure on the surface of the mold or tool laminate, as previously mentioned, is formed by pumping and evacuating the air from beneath the vacuum bag, which is sealed along the master pattern or tool edges using zinc chromate, natural, or even butyl rubber sealants.

Oven cure vacuum-bagged mold and tool laminates require the use of disposable films or reusable elastomers that can withstand the oven processing temperatures.

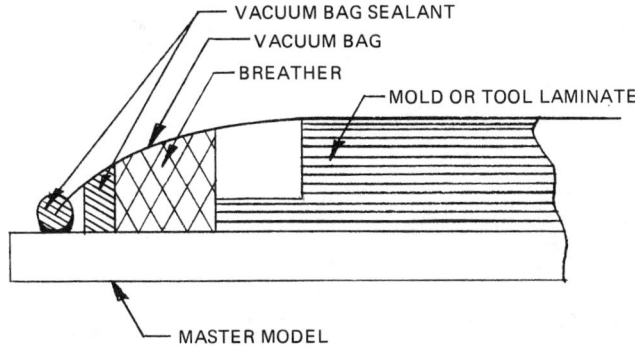

Fig. 8.3. Double-Sealed Vacuum Bag Edge.

Autoclave use, such as debulking (densifying, consolidating, and squeezing with pressure at moderate temperatures) of the laminate, as well as compacting the laminate at room temperature, can be accomplished with less expensive vacuum-bagging materials. The final cure of the laminate, however, must be performed using higher temperature bag materials such as nylon, kapton, or equivalent temperature-resistant materials. In this case, the vacuum bag is sometimes double-sealed along the edge (Fig. 8.3).

In some cases, a double bag, or two bags placed together, are used over the laminate. This is to assure that if a leak occurs in one bag, the other will maintain the vacuum pressure (Fig. 8.4).

This extra precaution is also sometimes used to assure that bag leaks will not be the cause for the loss of a costly laminate material such as graphite-fabric-reinforced epoxy or the labor required to build the tool.

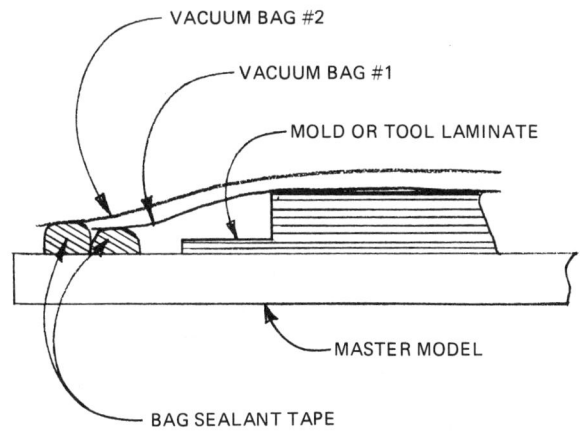

Fig. 8.4. Double Vacuum Bag.

258 ADVANCED COMPOSITE MOLD MAKING

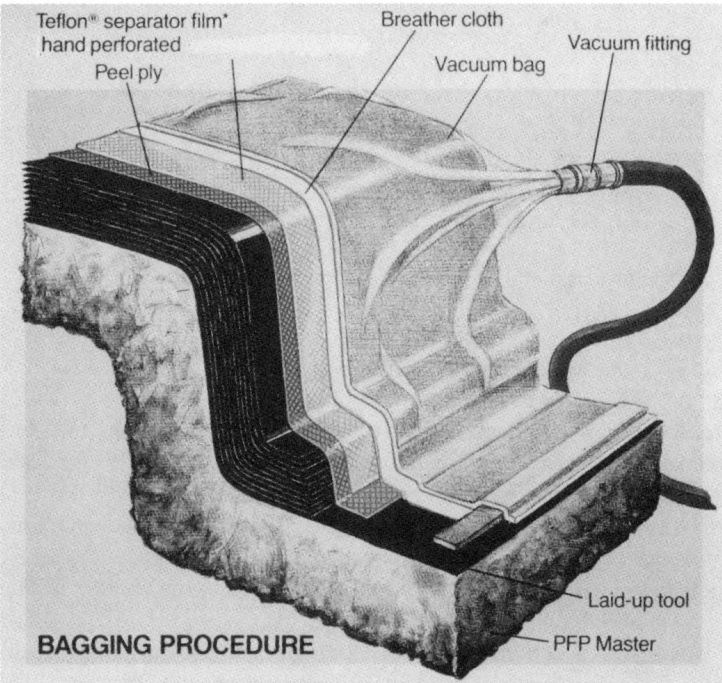

Fig. 8.5. Typical Bagging Procedure. *(Courtesy of the Toolrite Div. of Fiberite Corp.)*

There are two bagging methods usually used to form a mold or tool laminate: 1) skin or surface bagging, and 2) envelope bagging.

Skin or surface bagging is used only when vacuum integrity of the master pattern or mold surface has been assured. The surface may be a plastic-faced plaster model, wood model, plaster master, laminate, metal form, or any other contour which needs to be duplicated (Fig. 8.5).

If the surface is porous, or vacuum integrity cannot be assured, envelope bagging the entire assembly or item is employed.

Whatever the bagging method used, the bag becomes the sealed-edge membrane or nonporous blanket that separates the lay-up from the atmosphere so that the pressurizing gases of the atmosphere or autoclave (such as carbon dioxide, nitrogen, etc.) can compact, debulk, and compress the lay-up, allowing volatiles, entrapped air, excess resin, or other undesirable constituents to be removed. The bag, when used in an autocalve, is usually vented to the atmosphere through the clave wall, since autoclave pressures far surpass atmospheric ones, and the increased pressure will drive undesired by-products, volatiles, or moisture into the atmosphere (Fig. 8.6).

Fig. 8.6. Porous Release-Coated Fabric Material Being Processed. *(Courtesy of the Keene Laminates Div. of Keene Corp.)*

Whether surface, skin, or envelope bagging, there is no size limitation for the mold, tool, or part to be processed.

As mentioned earlier, the pressure of the vacuum bag against the laminate is formed by a vacuum at atmospheric pressure, the increased pressure of the autoclave, and, in some rare instances, the pressure that is formed by the pressurizing of the space above the bag, but within the mold confines. This last method is defined as pressure bag molding.

Pressure bag molding is between vacuum bag and autoclave molding, since it is only in a chamber that pressures above atmospheric can be developed. The choice of pressure molding methods used to produce a laminate mold and/or part may be picked from the following methods:

- Contact molding pressure only (placing laminations by hand with no vacuum bag assist).
- Vacuum bag molding.
- Pressure bag molding.
- Autoclave or pressure vessel molding.

260   ADVANCED COMPOSITE MOLD MAKING

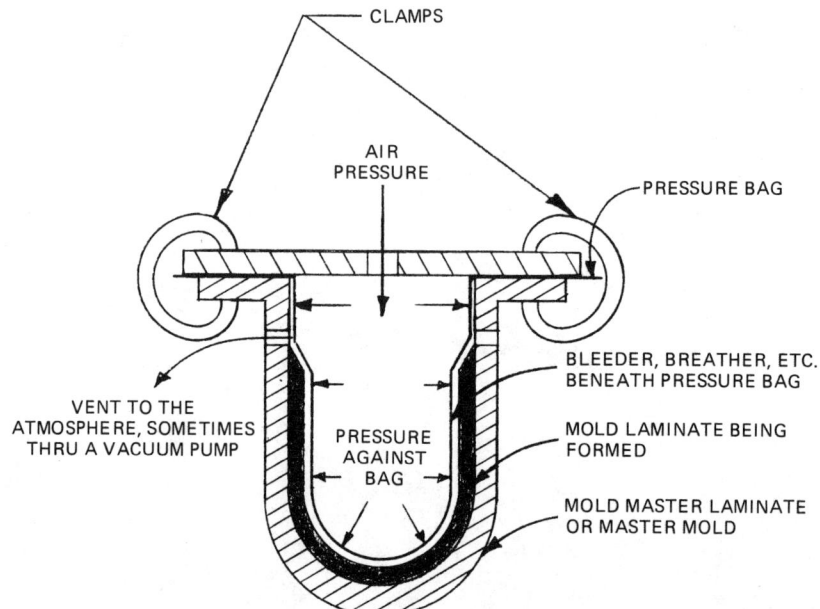

Fig. 8.7. Pressure Bagging.

When using pressure bag molding, a laminate can be pressurized within the mold master, laminate copy, or the master model itself, if the master or laminate can withstand the pressure forces. Pressure chambers or vessels, even autoclaves without internal heat, can be used to compact and cure a laminate mold at room temperature.

When producing a laminate mold in a self-contained master or model, the pressure bag is installed as illustrated in Fig. 8.7.

Pressures in excess of atmospheric (up to approximately 80 psi) are used.

At this point, in order to understand fully what is supposed to be beneath the bag as the laminate material and bag stack is assembled, it is necessary to present the suggested methods for establishing the vacuum bagging stack sequence of a wet lay-up or prepreg mold or part laminate.

Resin bleeding is used to remove excess resin from the mold or tool laminate, thereby leaving a stronger, more durable cured structure. The use of edge resin bleeding or vertical resin bleeding of the laminate is demonstrated by Figs. 8.8 and 8.9.

The suggested usual part bagging techniques are presented here for those unfamiliar with general, conventional, disposable vacuum bag techniques. Once these basic methods are understood, the mold or tool maker can improvise and adapt the process to any bagging method, including those required to form the laminate mold or tool.

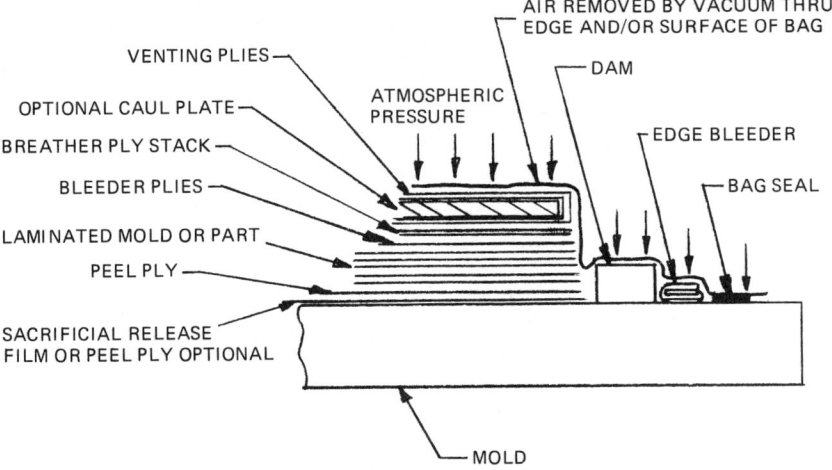

Fig. 8.8. Vacuum Bag Stack Sequence for a Vertical-Bleeding Laminate.

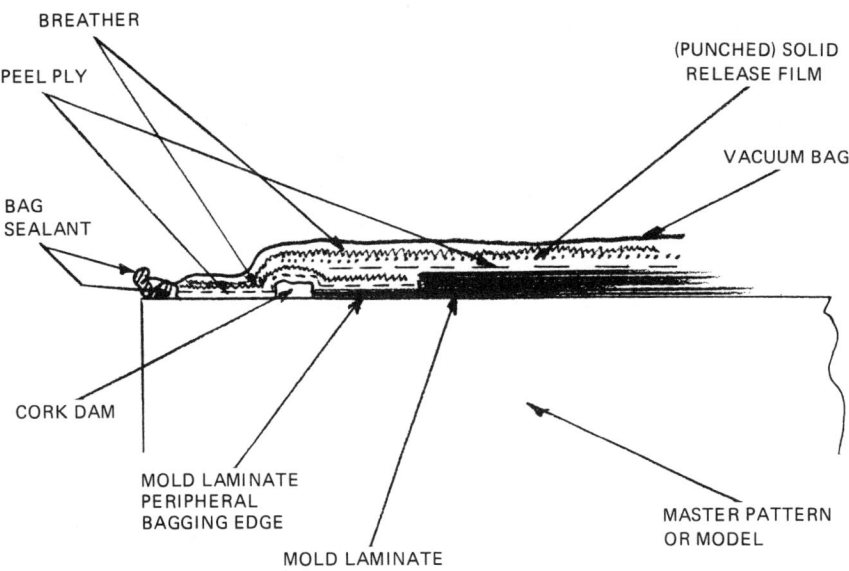

Fig. 8.9. Vacuum Bag Stack Sequence for an Edge-Bleeding Laminate.

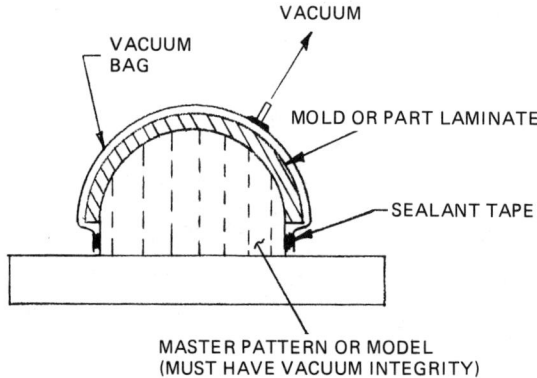

Fig. 8.10. Vacuum Bag on the Surface of a Master Pattern or Model.

8.1.1. Skin or Surface Bagging

Skin or surface bagging can take place on the surface of a master pattern or laminate mold copy if the surface being duplicated has vacuum integrity. The method is demonstrated by Figs. 8.10 and 8.11.

One of the most important considerations in the placement of a vacuum bag is the careful elimination of "bridges" in the disposable elastomer, film, or even reusable bag. Figures 8.12 and 8.13 depict the bridging of a bag and a suggested method of eliminating it.

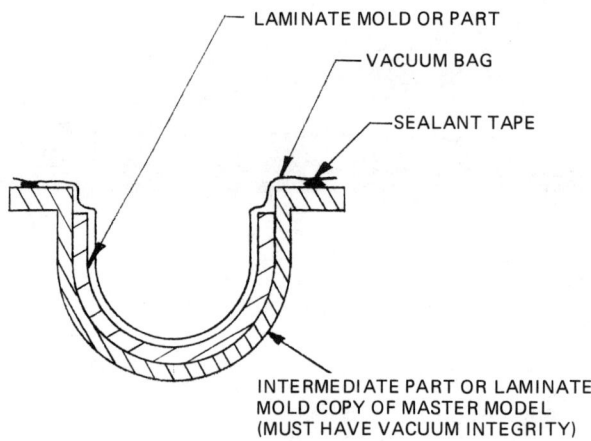

Fig. 8.11. Vacuum Bag on the Surface of a Laminate Mold Copy.

CONVENTIONAL OR PERMANENT REUSABLE VACUUM BAGS 263

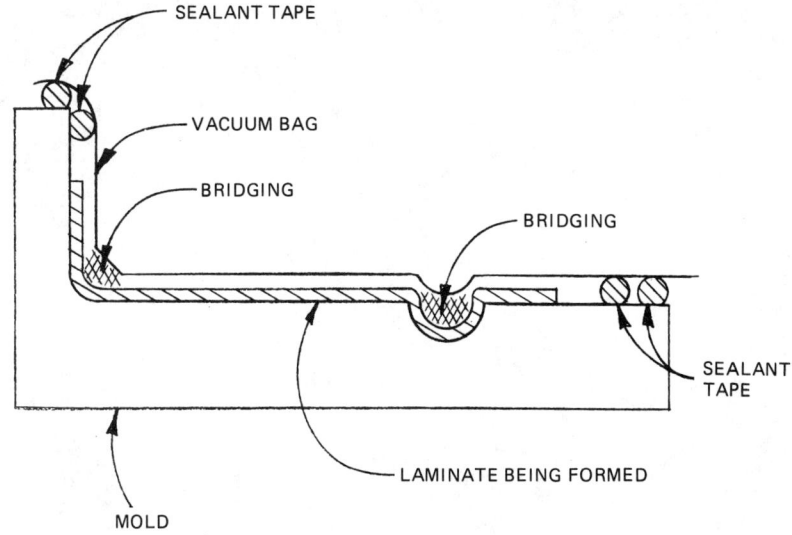

Fig. 8.12. Bag Bridging.

The bridging creates a lack of pressure through the bag in radius corners. This will, in turn, cause resin rich areas, weak and oversize laminates, and other characteristics of poor physical integrity in the laminate mold or tool. Careful place-

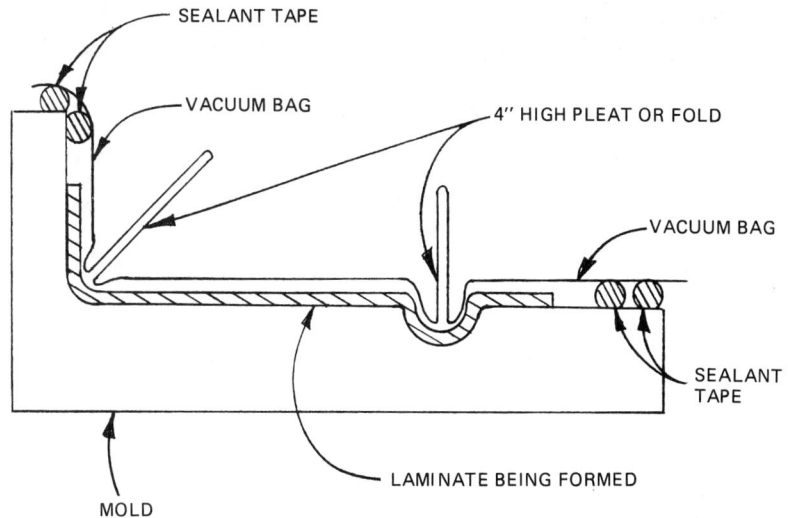

Fig. 8.13. Elimination of Bag Bridging.

264 ADVANCED COMPOSITE MOLD MAKING

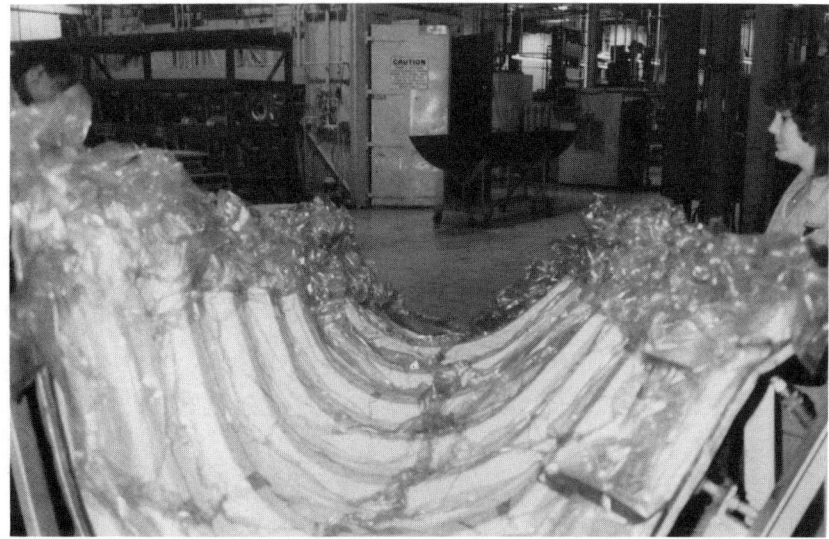

Fig. 8.14. Typical Disposable Film Pleated Bag. *(Courtesy of Grumman Aircraft Systems)*

ment of the vacuum bag and the introduction of pleats or folds will eliminate the problem (Fig. 8.14).

Another way to guard against radius bridging is to use bag materials with increased elongation. The cutting of breathers and bleeders for placement beneath the bag, and even the proper cutting of the laminate materials used to form the mold or part in radius corners must be performed in order to eliminate a corner bridge.

Another bag-forming aid, beside the pleat or fold, is the rabbit ear, or tab. A rabbit ear, or tab, is formed like a pleat or fold, but is approximately 4 in. high and ends at the bag edge (Fig. 8.15).

The inclusion of rabbit ears assists in relaxing the vacuum bag, eliminating tugging, pulling, or bridging of the bag. The rabbit ear or tab is placed at the edge of the mold or tool directly across from a contour change in the mold or part laminate being formed (Fig. 8.16).

Whether being used for skin bagging or envelope bagging, the disposable vacuum bag must have the corners, edges, and drastic changes in contour carefully fitted with a fold, which is referred to as a pleat or rabbit ear when it is extended out to the bag edge. To insure that a quality installation is performed, the bag edges of the vacuum bag material, whether a disposable elastomer or disposable film (such as nylon) must be cut to result in parallel, square, and straight corners and sides. This is more essential with disposable film since this

CONVENTIONAL OR PERMANENT REUSABLE VACUUM BAGS 265

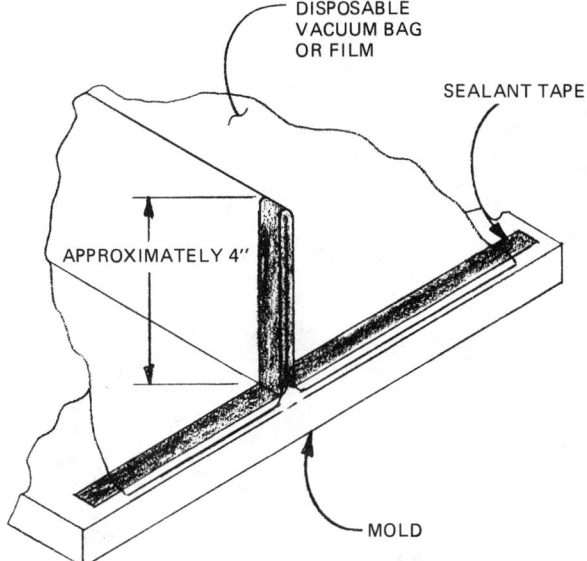

Fig. 8.15. Typical Rabbit Ear or Tab.

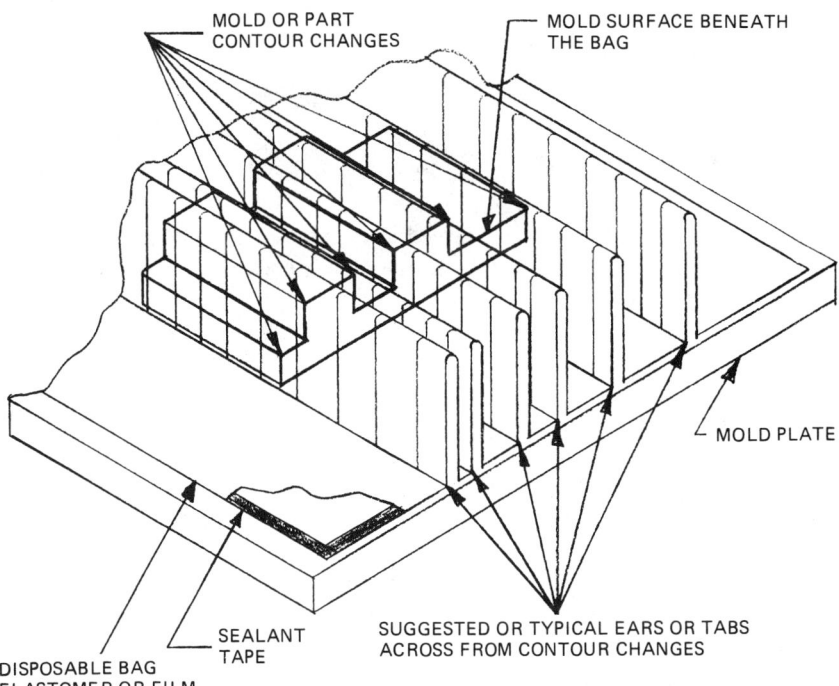

Fig. 8.16. Placement of Ears or Tabs.

266 ADVANCED COMPOSITE MOLD MAKING

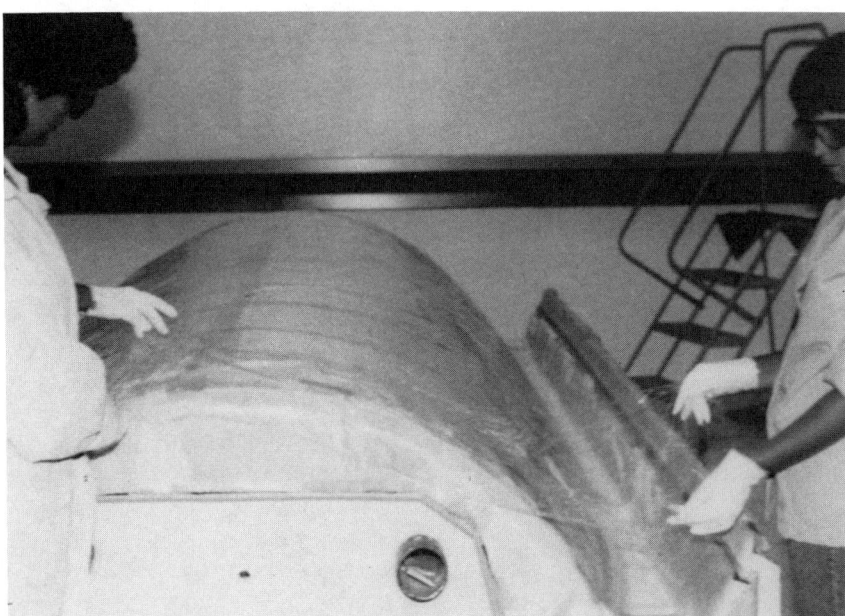

Fig. 8.17. Application of Disposable Film Vacuum Bag. *(Courtesy of the Ren Plastics Div. of Ciba Geigy Corp.)*

method requires pleats or ears, and the disposable elastomer, because of its ability to stretch, does not (Fig. 8.17).

An ear, pleat, or fold must be formed at each tool corner, in addition to any that must be formed along its edge. Corner ears are usually formed about 3 in. in from each tool corner.

When forming a pleat or ear, the following basic suggestions should be considered.

Pleat or ear height is approximately 4 in. This means that if a bag has 4 ears, the overall length of the bag side will have to be increased by an additional 32 in. in length (Fig. 8.18).

Therefore, if a bag side is 20 in. in length and has 4 pleats, the total bag side length will become 20 in. + 32 in., or 52 in.

Whether forming a corner ear or edge ear, bring together four fingers of your hand, and use the fingers to measure approximately 4 in. in height (8 in. in length) of each ear, and press the base of each side of the bag ear material down to the sealant tape (Fig. 8.19).

Now that the ear has been formed, and the base of the ear tacked in place against the sealant tape, the ear may be closed and sealed with the sealant tape

CONVENTIONAL OR PERMANENT REUSABLE VACUUM BAGS 267

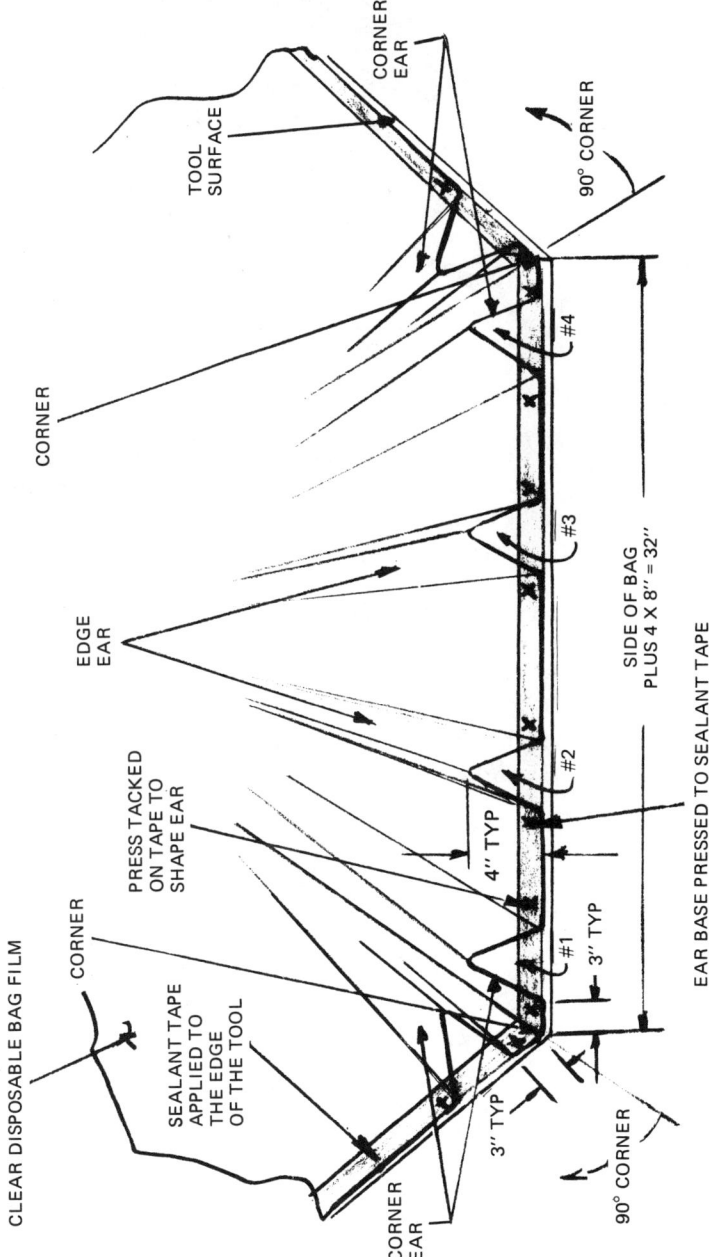

Fig. 8.18. Adjusting the Bag Length when Installing Pleats or Ears.

268 ADVANCED COMPOSITE MOLD MAKING

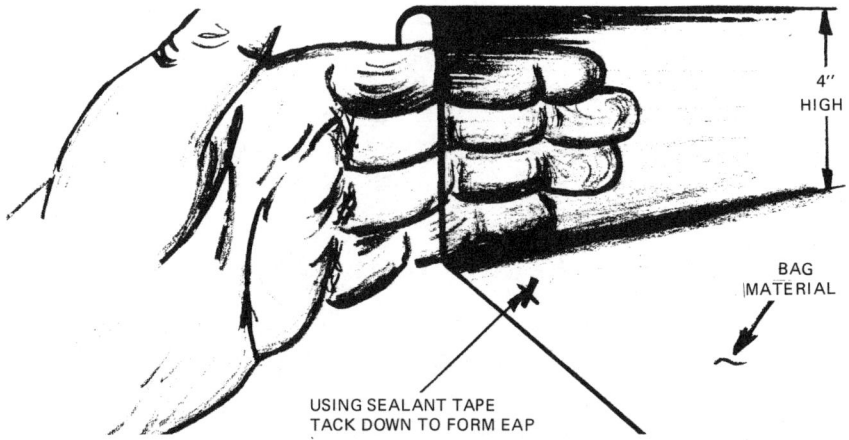

Fig. 8.19. Forming a 4 in. Ear "By Hand."

as follows: Unroll an appropriate amount of sealant tape, grasp the sealant tape and stretch tear the tape by twisting and pulling it apart in the appropriate location (Fig. 8.20). (Cutting sealant tape with a knife or sharp instrument is not desirable).

Using the same four fingers that measured the bag ear height, fold the sealant tape over the fingers and mark the tape tear location. After measuring the length, use the grip fingers of each hand, and grasp, twist, and break the sealant tape to this length (Fig. 8.21).

Now, using the same four fingers, apply the sealant tape to the ear edge of the bag (Figs. 8.22 and 8.23). *Note:* Careful application of the sealant tape at this point is essential.

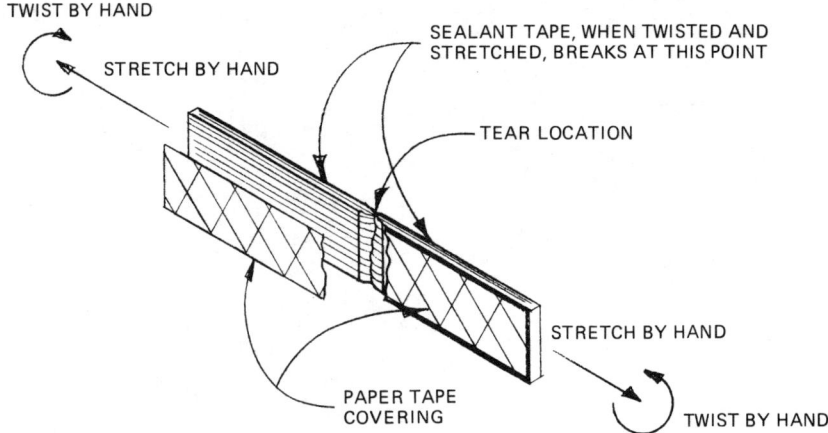

Fig. 8.20. Breaking, Not Cutting, Sealant Tape.

CONVENTIONAL OR PERMANENT REUSABLE VACUUM BAGS 269

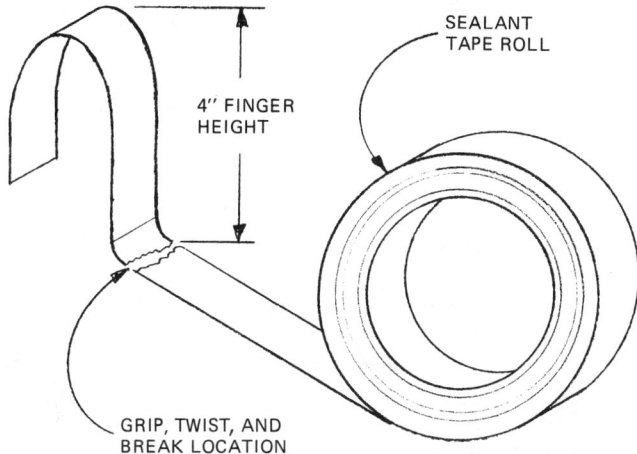

Fig. 8.21. Forming the Sealant Tape to Insert into an Ear.

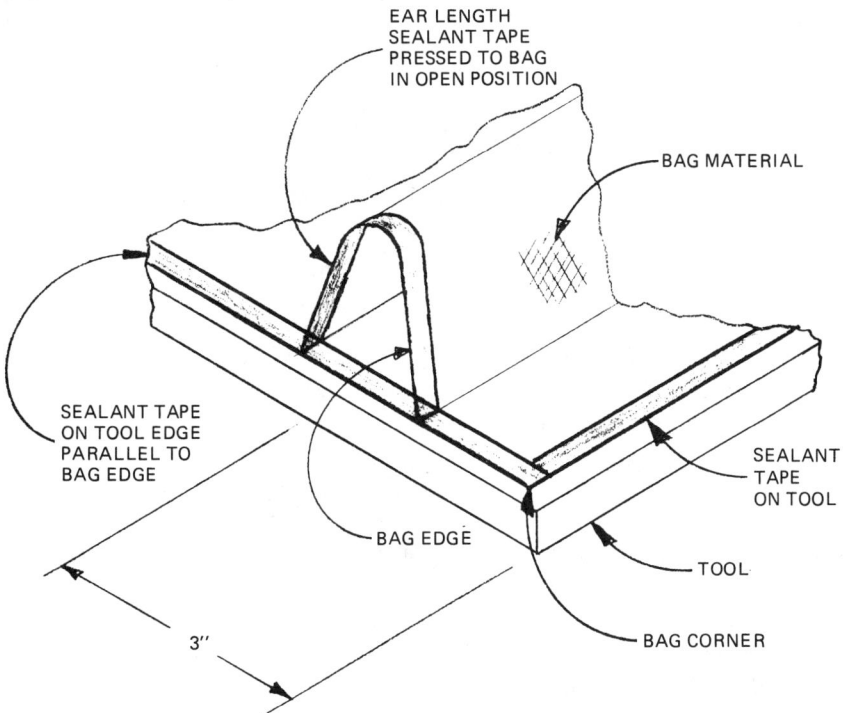

Fig. 8.22. Sealing the Ear (Ear Open).

270 ADVANCED COMPOSITE MOLD MAKING

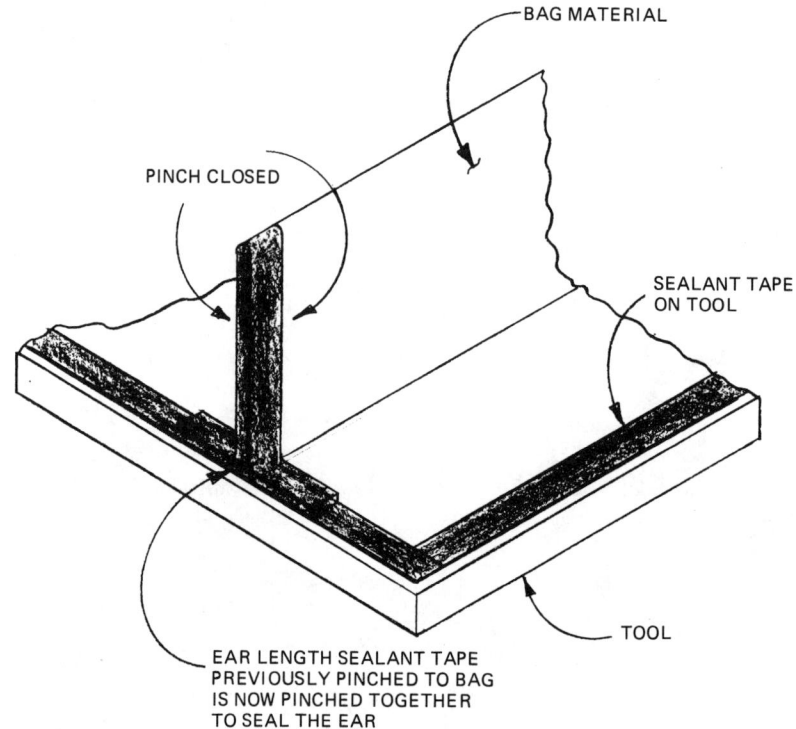

Fig. 8.23. Sealing the Ear (Ear Pinched and Sealed Closed).

Sometimes, a temperature-measuring device must be installed into the laminate lay-up, between the lay-up and the mold, or on the mold surface. The temperature of these locations must be tracked and monitored so that proper cure of the mold, tool, or part is assured.

When the thermocouples (used to track temperatures of laminate mold, tool, or part materials) are installed through the bag edge, the sealant tape and bag materials sandwich the thermocouple wire, which is inserted as shown in Figs. 8.24 and 8.25.

Tear the sealant tape into lengths that stagger over each other. Two short lengths applied over two longer lengths, which are in turn applied against the tool surface, form a strong bond.

A *double bag seal* encapsulates the bag material edge and thermocouple wire and also encapsulates the wire down to the tool surface.

CONVENTIONAL OR PERMANENT REUSABLE VACUUM BAGS 271

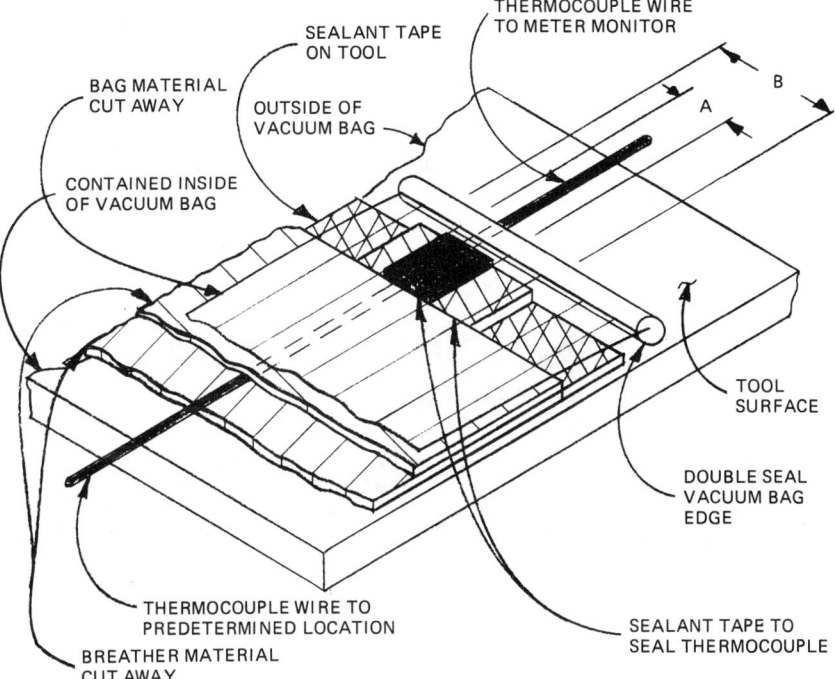

Fig. 8.24. Installation of a Thermocouple Wire Through the Sealant Tape.

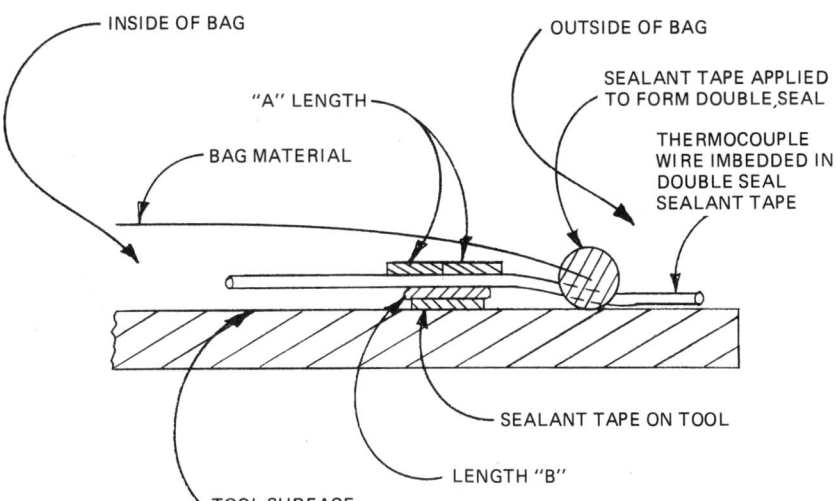

Fig. 8.25. Section of a Vacuum Bag Assembly Showing Thermocouple Wire Installation.

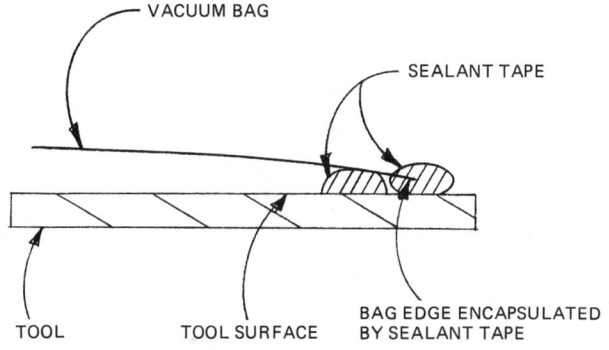

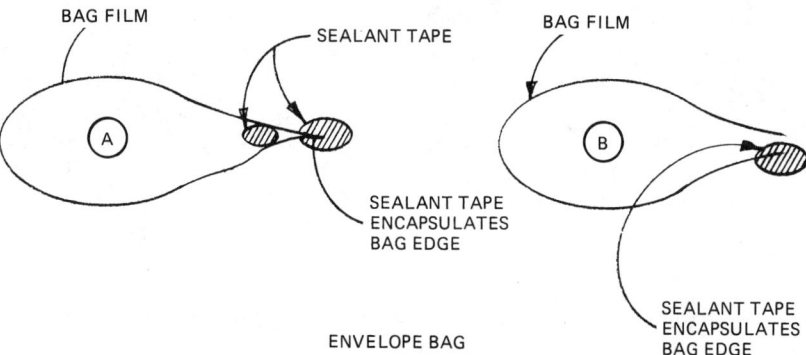

Fig. 8.26. Optional Double-Sealed Bag Edge.

8.1.2. Envelope Bagging Process

Envelope bagging a mold, tool, or part laminate is a positive way of assuring that no vacuum leaks can occur in the contour surface upon which the laminate lay-up has been applied. Like skin or surface bagging, the bag edge can be double sealed, to assure that no bag edge leaks can occur.

Figure 8.26 demonstrates the optional double sealing of a bag edge.

The envelope, as its name implies, is an edge-sealed pouch-type container, and is formed by a preformed film tube of a desired diameter or a piece of bag material folded and doubled over to form a bag that is sealed along the edges. In either case, the bag should be substantially larger than the laminate being formed, to avoid bag bridging at high points on the contour (Fig. 8.27).

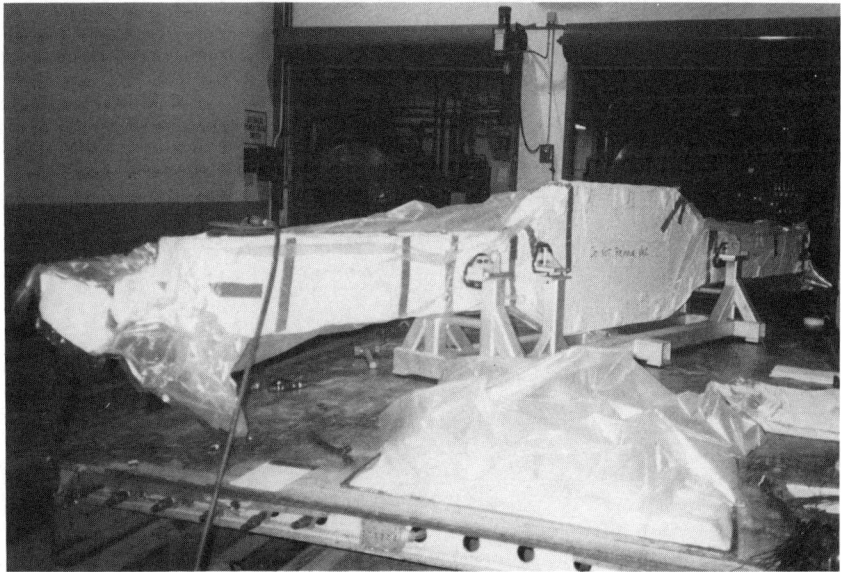

Fig. 8.27. Envelope-Bagged Mold Form. *(Courtesy of Grumman Aircraft Systems)*

The laminate mold, tool, or part is completely wrapped in a bleeder/peel ply/breather stack to allow air circulation around the laminate and to protect the bag from sharp mold or tool edges. The laminate is inserted into the bag, and the vacuum port or valve is inserted into the bag surface in the proper position. Finally, the bag is carefully sealed.

In this process, whether using a double-seal or single-seal bag, the bag edge is closed by starting in the center of the edge opening and working out to the ends and up to the fold (Fig. 8.28). In the case of a female mold, Figs. 8.29–8.31 illustrate the use of a bladder bag to avoid the otherwise inevitable bridging that would occur.

The master model or pattern can also be placed on a plate. Mold or tool laminate material is applied to the model surface, and a vacuum bag is then placed over it. Finally, the form is bagged down to the plate. In this approach, the master model or pattern must be able to withstand near-atmospheric pressure (approximately 10–12 psi). A quick calculation can determine if this is possible, so that the master pattern or model will not be damaged (Fig. 8.32).

After sealing the bag, draw a slight vacuum, so that the wrinkles and bridges can be adjusted out of the bag. Start from the bag edges and shift the bag around towards the laminate lay-up.

274 ADVANCED COMPOSITE MOLD MAKING

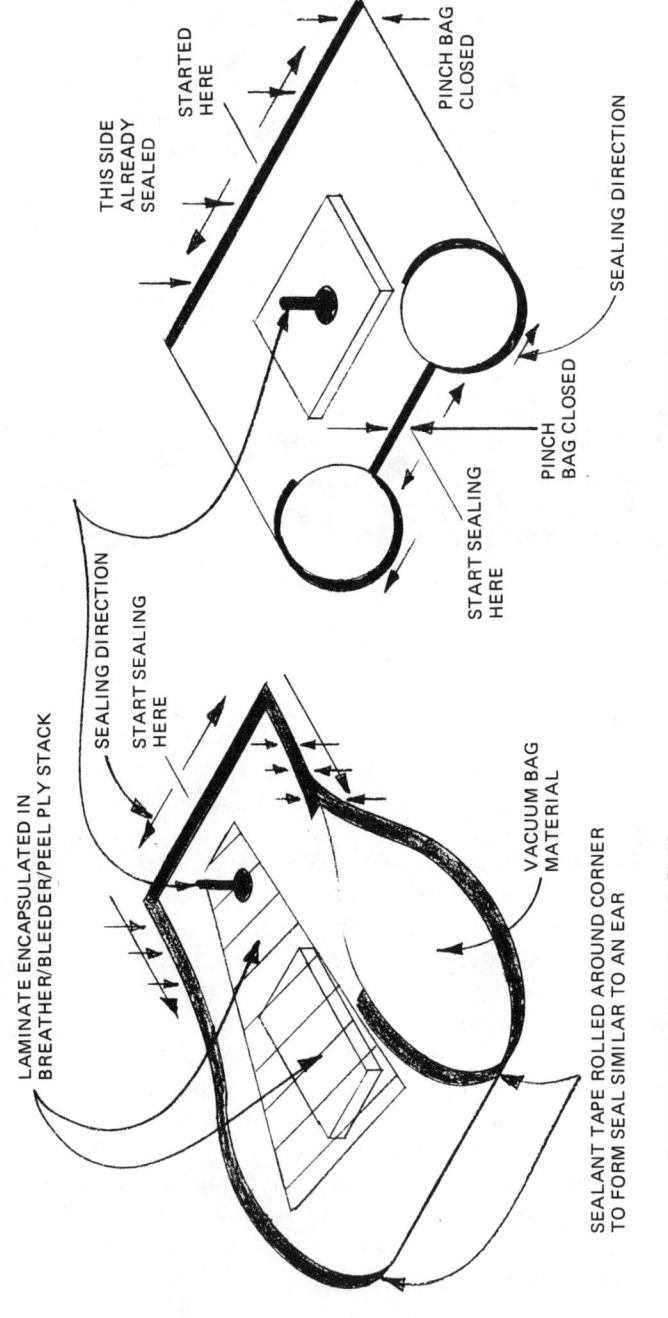

Fig. 8.28. Sealing an Envelope Bag.

CONVENTIONAL OR PERMANENT REUSABLE VACUUM BAGS

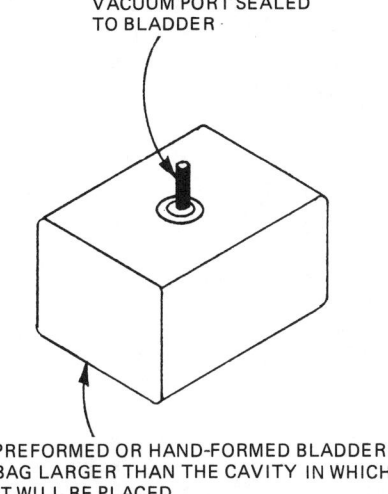

VACUUM PORT SEALED TO BLADDER

PREFORMED OR HAND-FORMED BLADDER BAG LARGER THAN THE CAVITY IN WHICH IT WILL BE PLACED

Fig. 8.29

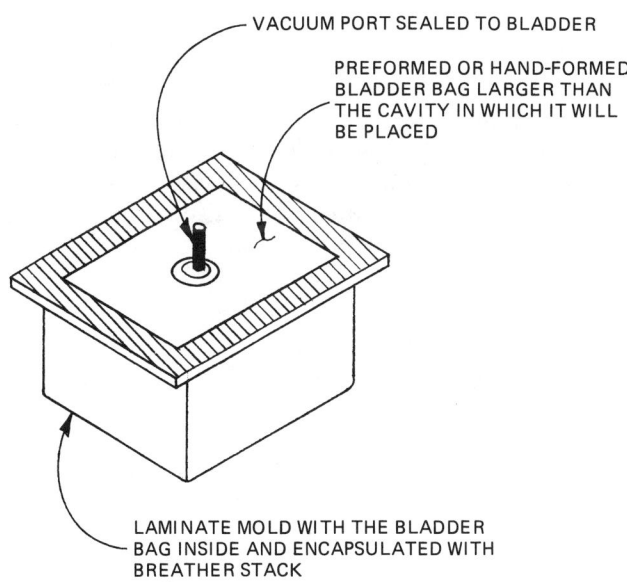

VACUUM PORT SEALED TO BLADDER

PREFORMED OR HAND-FORMED BLADDER BAG LARGER THAN THE CAVITY IN WHICH IT WILL BE PLACED

LAMINATE MOLD WITH THE BLADDER BAG INSIDE AND ENCAPSULATED WITH BREATHER STACK

Fig. 8.30

Figs. 8.29–8.31. Envelope Bagging Inside a Female Mold.

276 ADVANCED COMPOSITE MOLD MAKING

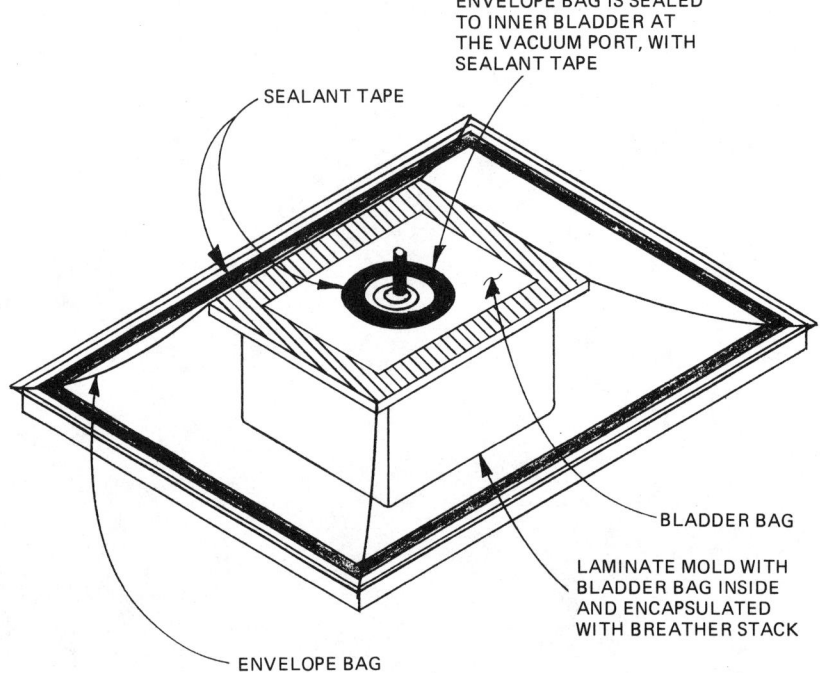

Fig. 8.31

This final step is used to adjust *all* vacuum bags before pulling a full vacuum, and even during the tool proofing procedure, or vacuum leak test.

Now that we have reviewed the basic steps in skin or surface bagging and for envelope bagging, let us proceed to study the actual bagging procedure used to form a laminate mold or tool.

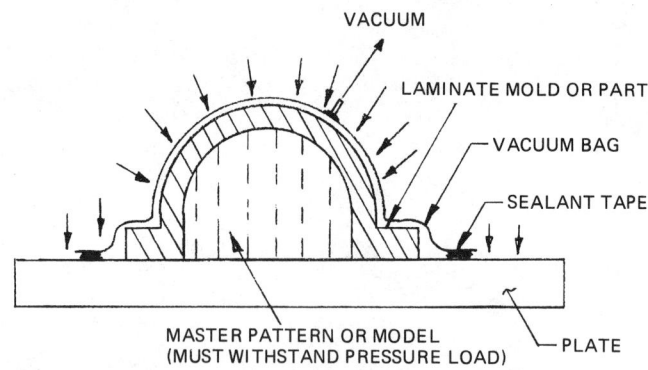

Fig. 8.32. Envelope Bagging a Mold or Tool Pattern Down to a Plate.

8.1.3. Bagging of Room-Temperature and Low-Temperature-Cure Mold and Tool Laminates

This section will present an approach to surface or skin bagging room-temperature and 200°F cure prepreg mold and tool materials, and wet lay-up materials. It will also explore the use of peel plies, peripheral bagging, and detail disposable elastomeric bagging techniques.

The basic lay-up procedure of a laminate mold is covered in Chapter 6. This chapter will detail the procedures for vacuum bag laminating as it is combined with the step-by-step laminating process.

Also included in this subsection will be suggested laminated mold and tool fabrication techniques, some of which could be adopted as mold and tool designs.

Mold and Tool Fabrics. It is suggested, at this point, to remember and note that, when preparing a wet lay-up, room-temperature-cure prepreg, or heat-cure prepreg laminate mold or tool, one of the following typical fabric or prepreg material choices should be used. Material suppliers should be consulted for additional suggested configurations.

It is also very important to note that these laminate lay-ups are formed from rectangular ply pieces, cut in sizes up to, but not larger than, 18 in × 24 in. All plies are butt-seamed, staggered seams over each other on successive plies, using picture framing techniques, 0° and 90° plies for simple, mild contours, and 0°, 45°, and 90° plies for complex contours. Plain, crowfoot, 8H satin (or even triple plain weave) fabrics may be used. The total laminate thickness should be approximately .250 in., although .400–.500 in. is more appropriate for molds and tools used for automatic tape-laying applications (Figs. 8.33–8.38).

Here are some helpful hints to aid in laying-up fabric in corners: 1) Always prepare the plies so that they can slide over themselves under the bag pressure.

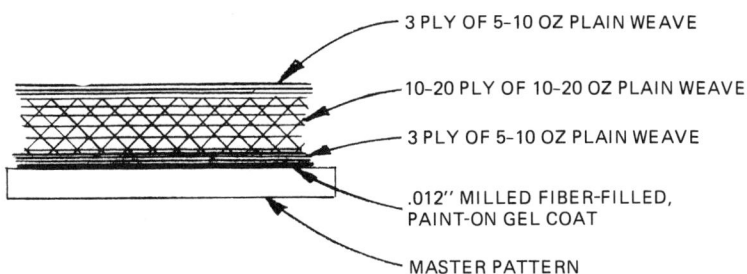

Fig. 8.33. Wet Lay-up or Room-Temperature-Cure Prepreg, Vacuum Bag, Free-standing Post-cure Ply Sequence for Fiberglass/Epoxy.

278 ADVANCED COMPOSITE MOLD MAKING

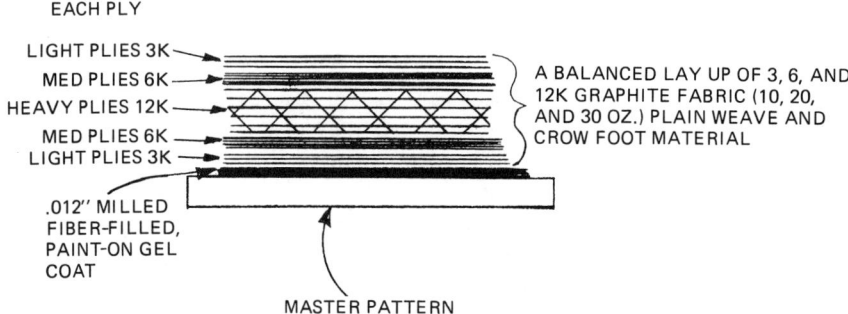

Fig. 8.34. Wet Lay-up or Room-Temperature-Cure Prepreg, Vacuum Bag, Free-standing Post-cure Ply Sequence for Graphite/Epoxy.

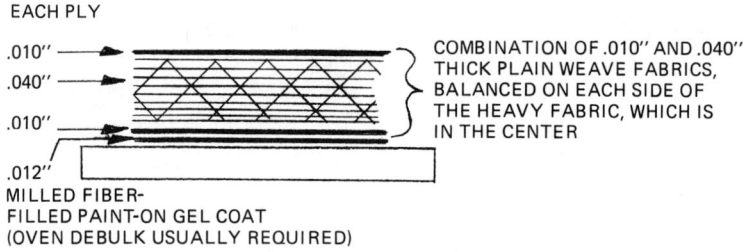

Fig. 8.35. Prepreg, Vacuum Bag, Oven Only, Ply Sequence for Fiberglass/Epoxy.

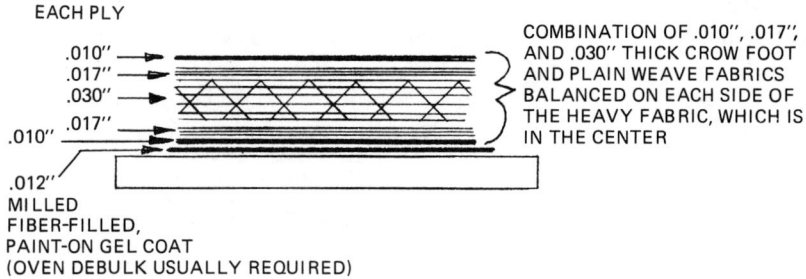

Fig. 8.36. Prepreg, Vacuum Bag, Oven Only, Ply Sequence for Graphite/Epoxy.

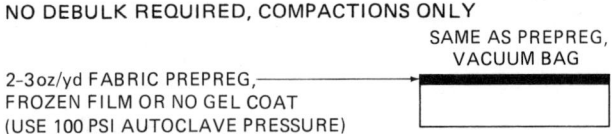

Fig. 8.37. Prepreg, Autoclave, Vacuum Bag Only, Ply Sequence for Fiberglass/Epoxy.

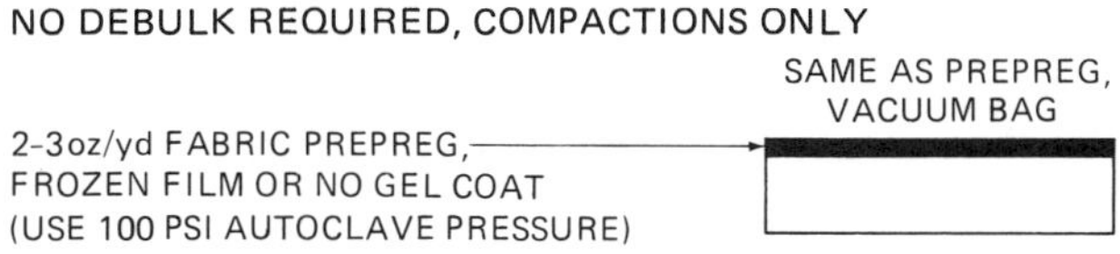

CONVENTIONAL OR PERMANENT REUSABLE VACUUM BAGS 279

NO DEBULK REQUIRED, COMPACTIONS ONLY

2-3oz/yd FABRIC PREPREG, ⎯⎯⎯⎯⎯⎯→ SAME AS PREPREG, VACUUM BAG
FROZEN FILM OR NO GEL COAT
(USE 100 PSI AUTOCLAVE PRESSURE)

Fig. 8.38. Prepreg, Autoclave, Vacuum Bag Only, Ply Sequence for Graphite/Epoxy.

Do not place plies into a radius without precutting, since they will "bridge" the radius (Fig. 8.39). 2) Do not fill a corner with a resin-rich fillet mixture, since the cured mold will warp or twist as a result of shrinkage. Prepare the layup as shown in Fig. 8.40A and B. 3) It is important *not* to use any fillet materials, such as expandable epoxy foam adhesives, that will expand in a radius corner, since the amount of expansion cannot be controlled. Over repeated and extended periods of use, a filleted corner could fracture and the laminate mold might delaminate (Fig. 8.41).

Bleeding or not bleeding resin from a mold and tool laminate is a very important step to be considered in preparing a laminate for vacuum bagging. The process for edge or surface bleeding is examined in the following paragraphs.

Edge bleeding of resin system and trapped air. If a wet lay-up is prepared and is not bled across the entire surface, but is instead bled along the entire edge, the bag configuration shown in Fig. 8.42 is recommended.

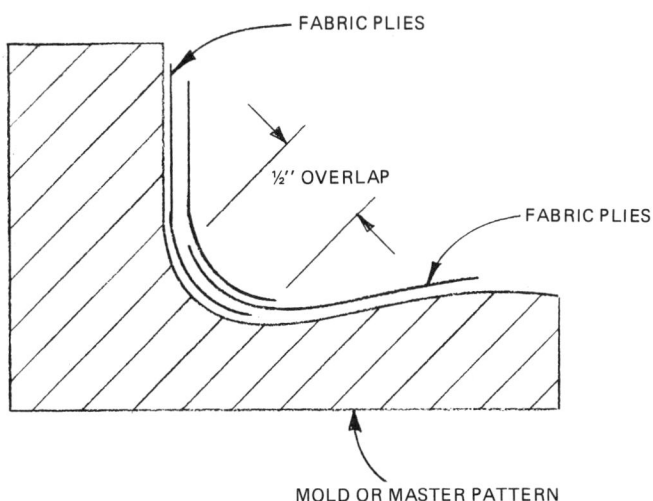

Fig. 8.39. Precut Plies for a Radius Corner.

280 ADVANCED COMPOSITE MOLD MAKING

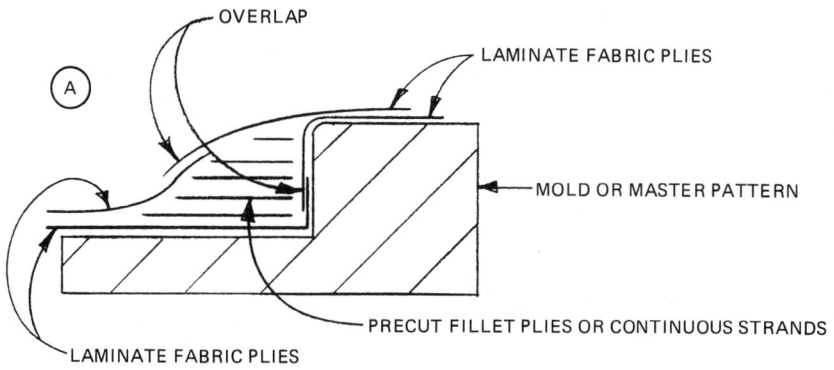

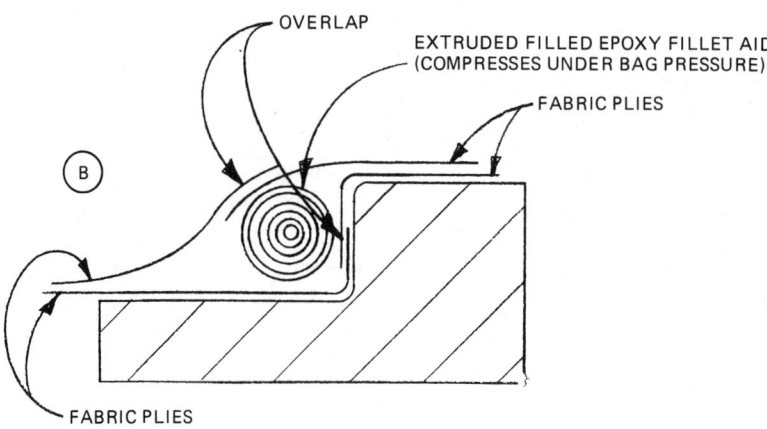

Fig. 8.40A and B. Laying Up a Radius Corner.

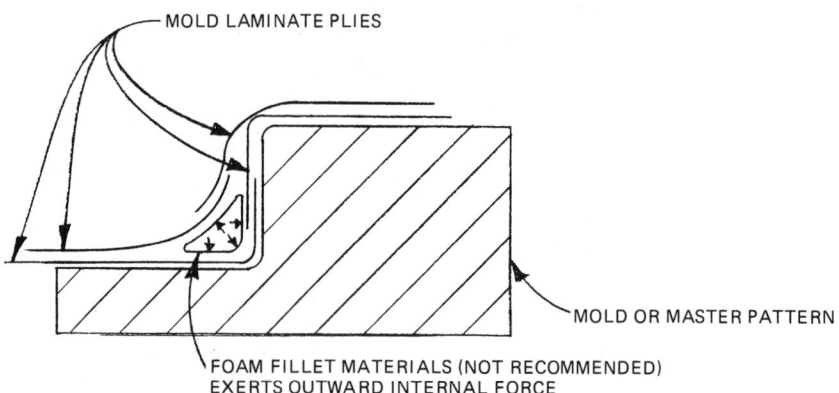

Fig. 8.41. Expanding Foam Filleted Corner.

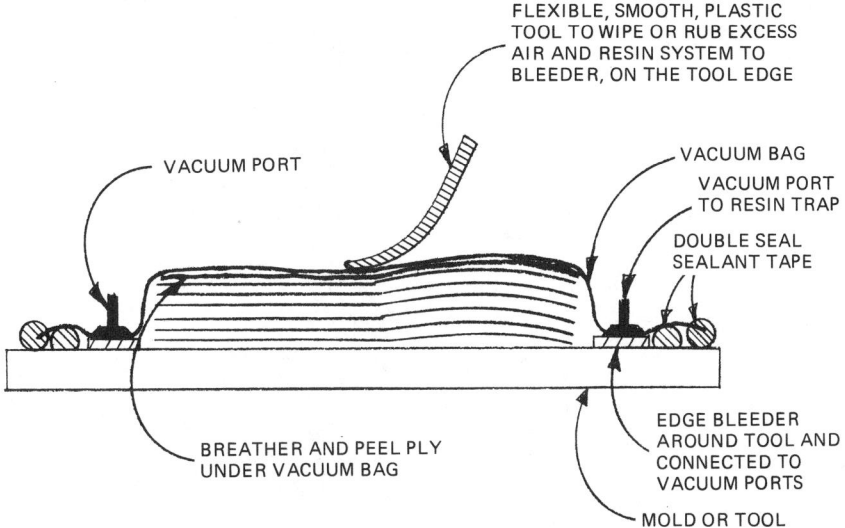

Fig. 8.42. Cutaway View of an Edge-Bled Lay-up.

Surface bleeding of resin system and trapped air. Surface bleeding is used for drawing excess air and/or resin up through and into a bleeder/breather stack which covers the entire laminate.

This process is used, for instance, when a wet-lay-up mold or tool is so large or complicated that bleeding up from and across the surface will be less expensive than edge bleeding. With surface bleeding, which is similar to the preparation of a no-bleed laminate, the laminate does not have to be physically wiped, combed, rubbed, or pressed by hand-held tools to remove air and resin.

Whether bleeding resin system and air, or just air, there must always be a perforated film layer separating the wet mold laminate from the bleeder and a solid film layer between the bleeder and breather. Bleeders can be formed from polyester felts that can withstand 350°–400°F or newer high-temperature felt materials capable of use up to temperatures of 1500°F and more. Plain weave fiberglass fabrics have even been used for these purposes.

Using colored release fabrics, peel plies, and colored perforated or non-perforated films and materials allows for immediate identification of the presence of the bag material or accessory material, so that the material will not be accidentally trapped and included in the composite mold or tool layup.

The configuration illustrated in Fig. 8.43 is suggested.

No-bleed system. A no-bleed system is usually used for controlled-resin-content prepreg laminate mold materials where the resin must remain within the laminate during and after cure. In this process, the resin is trapped around the periphery by a dam and vertically above the laminate. This system will now be

282 ADVANCED COMPOSITE MOLD MAKING

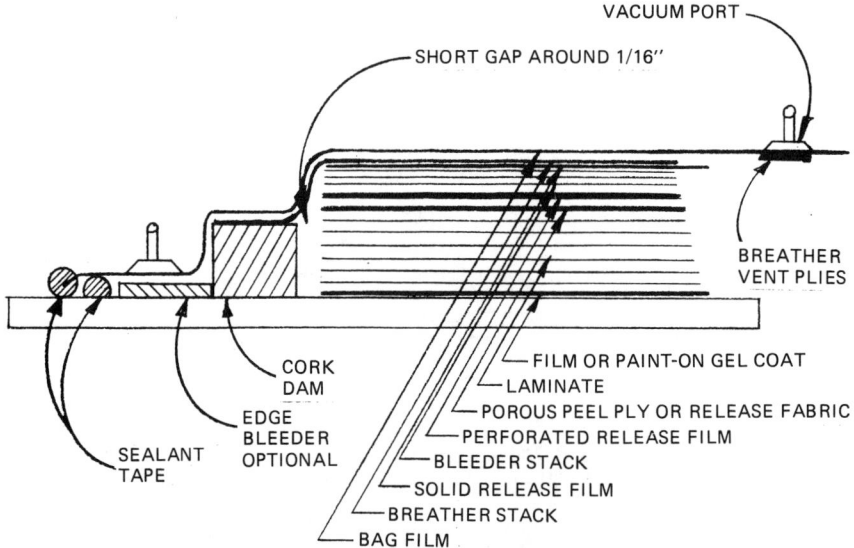

Fig. 8.43. Surface or Bleeding in a Vertical Direction.

used to demonstrate the preparation of a laminate and disposable vacuum bag system needed to form the mold or tool laminate.

In order to set up the vacuum bag sealant tape and establish the edge of the laminate mold or tool, it is necessary to add 1/2 in. to the mold edge as a trim line, which is the laminate material that will be removed so that the edge will be flush with the eggcrate substructure. Another 1 in. is allowed for the peripheral bagging edge. This process, which is termed peripheral bagging, and which will be described in detail shortly, is used to hold the laminate to the master mold when the eggcrate is being attached (Fig. 8.44).

When the no-bleed system is used to form the laminate mold or tool, the release film should be a nonperforated type and, if necessary, a pointed object (such as a scribe or nail) can be used to puncture the film once every square foot. This will allow air, but not resin, to bleed through to the breather stack.

The first two prepreg plies are now applied over the gel- or film-coated master model or tool pattern. Bushing pins, marking bushing locations, are surrounded by star cutting the rectangular mold laminate pieces so that they pass over and around the bushings. The bushings will be applied during the laminate process and then surrounded by the laminate material as it is built up. Then, air and wrinkles are swept out with a plastic squeegee or flexible roller, jabbing, or — if necessary — slicing the prepreg or wet fabric to remove the trapped air.

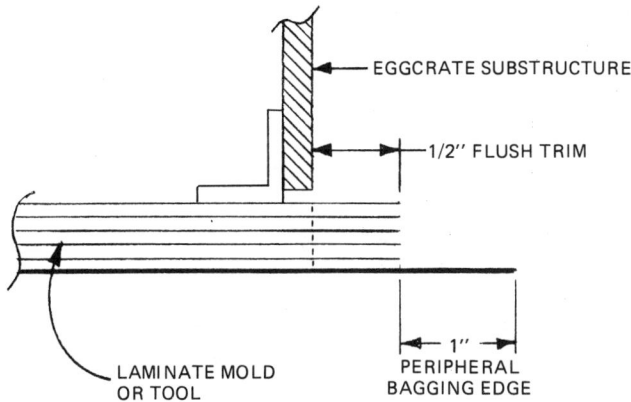

Fig. 8.44. Trim and Peripheral Bagging, Edge Compensation.

Picture framing, described in detail in Chapter 6, is employed next. The laminate plies are cut into 4 in. wide strips, butt seamed and staggered at the mold or tool end. Picture frame plies are the 3rd, 4th, and 5th plies.

Compaction or consolidation of the plies is required during lay-up, and is performed by applying sealant tape to the outside edge of the master pattern or tool mandrel. Usually, this step is performed every 4-6 plies, or however often previous experience dictates.

First, apply one layer of peel ply, butt-seamed pieces, followed by a ply of perforated release film, a ply of breather material, and finally, a less expensive room-temperature bagging film.

In order to prevent bag puncture, additional plies of breather material or a foamed flexible plastic are applied over any sharp protrusion, such as a bushing.

Compaction is performed under vacuum pressure, at room temperature, for approximately 15 min.

When the laminate (200°F cure) material cannot be completed within one day, the bag, under pressure, is left overnight. A room-temperature laminate that cannot be completed should be left under vacuum overnight, with the peel ply intact. On the next day, the peel ply should be removed and the laminate continued.

Whether utilized for a one-day or extended lamination, the peel ply provides a rough texture pattern that allows for continued bonding of subsequent plies.

Finally, the compaction bag set up is removed, and the remainder of the laminate mold process is completed.

284 ADVANCED COMPOSITE MOLD MAKING

Debulking (the application of pressure and heat) is sometimes performed to aid laminate consolidation by causing a slight resin softening and flow. This step is performed in an oven or autoclave at temperatures of room temperature ambient to 150°F for a time period of 15–60 min. The same bagging system described for compaction is usually used.

Before installing the final high-temperature bag system to be used to cure the room-temperature or 200°F prepreg laminate lay-up, thermocouples are installed. These are used to monitor the temperature of cure as well as the rise and fall of heat. Thermocouples are installed at the mold/laminate interface in pairs, in case one fails during the final cure.

Next, a cork dam, usually sealed with release-coated tape, is installed and butted to the outer edge of the thin peripheral bagging laminate edge. The dam is usually 1/4 in. high × 1/2 in. wide. Sealant tape is now installed beyond the dam, encapsulating the thermocouples, as previously illustrated in Figs. 8.24 and 8.25 of Section 8.1.

A suggested variant installation of a thermocouple through bag sealant material is to route the thermocouple wire through two layers of tape for a distance of approximately 6 in. before exiting from the tape. This will compensate for an accidental tugging or jarring loose of the wire from the sealant (Fig. 8.45).

Next, the peel ply layer is installed by extending it from the inside edge of the bag sealant tape, over the dam, and over the thin periphery. Then, it is butted to the periphery of the thick edge of the laminate lay-up. This will protect the peripheral bagging edge.

Finally, the peel ply is cut into manageable pieces, and butt seamed.

Next, two plies of breather material are laid down, covering the peel ply but also extending from the inside edge of the bag sealant tape, over the dam, over

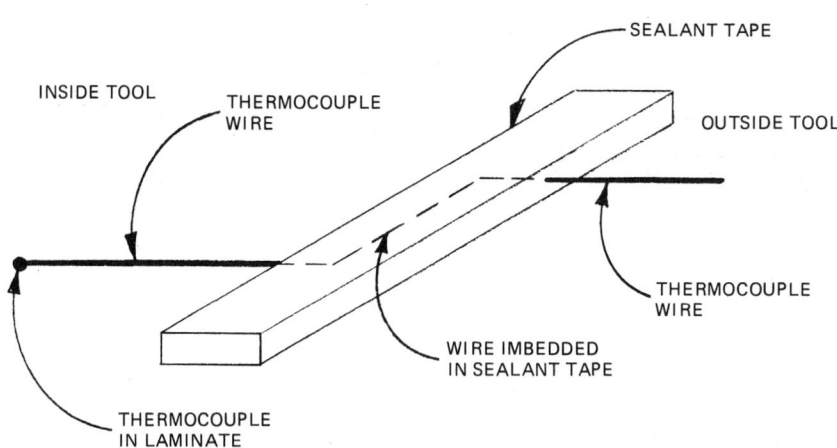

Fig. 8.45. Imbedded Thermocouple Wires.

the thin laminate periphery, and butted to the thick edge of the laminate layup. This is called the "edge breather."

Now, one layer of peel ply is applied over the *entire* laminate, extending over the dam and ending at the dam's top outer edge. One layer of solid release film is applied over the peel ply, also extending over the dam and ending at the dam's top outer edge.

The solid release film should be punched, using a sharp instrument such as a scribe, every square foot, to prevent a local bag sealoff.

Next, two plies of breather material are laid down over the solid release film and edge breather, extending to the inside edge of the bag sealant tape.

Now, the high-temperature vacuum-bagging film is applied. Another layer of sealant tape can then be installed (double seal) to assure that no bag leaks occur when the bag is pulled inward during the cure cycle.

At least two vacuum ports, chucks, or valves are recommended. The first one is installed through the bag at one corner of the layup and then a vacuum gauge is attached.

The second vacuum valve is then installed, directly across from the other, and a vacuum gauge and vacuum intake are attached. The vacuum chucks should be located over the edge breather, and additional plies of breather material should be placed under the vacuum chucks to prevent seal-off.

It is important that no part of the bag system bridges over the pad build-ups.

Next, the bag pleats or ears are installed, as previously described in this chapter.

Finally, vacuum is applied, and a test for leaks is conducted by turning off the vacuum and watching the leak rate. Both gauges should be at the same reading (22 in. Hg minimum) and within 1.0 in. Hg from each other. If the gauges do not read the same, a sealed off condition probably exists and extra breather plies should be installed. The room-temperature leak rate should not exceed 2.0 in. Hg within a 5 min. period.

To cure the room-temperature-tested laminate, the laminate is placed in an air circulating, vented oven or autoclave, capable of $\pm 10°F$. Inert atmosphere autoclaves are recommended for this. In this process, a room-temperature leak test is repeated within the autoclave.

After the test, vacuum is reapplied and the temperature is increased, as per the required cure schedule. The autoclave is pressurized when used, and the bag is vented to atmospheric pressure once the autoclave has reached a pressure of 20 psi. The heat rise should be controlled at a rate of $5°F/min$.

Cured molds or tools should be cooled within the autoclave or oven, under pressure, to at least $125°F$ under (85 psi in autoclave and atmospheric pressure in oven). The molds or tools are then cooled below $125°F$ before the pressure is released.

286 ADVANCED COMPOSITE MOLD MAKING

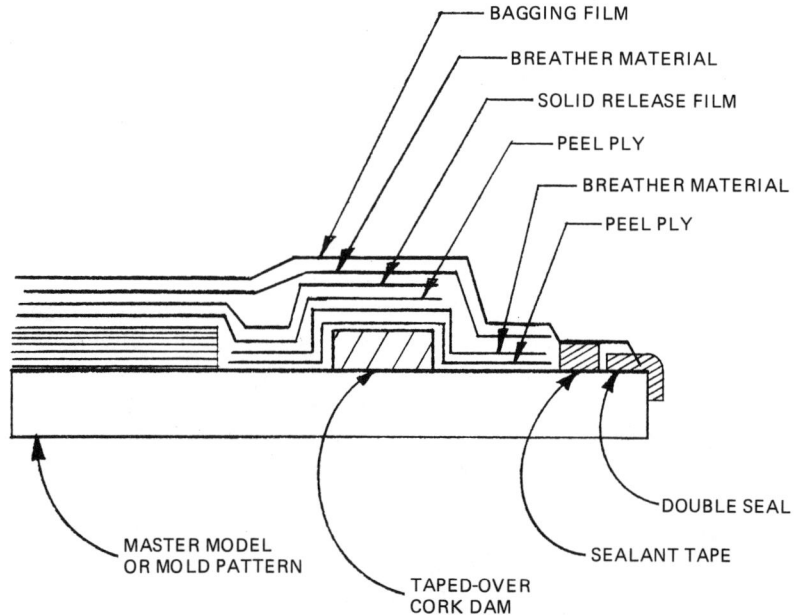

Fig. 8.46. No-Bleed Bag System.

The bag system is now removed, but the peel ply is not. *Note:* It is aboslutely imperative that the laminate lay-up is not lifted from its molded surface. The peel ply will protect the surface, to which the support structure must be bonded, from contamination. This peel ply will also aid in the peripheral bagging and eggcrate or substructure attachment step.

The peripheral bag configuration just described is shown in Fig. 8.46.

Peripheral bagging is the mold-making method used to hold the laminated mold or tool laminate against the surface from which it was formed prior to attaching the substructure. Usually, mold makers merely rely upon the weight of the mold or tool and gravity to hold the laminate against the surface from which it was formed, prior to attachment of the substructure. This method, however, does not demonstrate a reliable mold-making procedure, and is risky besides.

A better, safer, and more reliable method involves peripheral bagging, which can be used in the formation of wet lay-up and prepreg molds. In this method, it is assumed that the laminate has been previously prepared with a 1 in. wide flange around the periphery. This flange should be 3 plies thick, and will be the edge that is bagged to the master model surface prior to attachment of the substructure (Figs. 8.47, 8.48, and 8.49).

CONVENTIONAL OR PERMANENT REUSABLE VACUUM BAGS 287

Fig. 8.47. Peripheral Bagging Edge Around Tool. *(Courtesy of Grumman Aircraft Systems)*

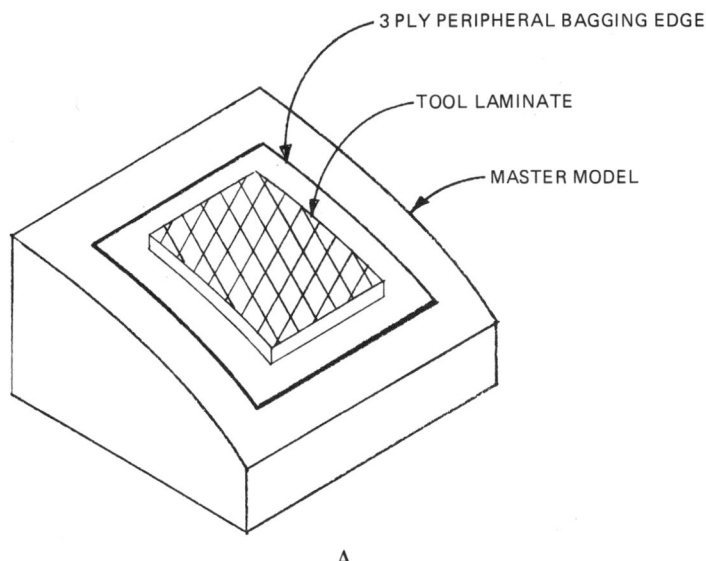

A.

Fig. 8.48. Peripheral Bagging.

288 ADVANCED COMPOSITE MOLD MAKING

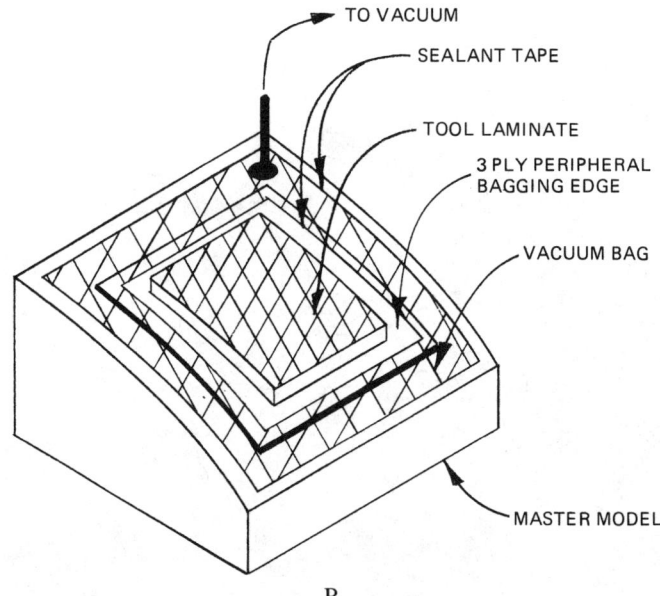

B.
Fig. 8.48. Continued

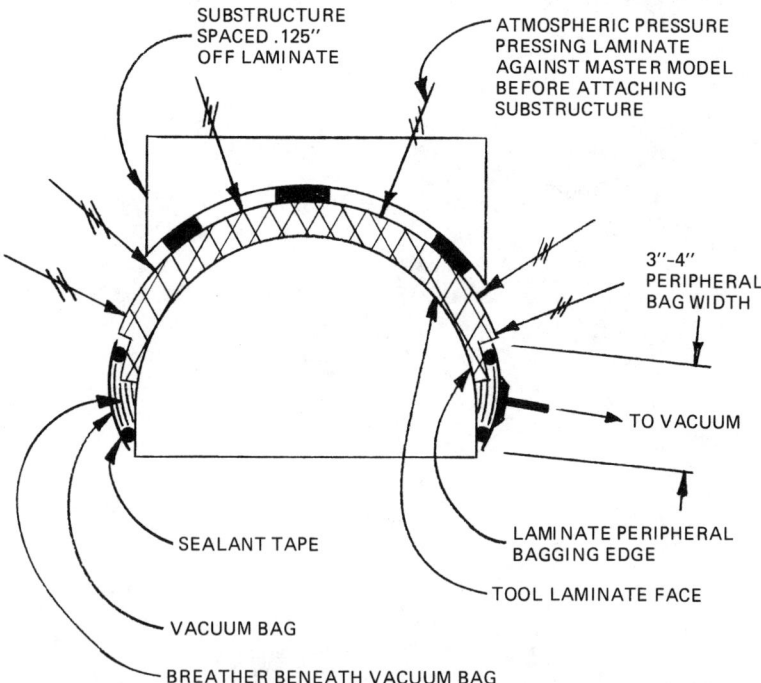

Fig. 8.49. Side View of Peripheral-Bagged Tool.

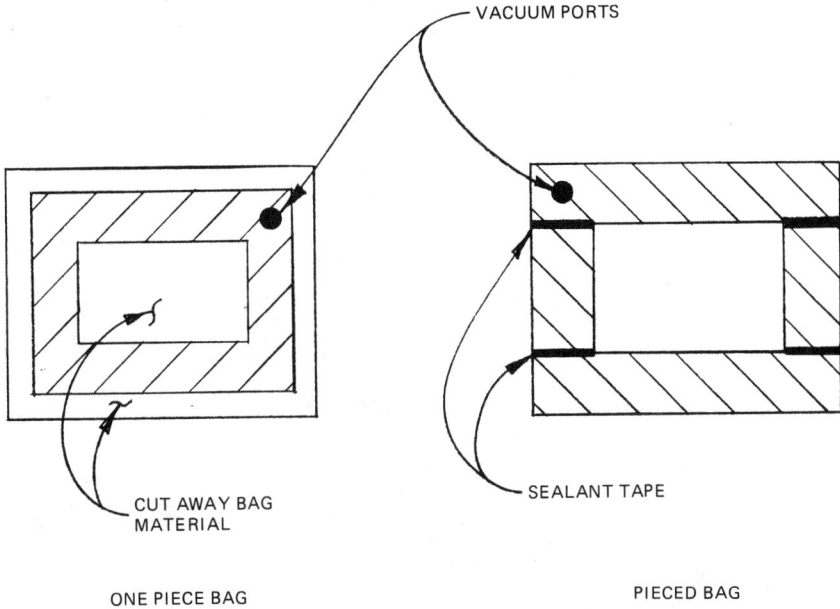

Fig. 8.50. Peripheral Bags (One- and Multiple-Piece).

Sealant tape is now applied around the thin, three-ply lip formed at the tool laminate periphery. Sealant tape is then applied along the mold master edge, forming a peripheral bag width of approximately 3–4 in. Finally, two plies of breather material are laid between the strips of sealant tape. If a tool is small enough, the peripheral bag can be formed from one piece of bagging film by cutting the bag into a picture frame. Larger molds or tools require bag pieces to be connected to each other using a sealant tape (Fig. 8.50).

With the peripheral bag in place and the vacuum pressure on, the peel ply is removed on the rear of the laminate, and the substructure is installed.

The substructure, which has been prethermally stabilized or post-cured, is secured in place with laminate tie-ins, while the peripheral bag pressure is maintained. If a low-temperature cure of the tie-in material is required, the vacuum is maintained during the cure.

Face bagging the cured laminate, prior to the "free-standing" post-cure, is a measure that will reveal surface flaws during the post-cure. This is important because a flaw in the surface that shows up during the formation of a part will be cause for the rejection of the part. The vacuum pressure usually collapses voids during the post-cure cycle, and this will alert the mold or toolmaker to their presence prior to the release of the tool.

Should these types of voids be present, they can be "dressed-up" or filled at this time (Fig. 8.51).

290 ADVANCED COMPOSITE MOLD MAKING

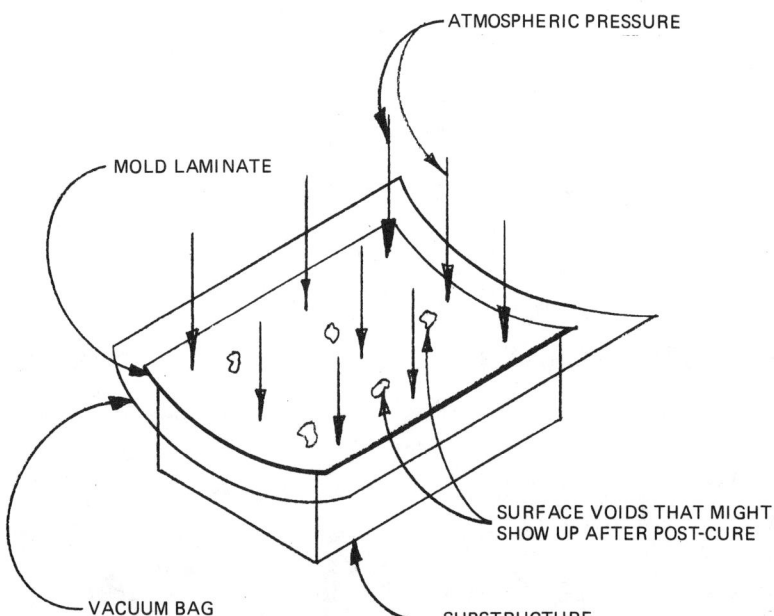

Fig. 8.51. Face Bagging During Post-cure.

8.2. SEMIAUTOMATIC REUSABLE VACUUM BAG SYSTEMS

Reusable vacuum bag systems are production part forming devices. The subject is presented here for those who are required to design and fabricate such a bag or bag system and as an explication of the theory behind their functionality.

For those who must design the mold or tool that will have a reusable bag as part of the assembly, or those designers who must alter a mold or tool to include a reusable bag, these sections should provide the design parameters required to devise, fabricate, and install the bag.

The materials used in the formation of the reusable bag can range from simple elastomeric materials such as latex rubber to exotic blends or formulations of specially catalyzed silicone or fluoroelastomers.

Conventional silicone rubber bagging materials — the most widely used materials — are prepared through chemical reaction, using peroxide catalysts. There are proprietary catalyst systems that have just recently been developed that produce an elastomeric bag rubber that is highly reversion-resistant. This system has an involved material chemistry and manufacturing process which only a very limited number of manufacturers have mastered. The process involves special compounding, blending, and calendaring of these special elastomer formulations, which yield bag materials capable of operating at temperatures of 450°F and higher.

By using this new material manufacturing process, silicone bag materials have been manufactured and are available to be formed into vacuum bags that will not transfer silicone by-products, oils, or other residuals and reaction by-products during use.

These materials can be supplied uncured for lay-up like prepreg mold laminate materials, reinforced with various fabrics made from fiberglass, kevlar, and other materials, or cured and supplied post-processed by heat stabilizing the cured sheet materials to temperatures of 450°F. This post-cure step assures that the materials (when used as vacuum bag materials) will not shrink, weaken, or take a compression set.

Reusable bag systems eliminate costly labor operations (such as hand-installing sealant tape around mold peripheries), reduce film bag labor installation costs, and eliminate mold and/or laminated part loss due to failure of a bag seal, pleat, or disposable bag material itself.

The savings attainable with reusable vacuum bags can be as much as 50% over conventional materials and processes. The production-use life expectancy of an elastomeric bag made from state-of-the-art materials can be in the 500–1000 part range before maintenance must be performed on the bag surfaces. In the formation of composite laminates or bonded assemblies, the reusable vacuum bag system will pay for itself quickly if the system is used to automatically compact, densify, consolidate, and debulk the composite lay-up or assembly.

The bags are formed from flat sheets bonded to frames or custom bonded patterns shaped to a size near to that of the finished contour. With the new laminate part-making procedures of co-curing structural stiffeners, inserts, laminate materials, and other inclusions all at one time, the reusable, preformed, or shaped bag has unmatched advantages.

Unlike the situation posed by the disposable film bag, reusable bags allow the mold, tool, or part maker to install various types of hardware into the bag surface when required. Thermocouples, vacuum ports, stiffeners, metal support frames, hoist brackets, and internal heaters are just some of the added accessories that can be installed in or attached to a reusable bag, which may not be possible when a conventional disposable film bag is used.

Edge seal or closure systems, made as part of the bag or tool, become a reusable permanent part of the vacuum bag. For this reason, and the fact that reusable seal materials withstand cure cycle temperatures better, disposable sealant tape cannot compete and is not a good substitute. Disposable sealant tape softens upon use. Usually, leaks occur in the seal or the conventional bag film. Disposable, conventional bag materials do not conform as well as reusable elastomers. Leaks can occur in spots and failures can ruin the part, mold, and sometimes the autoclave interior.

When using reusable bags, through-the-bag vacuum port or valve locations must be reinforced, and not used as bag lifting supports. Bag supports or lift locations can be laminated into the bag surface. Bag surfaces must be inspected and questionably thin areas patched with uncured or RTV rubber materials.

There is no limitation to the size that a reusable bag can be fabricated to, since seams are easily and smoothly formed to any contour or shape. Contoured bags can even include intensifiers, pressure pads or blocks, rails, and other inclusions that have been molded and placed beneath the bag surface.

Elastomeric, unreinforced reusable bags make better quality parts because the bag material has an elongation factor up to 700%, and as such can form to the required contour and provide intensified pressure in radius corners. This cannot be attained by the use of conventional bag films.

Another important feature of reusable bag systems is that, if they are carefully designed, they can be stored by rolling them up and racking them in tubes for easy handling. Reusable bags should not be stored until they have been solvent cleaned and dried. Storage should be away from sharp objects and tool sections.

Other forms of reusable vacuum bags include cast bags and sprayed elastomers that are applied over a release-coated object or actual part used to simulate the part thickness during bag fabrication. Reusable vacuum bags of this kind must have adequate space beneath them to allow for the expansion of mold and part surfaces, allow for insertion of peel ply, bleeder, and breather layers, caul plates, pads, and other necessary bag accessories.

8.2.1. Permanent Reusable Elastomeric Rubber Bag Fabrication Techniques

There are a number of manufacturing methods available to form reusable vacuum bags. The following methods are the most common:

- Uncured sheet rubber: reinforced and unreinforced silicone
- Cured and post-cured sheet rubber: reinforced and unreinforced silicone
- Spray-on: silicone elastomer
  latex elastomer
  polysulfide elastomer
- Paint-on: silicone elastomer
  latex elastomer
  polysulfide elastomer
- Cast: silicone RTV (room temperature vulcanizing) elastomer
  latex elastomer
  polysulfide elastomer

Precured, post-cured, unreinforced sheet rubber bags are recommended above all others. This type of construction will produce the highest quality parts, be the least expensive to produce, and will last longer than any of the other forms of elastomeric bags.

We will cover the fabrication techniques of this type first.

Reinforced and Unreinforced, Uncured and Cured/Post-cured, Latex and Silicone Sheet Rubber Vacuum Bags. When practical, the use of precured silicone sheet rubber that has been post-cured to temperatures in excess of 400°F will yield the most reliable reusable bag. In choosing, a careful examination of the rubber formulation and bag material manufacturing technique will assure the correct selection. If, for instance, a silicone material is used, the choice of peroxide catalyzed elastomers should not be made.

Silicone rubber is suggested for use in forming reusable bags. Recently, however, latex and other elastomers have been used to form lower-temperature reusable compaction or consolidation bags.

The following subsection will be devoted to the fabrication of seams for silicone sheet materials, since they have proven to be the most reliable bag materials to date.

Seams. Forming seams in precured silicone sheet rubber materials is a simple task that allows for the fabrication of any size vacuum bag. The seam edge must be straight and butted, with a doubler strip usually 1/2 to 1 in. wide that has been bonded along and over the seam, using a compatible silicone RTV or heat-cured adhesive recommended for the bag material chosen.

For instance, when forming and bonding a silicone rubber bag material seam, an RTV adhesive is used to coat the seam area and doubler strip. RTV adhesives that cure by an alcohol reaction to air, rather than an acetic acid cure, will not affect metal frame or plate surfaces, especially if bonding to aluminum. RTV adhesives are applied by spatula and are cured for 24 to 48 hours at room temperature. Because the bag material is usually stronger than the adhesive, a bond-line RTV thickness of .020 to .060 in. can be used.

Uncured silicone sheet rubber may also be used to form the doubler strip and bond the seamed sheet rubber together.

Cohesive or adhesive failure of the bonding material will usually occur, so it is better to have more adhesive in the bonded seam than not enough. Using a conventional vacuum bag to hold the seamed material in place during the adhesive cure will assure that the material being seamed will not shift and air, trapped in the adhesive, will be degassed during the vacuum cycle (Figs. 8.52 and 8.53).

294 ADVANCED COMPOSITE MOLD MAKING

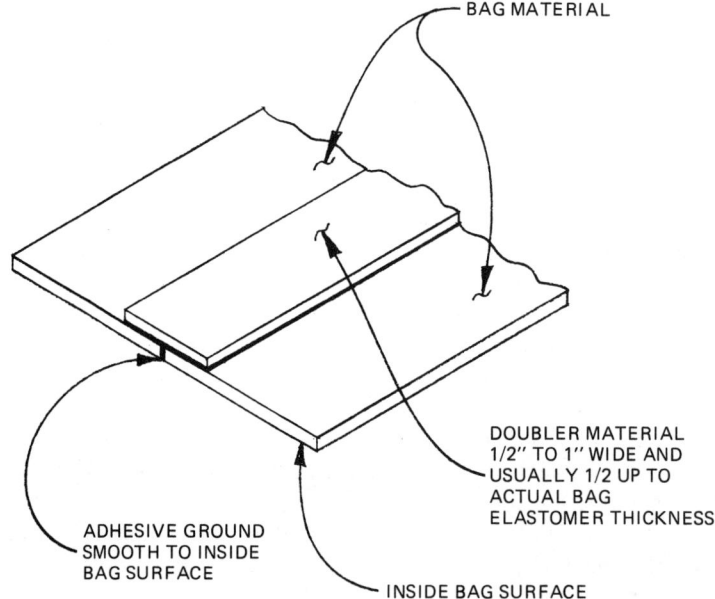

Fig. 8.52. Typical Seam with Doubler (Silicone Rubber or Other Elastomer).

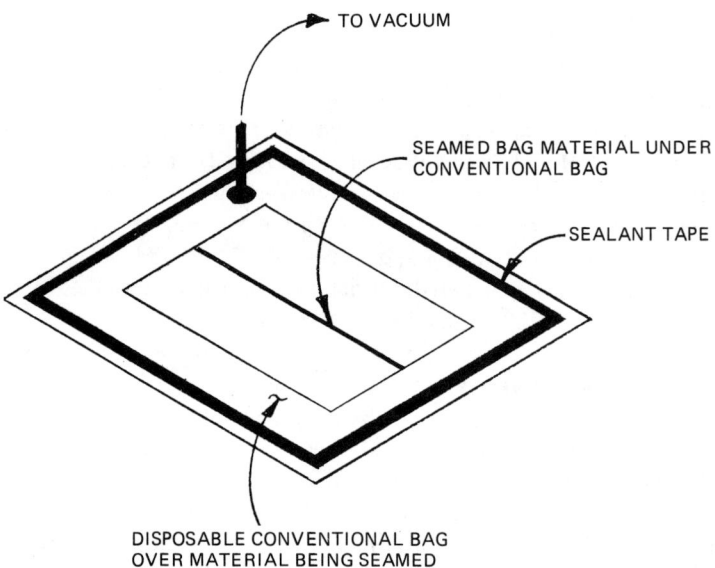

Fig. 8.53. Conventional Vacuum-Bagged, Reusable Vacuum Bag Seam.

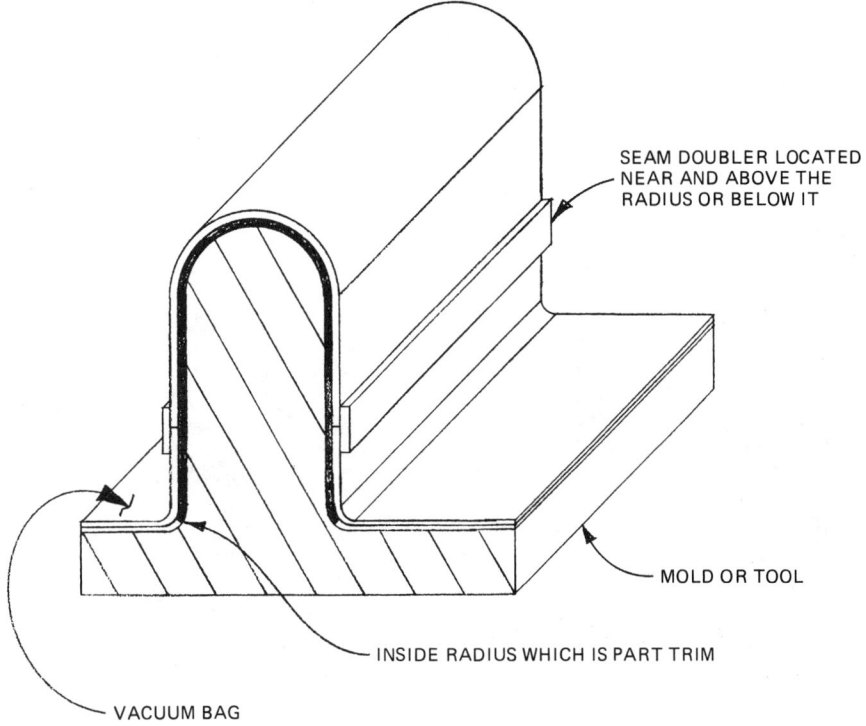

Fig. 8.54. Locating the Doubler Near the Inside Bag Radius.

Whether using a seam in a flat reusable vacuum bag, or a custom contoured bag, the location of each seam should be carefully thought out ahead of time. It is important to try to locate as few seams as possible over an inside radius. The reason for this is that the bag is expected to stretch into the radius corner, and if a doubler is present, the ability to stretch into the corner is reduced. If a seam must be placed on the inside radius, try to locate the trim line for the part at a location higher than the radius corner (Figs. 8.54 and 8.55).

On occasion, flexible high-temperature open-weave fabrics have been used to form the doubler, with the scrim or box-weave type material being impregnated with the pourable or RTV adhesive. Dacron or other temperature-resistant materials are suitable to form the reinforced elastomeric seam (Fig. 8.56).

Although very difficult for large bags, the uncured flush elastomeric seam material technique has been used successfully for some applications. Here, a pressure device cures the seam material in a space that is established between the two straight seam edges (Fig. 8.57).

296   ADVANCED COMPOSITE MOLD MAKING

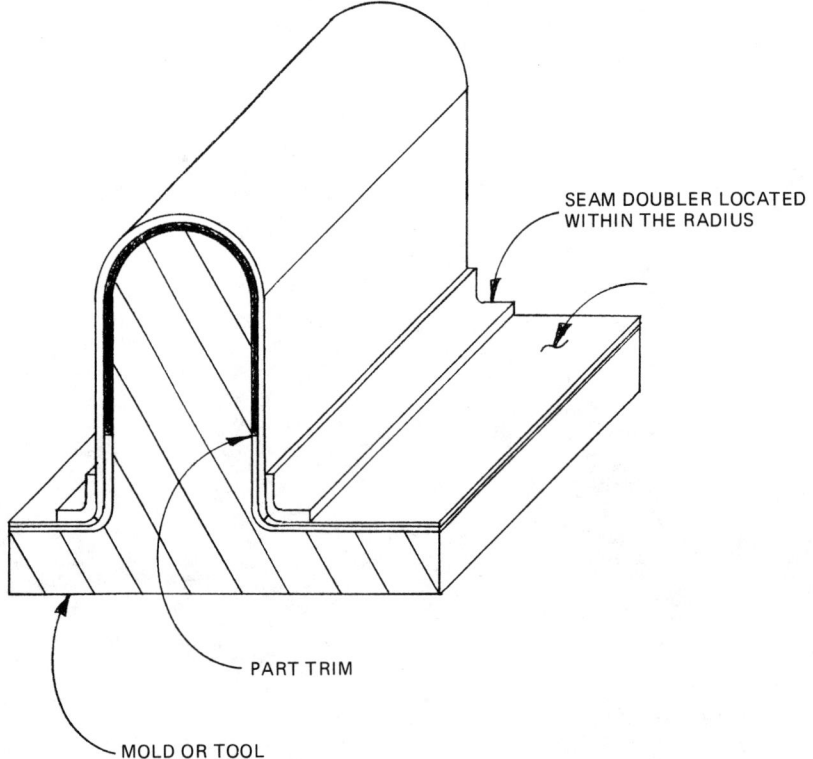

Fig. 8.55. Locating the Doubler Within the Inside Bag Radius.

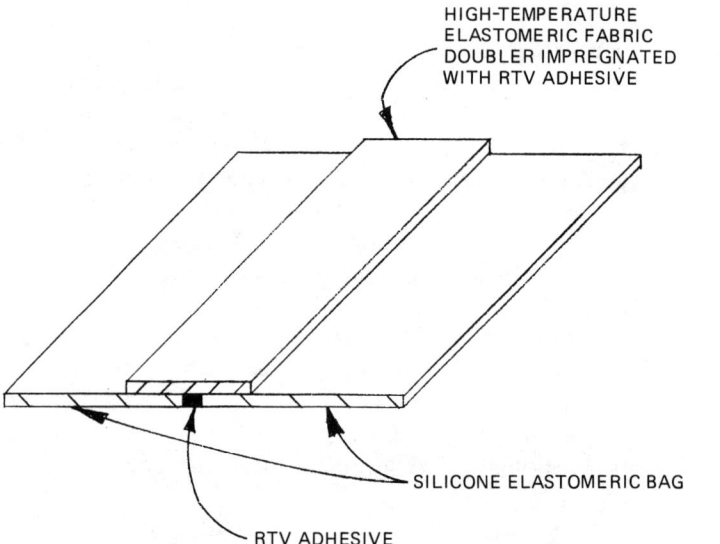

Fig. 8.56. Silicone Rubber, RTV Elastomeric Fabric Doubler Seam.

CONVENTIONAL OR PERMANENT REUSABLE VACUUM BAGS 297

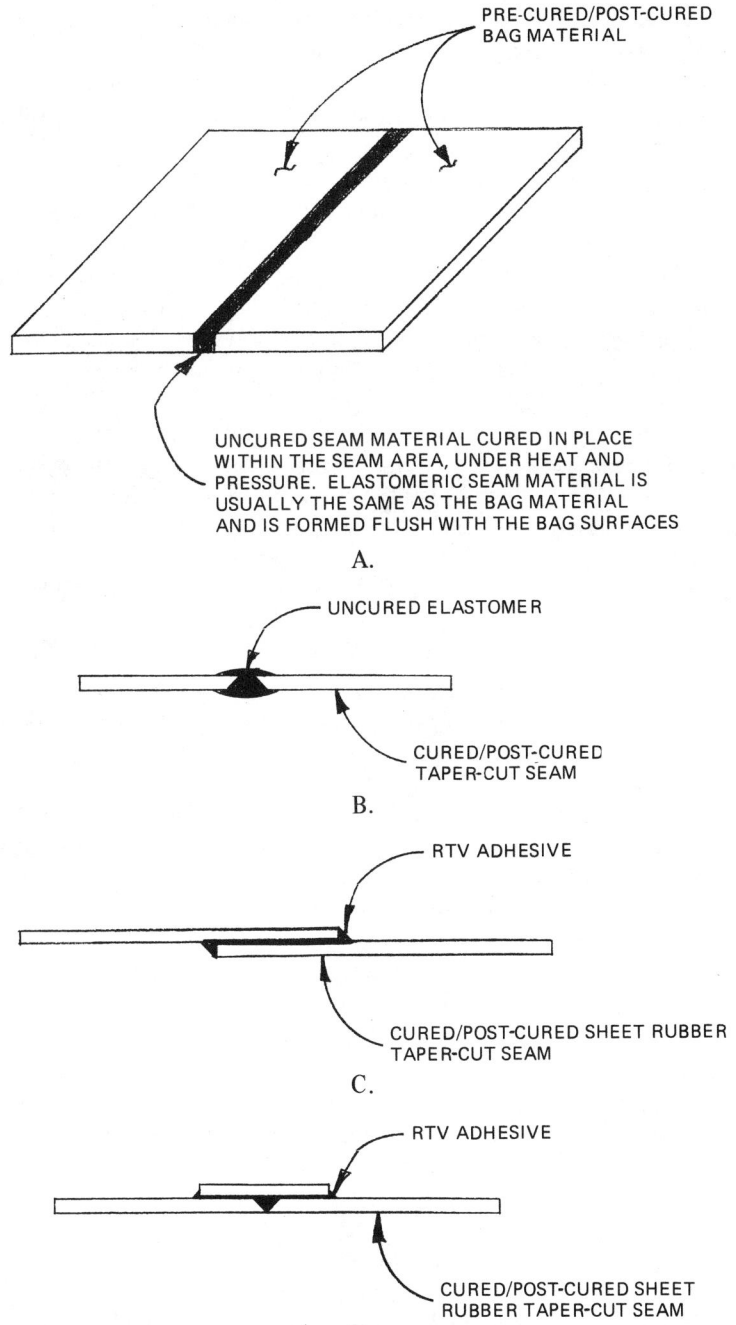

Fig. 8.57A–D. Flush, Heat/Pressure-Cured Elastomeric Seam (Three Additional Configurations also Shown).

Uncured elastomer material, calendared (rolled and squeezed to a predetermined thickness) to a thickness of about one half of the bag material thickness has been used to bond seam doublers in place and bond bag edges to a frame. Like the flush seam configuration just demonstrated, the uncured adhesive must be cured under heat and pressure.

Forming the reusable bag pattern. Now that seams and their formulation has been discussed, it is time to consider the steps required to form the bag pattern. Simple and flat bag patterns are basic and self-explanatory. Custom-contoured patterns are more complex, and we will deal with these in this section.

Bag patterns have been formed in a variety of ways — by using corrugated paperboard stock, sheet metal, plastic films such as polyester and vinyl, kraft paper, and textile fabrics.

Kraft paper is a very common pattern or template material and is used in the transportation, marine, aerospace, and aircraft industries to form patterns for various purposes. Here, it will be used to form the demonstration pattern, which will be the template from which precured/post-cured or uncured sheet rubber material can be cut.

Before starting application of the kraft paper, a space beneath the bag of approximately 1/2 to 1 in. must be provided to allow the bag to draw down into a recessed cavity or over a protruding surface, such as a dome. In either case, a space must be provided beneath the bag to allow for peel plies, separator films, bleeder or breather layers, intensifiers, caulplates, pads, etc. A custom-contoured bag that is too tight may seal off air flow or disturb lay-up materials beneath it during installation.

Styrene or polyurethane foam can be used to form the gap required. Sheet foam is cut and fitted in place so that the pattern can be formed over this surface. When possible, plan to make the reusable vacuum bag over a male pattern.

If the bag material is to be laid-up and formed over these foam surfaces, and these surfaces are used to hold the bag material and seams in place under vacuum during cure, the void compensation material should be capable of withstanding the atmospheric vacuum bag pressure. Radius intensifiers, triangular ramps, or reusable fillets placed in an inside corner or radius will eliminate a bridge or pressure fluctuation of the bag during use. A larger bag radius in a corner is more desirable. When possible, angles or corners that are sharp should be eliminated or heavily padded with a breather cushion, so that the bag will not be abraded or punctured during use.

Other considerations for preparing the pattern should be marked accordingly on the pattern. These include the edge space for edge bleeders, and provision for a flat surface at which the seal or closure will be installed. (A seal can be installed more readily on a smooth and flat edge surface.)

Next, the location and configuration of vacuum ports, whether in the mold, tool, or bag, must be determined and noted on the pattern.

When all preparation and notation has been completed, and the pattern material has been taped in place over the entire subject contour, flats and edge, the pattern is now ready to be slit and laid over the bag sheet rubber for the rubber cutting step.

The pattern is carefully cut or slit from the compensated mold or tool surface, laid flat and, depending upon the bag size, is recut or used as is to prepare the rubber bag material pieces. Overlaps, doubler strips, frame size areas, etc., are considered and marked on the pattern at this time (Fig. 8.58).

Forming, bonding, curing, and thermally stabilizing flat or contoured reusable vacuum bag materials.

A. Precured/postcured sheet rubber bags. To bond cured sheet rubber to itself or to various frame materials, the following process is suggested: First, cut the precured/post-cured sheet rubber into the required size pieces and shapes. It is now ready for bonding. Next, using clean, lint-free clothes, alcohol, chlorinated safety solvent, or ketone solvent, clean the surfaces to be bonded. Application of the RTV adhesive must always be made upon solvent-cleaned surfaces.

If the precured sheet materials are to be bonded to themselves or to a metal frame, the materials and process described in the subsection on seams must be used.

When rubber is to be bonded to rubber, as in the case of silicone, a primer is not necessary. Primers are adhesion promoters and are required when bonding rubber to metal or other hard, dense surfaces, such as hard plastic or other rigid surfaces.

Before applying the primer, prepare the hard surface for bonding by abrading the area where the RTV or heat-cured adhesive will be applied. Grit blasting, abrasive disk grinding, even chemical etching is recommended.

Now, blow away the abrasive residue, using dry, filtered air.

Solvent clean the bonding area when required. The bonding area is now ready for application of the primer.

Apply the primer by wiping a very thin coat over the hard and rigid surface. Allow the primer to dry for 30 minutes or more, as described by the manufacturer's instructions.

Next, with a spatula or other appropriate wiping tool, apply the RTV adhesive to both surfaces being bonded. (A detailed description of the application of the adhesive is described in the subsection on seams).

Now, position the rubber material carefully in place over the rigid surface. A roller or other suitable hard device can then be used to flatten the bond line materials. A conventional disposable vacuum bag, as described in this subsection on seams, can be used to maintain pressure over the bonded areas for the room-temperature cure of 24 to 48 hours.

It is important not to promote the cure of the RTV adhesive with heat or other surrounding environmental changes unless absolutely necessary, since this

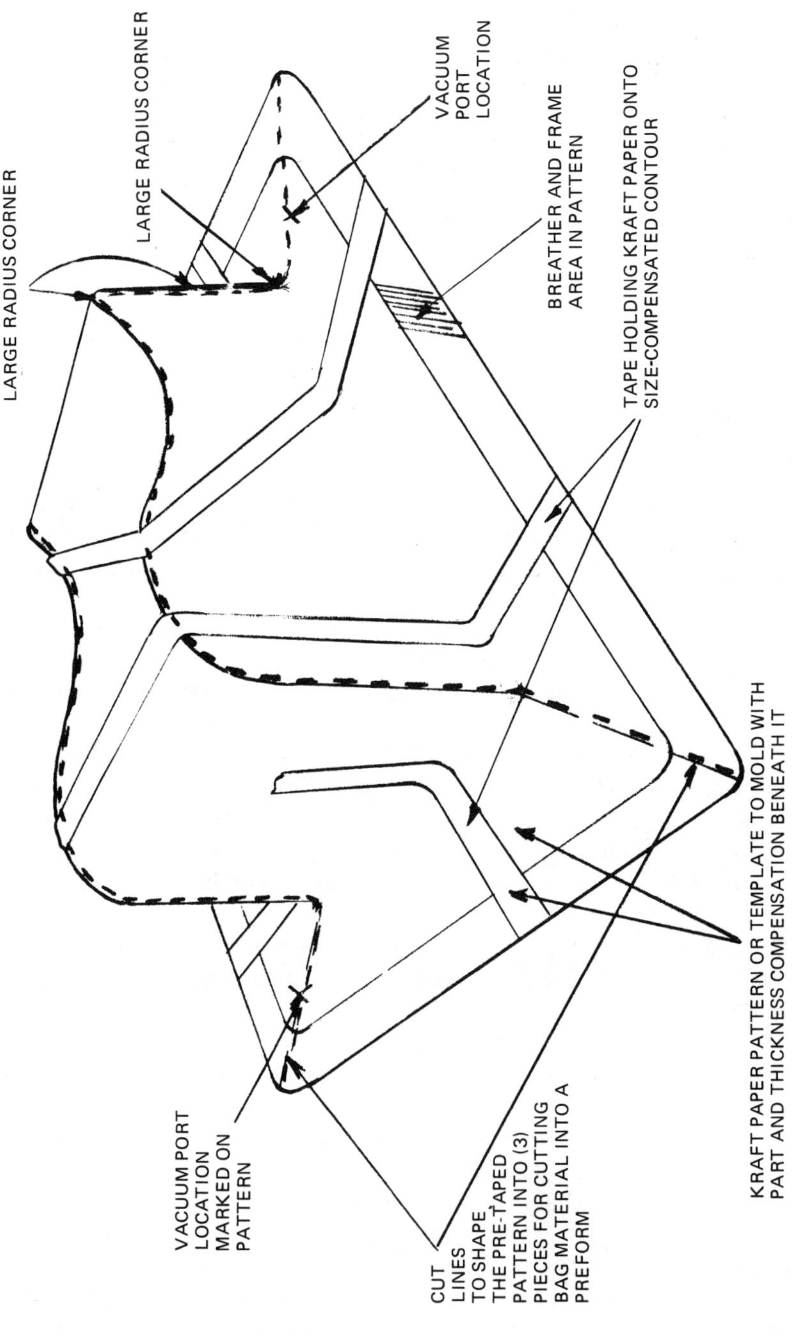

Fig. 8.58. Preparing a Kraft Paper Pattern.

CONVENTIONAL OR PERMANENT REUSABLE VACUUM BAGS 301

may reduce the physical properties of the cured bond line.

RTV adhesives are not the only materials used for bonding precured silicone rubber sheet materials to themselves or other rigid materials. Uncured silicone rubber bag materials or uncured calendared sheet rubber adhesive, as described in the subsection on seams, can be applied as the adhesive to form the vacuum bag.

B. Uncured sheet rubber bags. Besides using the uncured sheet rubber as an adhesive to bond precured sheet rubber to itself or a rigid material frame, the uncured rubber can be used to form the bag.

Uncured rubber is available reinforced with fiberglass, aramid, and graphite fabrics. Reinforced bags are usually used as bonding fixture bags, not composite laminate part-forming bags. Unreinforced bags are recommended over reinforced ones, since the unreinforced material has the ability to stretch up to 700% over its original size (Fig. 8.59).

In the formation of a vacuum bag from uncured sheet rubber, the mold or tool master, part, etc. upon which the uncured rubber is laid up must be capable

Fig. 8.59. Elastomeric Reusable Bag Material Being Compounded and Processed. *(Courtesy of the Keene Laminates Div. of Keene Corp.)*

of being exposed to the vulcanizing or cure temperature of the uncured sheet rubber.

Metal or other rigid surfaces to which bonding will take place should be treated as described for applying RTV earlier in this chapter.

Then, after the surfaces to be bonded have been prepared and primed, apply a release film or mold release to all areas to which the uncured sheet will not bond. A suitable mold release can be mixed from a solution of liquid soap-type dish detergent. A dry fluorocarbon spray release, or other release coating supplied by a mold release manufacturer may also be used.

Now, lay up the cut, uncured rubber pattern pieces over the released, thickness-compensated surface. Compensation for breather, bleeder, and other bag accessories, beneath the bag, is accomplished by the same process used for precured/post-cured bag fabrication.

Next, apply release fabrics, or dry and clean nylon fabric, or other conformable release materials to flatten out the cut, uncured pattern pieces of rubber. Simply place the fabric over the uncured rubber, and, starting from the mold center and working out to the edges, roll or iron out the bubbles. Do not allow any wrinkles or bubbles to remain in the lay-up. Carefully slice the lay-up and release the air when necessary.

Overlapping joints of uncured sheet rubber are allowable, and the overlap, from 1/2 to 1 in. wide, should be feathered, ramped, blended, or "skived." Using a spatula, smooth the overlap flat and into the layer beneath the overlapping piece.

If two plies of rubber are used to form the bag, carefully place the seams so that they do not overlap. The finished thickness of a bag usually is from .040 to .100 in. thick.

Usually, the uncured bag lay-up must be enclosed under a disposable vacuum bag and cured in an autoclave at 300° to 350°F. Pressures of 45 to 85 psi are used, and the cure time ranges from 30 to 60 min (Fig. 8.60).

The variations occur because material formulations vary from one manufacturer to another. Disposable vacuum bag curing of uncured rubber sheet assures that air will not inhibit the bag material surface during cure and cause a sticky surface to result after the cure step.

After the initial cure, the vacuum bag should be post-cured at 400°–450°F, depending upon the manufacturer's recommendations. Post-cure time is usually 3 to 4 hrs. Do not allow any bag surfaces to touch each other during post-cure, or they will either bond together or not cure properly.

General precautions. When processing uncured sheet rubber as a bag material or adhesive, or when using RTV rubber as an adhesive, the following general precautions should be taken:

CONVENTIONAL OR PERMANENT REUSABLE VACUUM BAGS 303

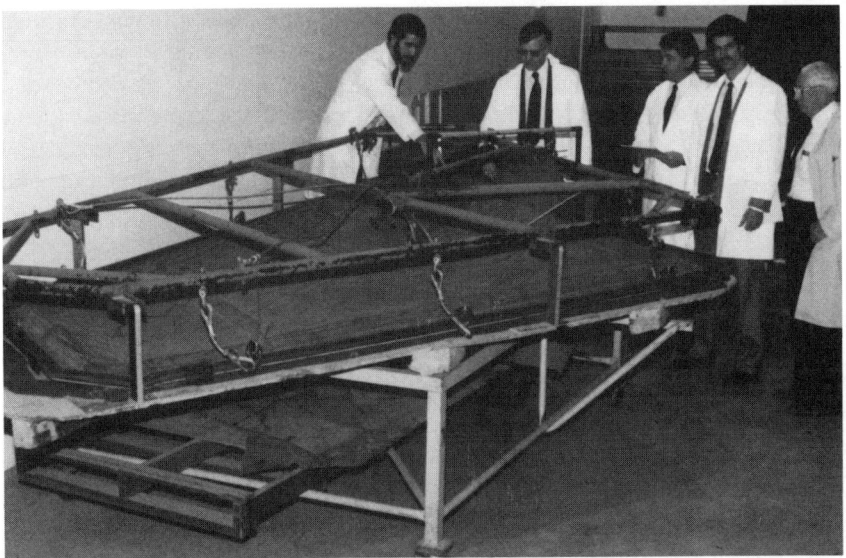

Fig. 8.60. Reinforced Elastomeric Reusable Vacuum Bag and Support Frame. *(Courtesy of Grumman Aircraft Systems)*

- Primers should not be used for rubber to rubber bonds.
- RTV adhesives cure by exposure to air and humidity in the air. Do not totally trap the adhesive in a confined area and expect it to cure. Covered adhesive bond lines, in fact, could take up to 72 hrs at room temperature to cure.
- One and two component RTV adhesives should not be heat cured.
- Do not allow RTV adhesives to come in contact with uncured sheet stock, since the rubber sheet material will be inhibited during cure.
- For the same reason, do not allow petroleum products to come in contact with uncured rubber stock.
- Outgassing of rubber during post-cure will cause reversion of the rubber sheet materials, thus causing the cured rubber bag to soften and become sticky.
- Adhesive from various kinds of tape will inhibit the cure of some uncured rubber stocks or RTV adhesives.
- Sample test combinations of bag materials and adhesives for use in forming vacuum bags.

Spray-on, Paint-on, and Cast Vacuum Bags. Although sheet rubber is the most reliable material for precured/post-cured and uncured reusable vacuum bags, there are other materials and processes available that can be used to produce

304 ADVANCED COMPOSITE MOLD MAKING

low-production-volume parts, inexpensive — yet still unique — reusable vacuum bags, and assemblies. These materials and processes include the spray- or paint-on application of latex, polysulfide, and even silicone elastomeric compounds.

In addition to the spray- or paint-on application of the above materials, the use of cast RTV elastomers is another available vacuum-bagging material and process choice. When casting, RTV silicones and polysulfides are by far the best choice of elastomer.

Spray-on and paint-on elastomeric vacuum bags. In a manner similar to the preparation methods used to form other types of reusable vacuum bags, the spray-on/paint-on method requires the installation in the mold or tool of a simulated or actual part, plus some types of thickness compensation, such as felt or precut sheets of foam, to simulate the buildup of peel ply, breather, bleeder, and other vacuum bag accessories. This procedure is required to allow a space of approximately 1 in. beneath the bag for accessories, shrinking, and stretching of the bag elastomer.

Another use for paint- or spray-on elastomers has been the application of the limited life elastomer as a protective layer or shield, covering sharp mold or tool parts that must be used in the final-assembly forming process (Fig. 8.61).

Whether using spray- or paint-on elastomers, for forming elastomeric shields or bags, the bag material must be applied over a solid release film that has been carefully positioned, butt seamed, and taped on the seams with polyester tape. The taped seams will prevent wrinkling of the applied bag elastomer — a crucial step, since wrinkles can be a cause for the bag to fatigue and subsequently fail during use, especially in the folded or wrinkled area.

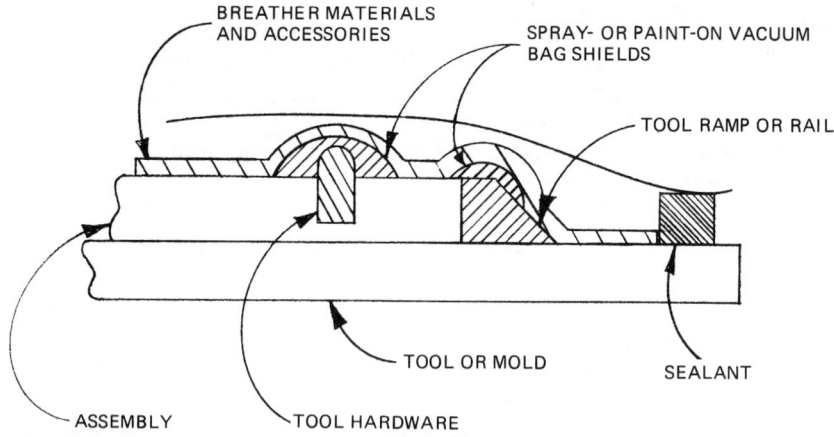

Fig. 8.61. Paint- or Spray-on Elastomers as a Vacuum Bag Shield.

Although polysulfides and latex elastomers have been mentioned as candidate bag materials, they are usually only used for room- or low-temperature molding or bonding applications. Silicones have proven to be the most reliable materials for use at elevated temperatures.

Important note: At this point in the process, thermocouples must be installed in all vacuum bags used at temperatures above ambient room temperature. Their installation is covered in Section 8.3.1, pages 306 and 307.

When spraying- or painting-on bag elastomers, the best rule to follow is to dilute the bag elastomer as little as possible before application. Reactive diluents are better to use than solvents that evaporate, as they take part in the chemical reaction.

Diluting elastomers causes the material to weaken during use. Thus, this material may shrink during application or use. *If* practical, fiberglass fabric or other suitable reinforcements should be used to support the applied elastomer, eliminating some of the problems described above. The fabric materials are applied in layers and impregnated during the bag lay-up process. Paint-on or spray-applied coatings can be either heat-cured or RTV types, but in either case, the build-up process is performed by gradual application of coats that are from .015 to .020 in. thick, until the final cumulative thickness is attained. This final thickness, similar to pre- or uncured sheet, should be from .040 to .080 in. thick.

Cast vacuum bags. Vacuum bags and/or portions of reusable vacuum bags have been formed using two-part RTV and heat-cured elastomers. Again, the silicones are the best choice over polysulfides or other pourable elastomers such as polyurethanes.

In all cases, it is suggested that the elastomer be evacuated, degassed, or de-aerated prior to pouring. Ten to twenty mintues of degassing time after the initial air surge from the mixed resin is optimum.

The same master model preparation used for all other bag-making processes is used for casting. Casting will result in the greatest degree of shrinkage experienced. For this reason, it is suggested that the thickness compensation installed beneath the bag also allow for a shrinkage of approximately .002 to .010 in./in., upon curing. Of course, this may vary from manufacturer's material to manufacturer's material. The CTE, usually in the 80×10^{-6} range, will not affect part forming, since the cast vacuum bag will conform to the part material and bag accessories applied beneath it.

When pouring the degassed elastomer, carefully position the flowing material so that a continuous flow with no ripples in the material will occur. Ripples in the flowing material can cause air to be trapped in the casting. Trapped air bubbles will expand during part formation and can cause surface discrepancies on the part, or damage to the bag.

306 ADVANCED COMPOSITE MOLD MAKING

RTV compounds have proven to be the most stable materials for use in forming cast bags. However, two-component, low-temperature compounds can also be used. In this case, the heat cure will increase shrinkage, so it is recommended that, when heat curing, the degree of shrinkage should be allowed or compensated for.

8.3. FRAMES, CLOSURES, SEAL CONFIGURATIONS, AND ACCESSORIES FOR PERMANENT REUSABLE VACUUM BAGS

The time has now come to explain the techniques for the installation of through-the-seal or bag thermocouples, bag frame or cable supports, hoist brackets, and metal or plastic bag frames. Vacuum ports or valves for gauge attachment and pulling a vacuum within a bag will also be described, as well as methods of installing ports or valves through the tool, seal, or bag.

Then, unique seal or closure designs will be demonstrated through the use of such material combinations as extruded bag seal ribs, pressure sensitive reusable seal adhesives, conventional sealant tapes, and others.

8.3.1. Thermocouples

These wire devices are used to indicate the temperature of the part laminate or assembly being formed beneath the bag. The thermocouple device can also indicate temperatures in other locations beneath the bag.

The safest and easiest way to install the thermocouple is to first insert the thermocouple wire into a predrilled hole in the bag or seal. Provide an extra amount of wire on the tool so that, should the wire be snagged or pulled, it will not be broken or the bag or seal ripped or damaged. Then, using an elastomeric adhesive that is compatible with the seal or bag, bond the thermocouple wire firmly into the predrilled hole (Figs. 8.62 and 8.63).

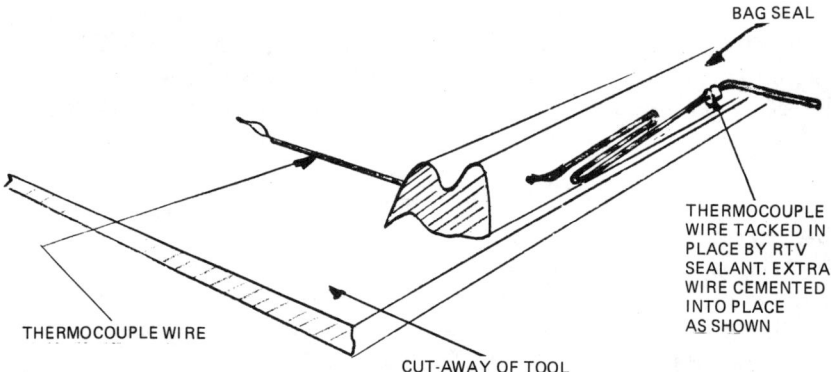

Fig. 8.62. Installation of a Thermocouple Wire Through a Seal.

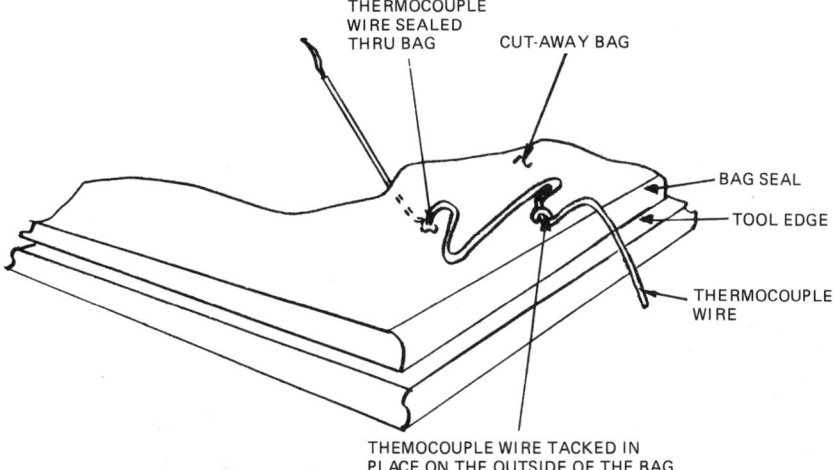

Fig. 8.63. Installation of a Thermocouple Wire Through a Bag (Only Used When Through-the-Seal Installation is Impractical).

8.3.2. Frames, Closures, Seal Configurations, and Accessories for Permanent Reusable Vacuum Bags

Frames used to support reusable vacuum bags can be simply described by comparing their appearance to a picture frame. Their function is to support the edge or periphery of the bag and to help to create a seal when the edges are pressed to the seal. Frames can be rigid or semirigid.

Rigid frames can be formed by shaping and bending light metal angles, such as aluminum, to the contour of the mold or tool periphery. Subsequently, the reusable bag rubber will be bonded to the contour frame. Whether the frame is flat or contoured, the metal angle material should be 1 to 1 1/2 in. high in cross-section.

Rigid frames can also be formed using room-temperature-cure/high-temperature-use epoxy resin systems, with fiberglass reinforcement. This allows a composites shop or other facility to totally fabricate a bag frame and bag, on site, without the presence of metal-working equipment. In this case, a dam can be formed around the mold periphery using a flexible polyethylene foam, and the laminate lay-up frame, which should be approximately 1/4 in. thick, can be formed by laying-up and shaping a laminate angle (Fig. 8.64).

Semirigid frames can be fabricated by using reinforced rubber along the bag edge. Some applications allow for the use of extrusions that are bonded to the bag edge and later locked into a mating surface seal, groove, or channel. High-temperature velcro materials have also been used to perform the frame lock-on function along the bag periphery.

308 ADVANCED COMPOSITE MOLD MAKING

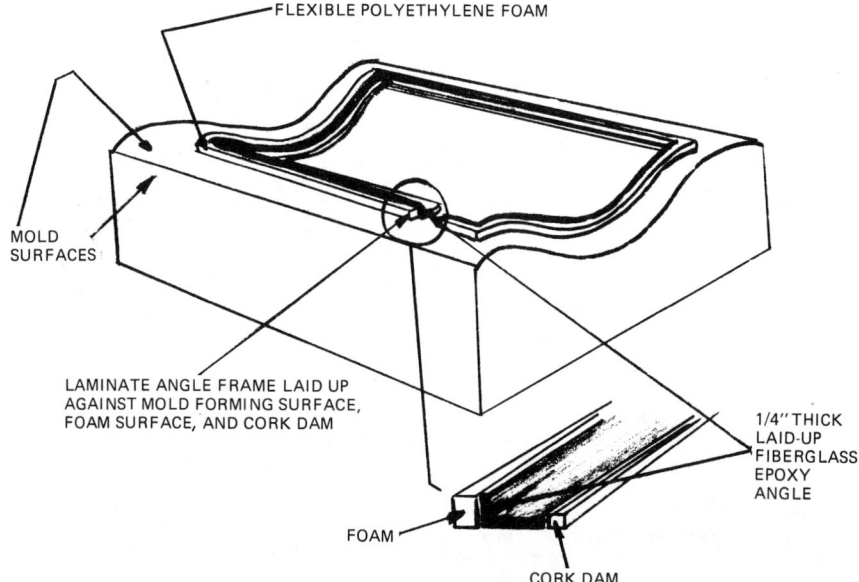

Fig. 8.64. Fiberglass/Epoxy Bag Frame.

Semirigid framed bags can be cleaned and rolled up after use. This makes them extremely desirable for storage in a small composite facility.

Seals or bag closures are the mechanisms that form the vacuum-tight seal along the bag edge and inside the frame surface. There are many ways to form bag seals on reusable vacuum bags, but the cross-section configurations illustrated in Figs. 8.65–8.74 are the most common.

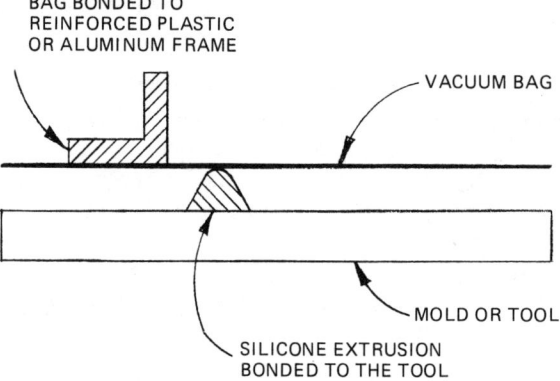

Fig. 8.65. Common Seal (Rigid): Bag Bonded to a Frame and Pressed Over an Extruded Rim Bonded to the Tool.

CONVENTIONAL OR PERMANENT REUSABLE VACUUM BAGS 309

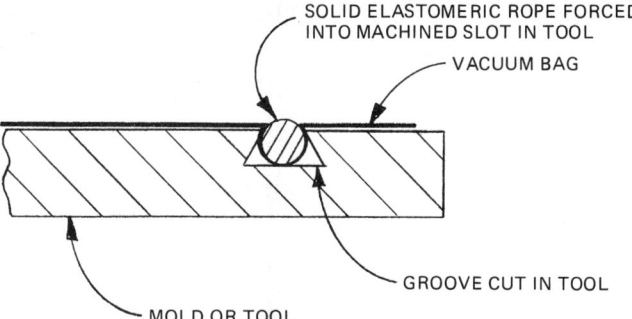

Fig. 8.66. Common Seal (Rigid): Bag Captured By a Solid Elastomeric Rope Pressed into a Groove Cut into the Tool.

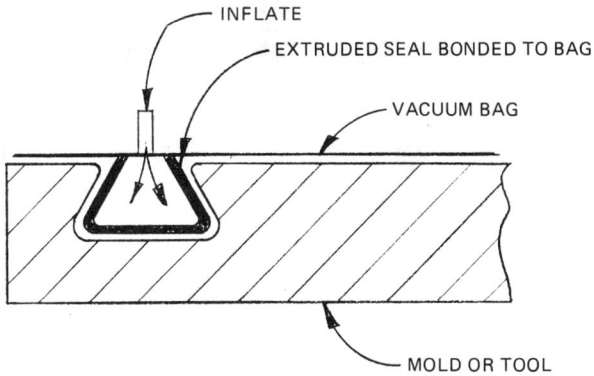

Fig. 8.67. Common Seal (Rigid): Bag Captured By an Inflatable Extruded Seal in Groove Cut into Tool.

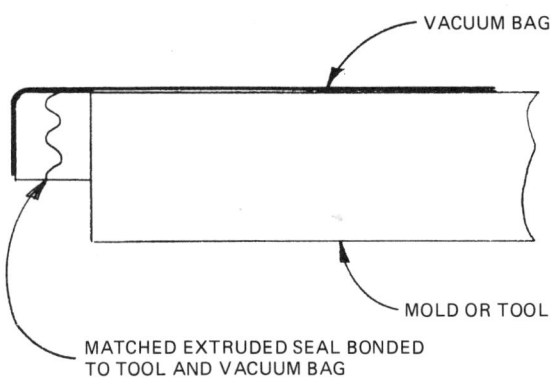

Fig. 8.68. Common Seal (Rigid): Interlocking Extruded Seal Halves Bonded to Bag and Tool.

310 ADVANCED COMPOSITE MOLD MAKING

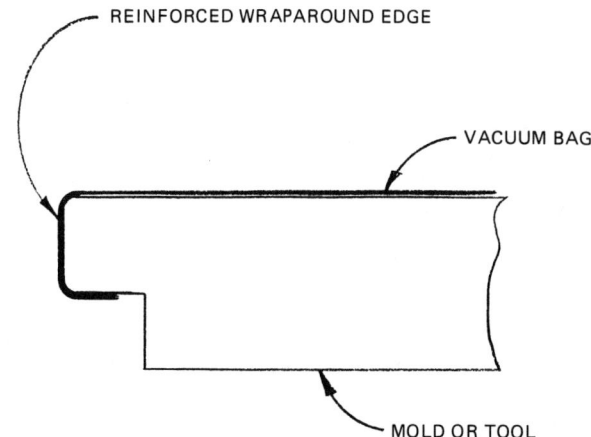

Fig. 8.69. Common Seal (Rigid): Reinforced Wraparound Bag Edge Attached Over the Edge of the Tool.

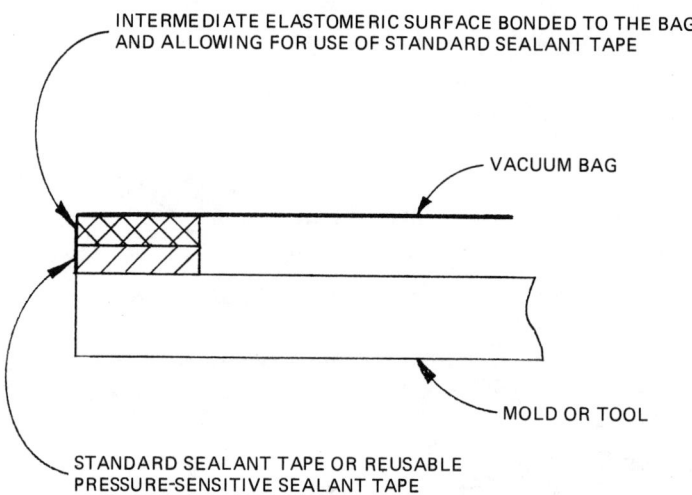

Fig. 8.70. Common Seal (Rigid): Bag Attachment to the Tool by Means of Sealant Tape on a Reinforced Surface of the Bag Edge.

CONVENTIONAL OR PERMANENT REUSABLE VACUUM BAGS 311

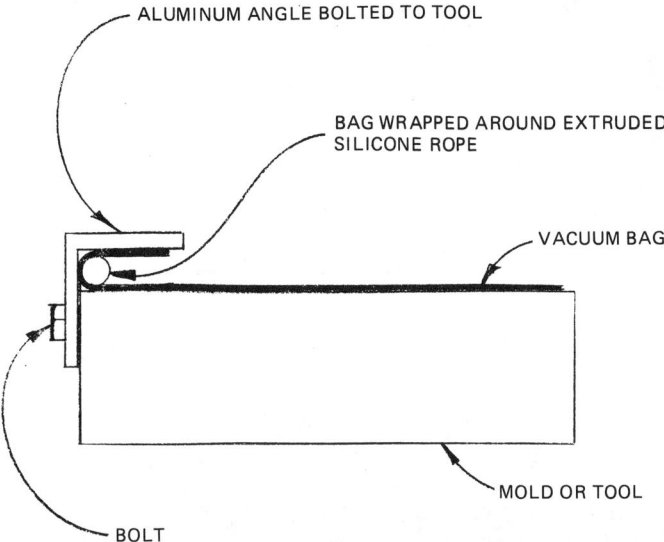

Fig. 8.71. Common Seal (Semirigid): Bag Edge Wedged Between a Silicone Rope and the Tool Edge by Means of an Angle Bolted to the Tool.

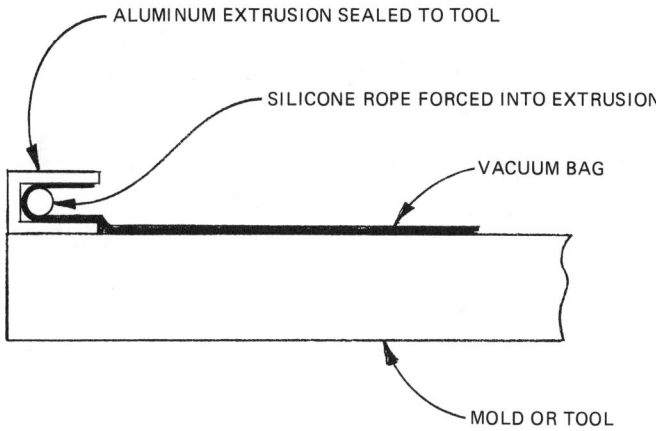

Fig. 8.72. Common Seal (Semirigid): Bag Edge Wedged, with a Silicone Rope, into an Extruded Channel Sealed to the Tool.

312 ADVANCED COMPOSITE MOLD MAKING

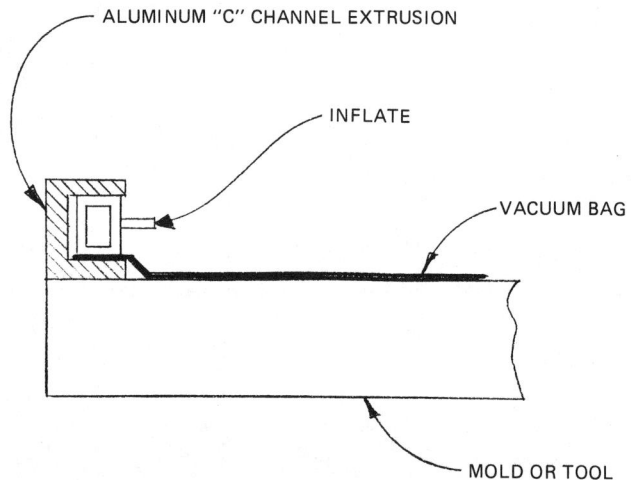

Fig. 8.73. Common Seal (Semirigid): Bag Edge Wedged, with an Inflatable Seal, into an Extruded Channel Sealed to the Tool.

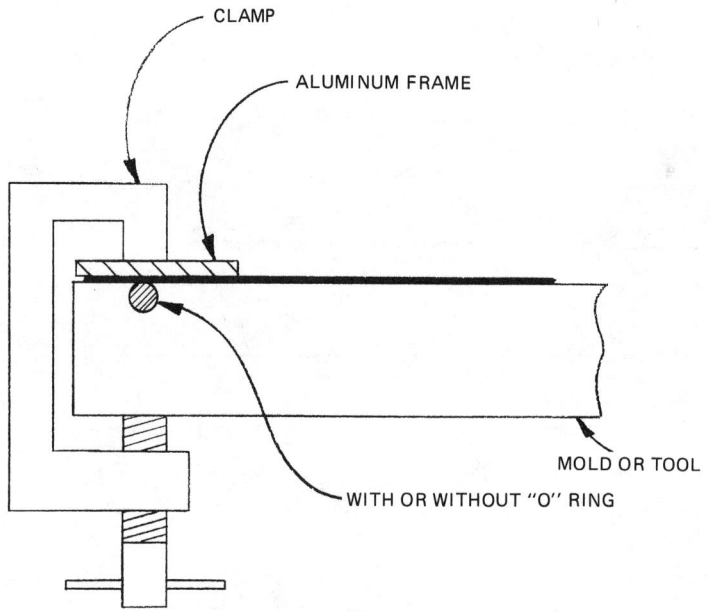

Fig. 8.74. Common Seal (Semirigid): Bag Edge Clamped to the Tool.

Cable supports and hoist brackets are usually used to lift rather heavy, large, and frequently used vacuum bags up and out of the work area. The brackets are installed into the rigid frame edge, using eyes as lift points.

If the reusable vacuum bag is so heavy that it must be supported while the frame is being used to lift and support the bag, cable supports are attached directly to the bag surface and frame, and these supports are used to raise and lower the bag (Fig. 8.75).

Vacuum ports installed throughout the reusable bag are another accessory that makes this type of bag more desirable than conventional disposable bags.

It is not always practical to design and locate the precise position where the vacuum port or valve should be located. With reusable bags and these types of vacuum ports, additional ports can be added or even relocated simply by carefully cutting and mechanically installing new ports, and, when necessary, sealing

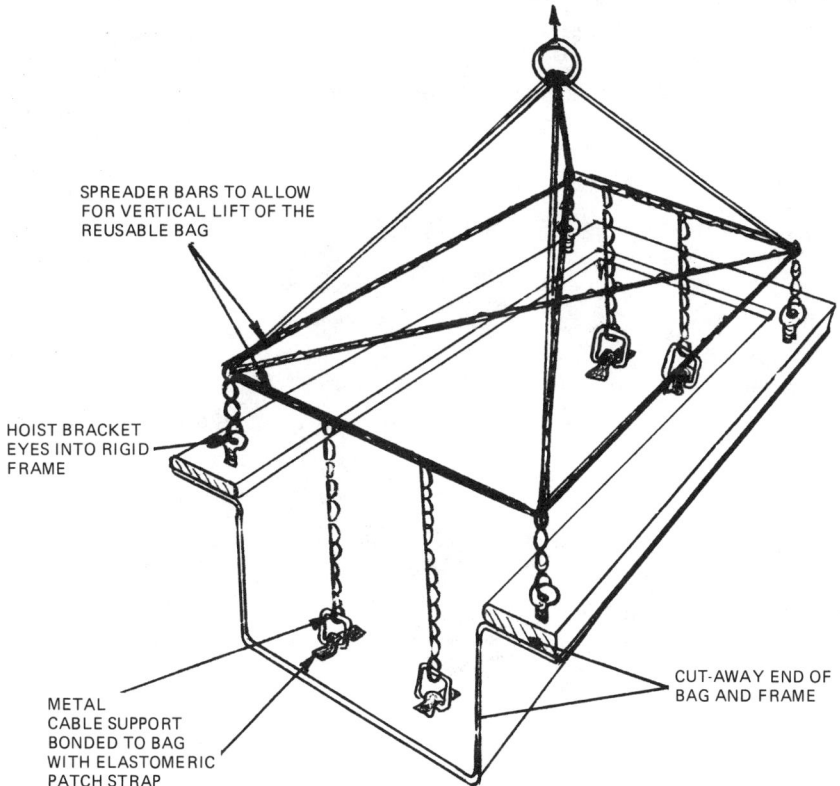

Fig. 8.75. Cable Supports and Hoist Brackets.

314 ADVANCED COMPOSITE MOLD MAKING

up an original location using an RTV adhesive and/or patch. See Chapter 9 for more details on reusable ports, fittings, accessories, and support tools.

8.4. CAUL PADS, INTENSIFIERS, AND CAUL PLATES FOR CONVENTIONAL AND PERMANENT REUSABLE VACUUM BAGS

Caul pads, intensifiers, and caul plates are just some of the names used to describe bag-side tooling aids, used to assist the composite laminate or bonding fabricator in forming more precise and accurate parts. In addition to their aid in precision forming, these bag-side tools may eliminate some additional manufacturing steps or substitute for a matched mold set or expensive matched tooling package (Fig. 8.76).

Figure 8.77 will illustrate the function and purpose of the caul pad, plate, or bag-side tooling aid.

As we have previously studied, the pressure required to form or bond a laminate part and that is acting on the surface of a vacuum bag is a result of the transfer of atmospheric pressure to the part material being formed through the vacuum bag material itself. If the surface opposite the mold surface must be finished during forming, then a tool aid adjacent to the bag must be used. Thus,

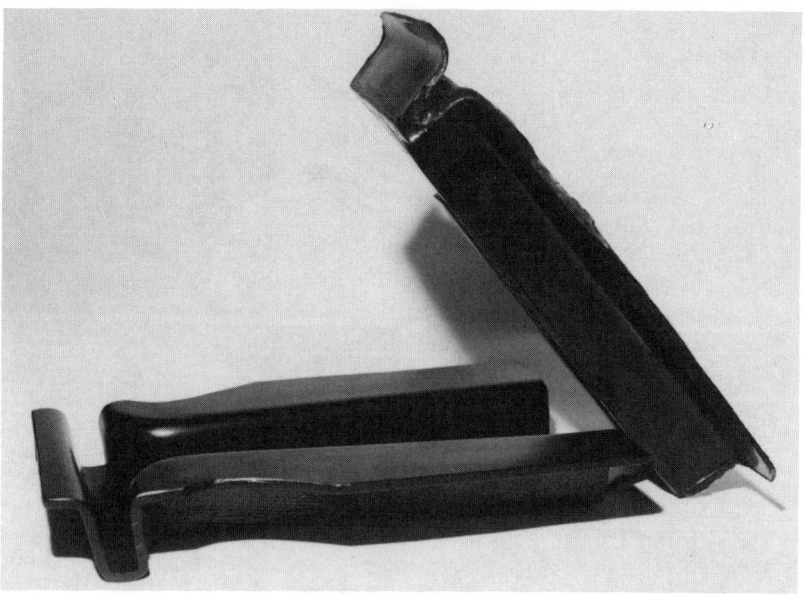

Fig. 8.76. Bag-Side Tool Over a Laminate Mold. *(Courtesy of A.C.E., Inc., the Keene Laminates Div. of Keene Corp., and the Ren Plastics Div. of Ciba Geigy Corp.)*

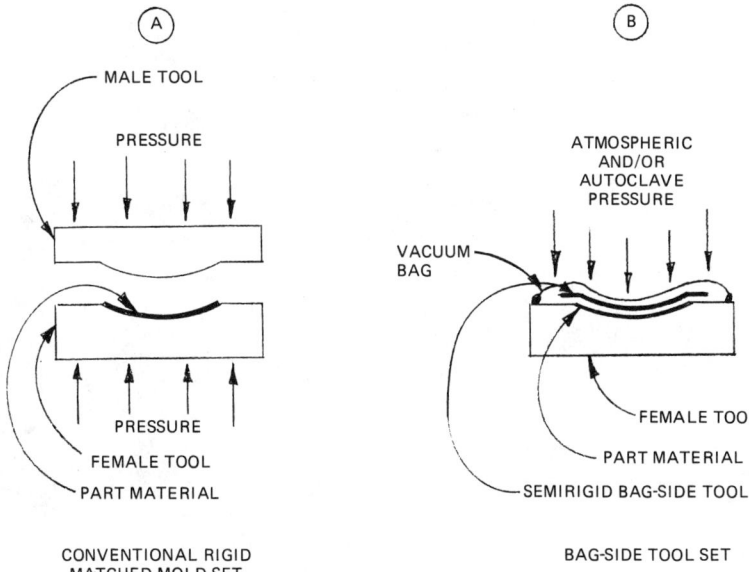

Fig. 8.77A and B. Function of a Bag-Side Tool.

the bag-side tool. This tooling aid, adjacent to the bag, can be semirigid (caul pad) or rigid (caul plate).

In use, the caul pad, plate, or bag-side tool is pressed against the part material being formed or bonded by the transfer of atmospheric pressure brought about by the use of a vacuum bag. Atmospheric and/or autoclave pressure, transferred through the bag and caul pad or plate, can eliminate one half of the tool set and the need for mechanical pressure applied by a hydraulic press or other physical means. The use of a bag-side tool also allows for the fabrication of any size laminated or bonded assembly, finished on both sides, and when pressures in excess of atmospheric are required, the limiting factor becomes the size of the autoclave (Fig. 8.78).

Bag-side tools or tooling aids also allow for the use of more innovative molding methods such as co-curing details and laminate skins simultaneously with finished inside and outside surfaces. Secondary bonding operations are thus eliminated, since details can be bonded in place at the same time that the laminate skins are formed. This is expedited by the fact that the bag-side tools can be elastomeric, and elastomeric tooling can form complicated shapes that, in some cases, could not be formed using rigid, conventional tooling.

Prototypes and small production runs can also be easily accomplished, since this type of tooling eliminates the need for rigid and hand-matched mold sets, which are usually extremely expensive.

316 ADVANCED COMPOSITE MOLD MAKING

Fig. 8.78. Bag-Side Tool (Elastomeric Caul Pad). *(Courtesy of Grumman Aircraft Systems)*

Of all the elastomers used in the formation of elastomeric molds and bag-side tools, silicones are customarily recognized as the best of the flexible materials. The reasons are many: First of all, they possess the following qualities that lend themselves easily to use as a flexible material: adjustable hardnesses, temperature, chemical resistance, and a predictable CTE (Fig. 8.79).

Secondly, the use of silicones as materials for elastomeric tools also provides for easy molding of areas of a part that might have a slight undercut or zero draft.

Thirdly, when molding from the surface of a silicone elastomer, the use of a parting agent is not usually required.

Finally, tool repair and replacement is easy. The operator simply cuts away and replaces sections and portions as required. (For further information concerning materials for these applications, consult Chapter 4, Mold Materials.)

Now that we have reviewed the basic description of caul pads, caul plates, intensifiers, and elastomeric molding, it is time to explore each of these in detail.

8.4.1. Caul Pads

Caul pads as elastomeric bag-side tools increase the molding possibilities for composite laminate part molding and bonding far beyond the scope of metallic or nonmetallic rigid tooling. As mentioned in the introduction to this chapter, the silicones are the most efficient materials for producing elastomeric tools,

Fig. 8.79. Elastomeric Material Being Manufactured. *(Courtesy of the Keene Laminates Div. of Keene Corp.)*

because of their thermal and chemical properties. But sometimes, a particular application does not allow for the use of silicones; therefore, the use of a nonsilicone is required.

Nonsilicone materials for bag-side tools are available processed from butyl, acrylate, and slightly higher temperature materials such as fluoroelastomers. Butyl, the lowest temperature-of-use material of the group, continues to shrink under repeated use, so the acrylates have proven to be the best material of choice.

Acrylate elastomers can be applied, processed, and cured in an unreinforced or reinforced state (Fig. 8.80). Reinforcement is available in various forms. Wire screen, fiberglass, or graphite epoxy prepreg, as well as some dry fabric reinforcement materials, have been used. In all cases, the reinforcement is selectively placed, so that the bag-side tool can stretch over and into radius corners as required. This is illustrated in Fig. 8.81.

Taking advantage of the fluid pressure provided by the autoclave allows more design freedom in the fabrication of the bag-side tool, and thus of the part. The elastomeric bag-side tool, as the name implies, will follow the mating mold or tool surface contour. The simple materials and process can be modified to allow for a vent beneath the elastomeric tool to the autoclave or atmosphere, and gain the advantage of internal or "beneath-the-bag" pressure, instead of the basic fluid pressure only within the autoclave.

318 ADVANCED COMPOSITE MOLD MAKING

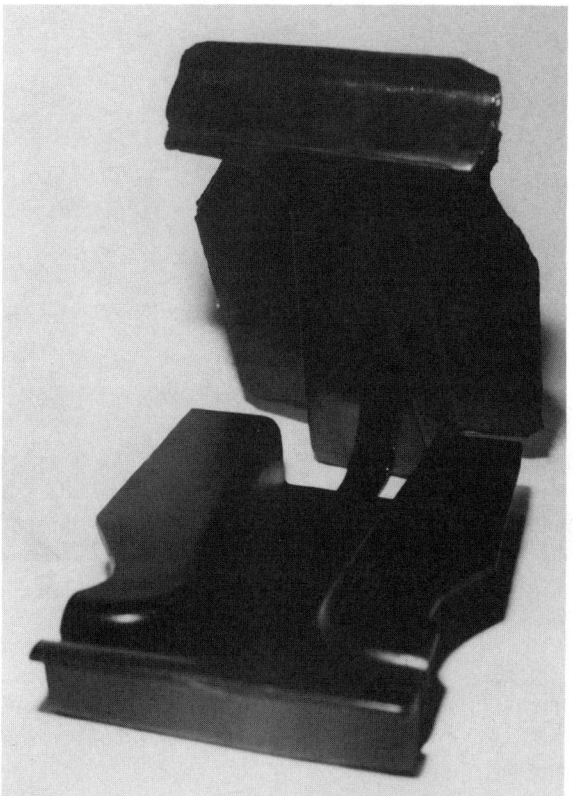

Fig. 8.80. Nonsilicone Caul Pad Over a Laminate Tool. *(Courtesy of A.C.E., Inc., the Keene Laminates Div. of Keene Corp., and the Ren Plastics Div. of Ciba Geigy Corp.)*

For instance, if it becomes necessary to form a stiffener or reinforcement on the rear of a tool or part during the cure of the face laminate, a co-cured manufacturing procedure must be followed. The necessary sequence of steps is illustrated by Figs. 8.82A and B and 8.83A and B.

Co-curing and venting of elastomeric bag-side tools evolves into a more complex form of the fluid pressure principle within an autoclave, by forming a rigid laminate in back of the caul pad and then allowing the atmosphere or autoclave pressure to assist in driving the elastomeric tool material into curves or radius corners. The theory is almost the same as envelope bagging a laminate during cure and allowing outside pressure to pass through vacuum ports in the bag to apply pressure within the tool (Fig. 8.84).

CONVENTIONAL OR PERMANENT REUSABLE VACUUM BAGS 319

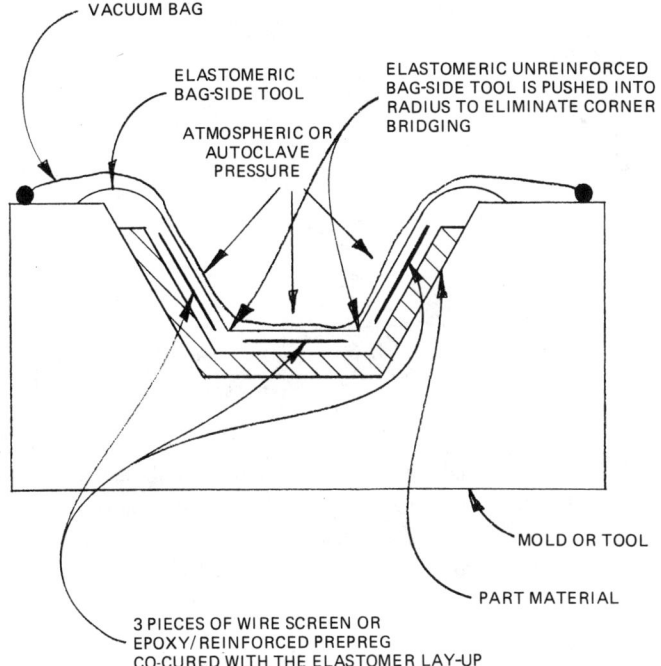

Fig. 8.81. Reinforcement and Lay-up of an Elastomeric Bag-Side Tool.

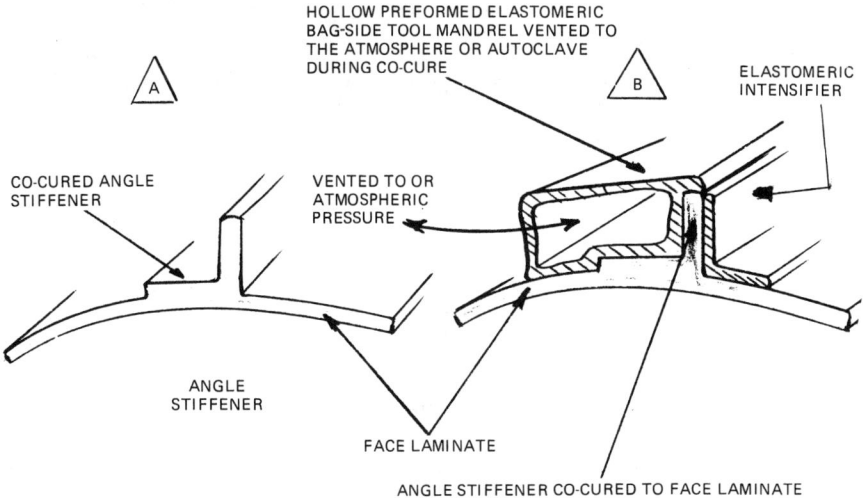

Fig. 8.82A and B. Co-cured Angle Stiffener.

320 ADVANCED COMPOSITE MOLD MAKING

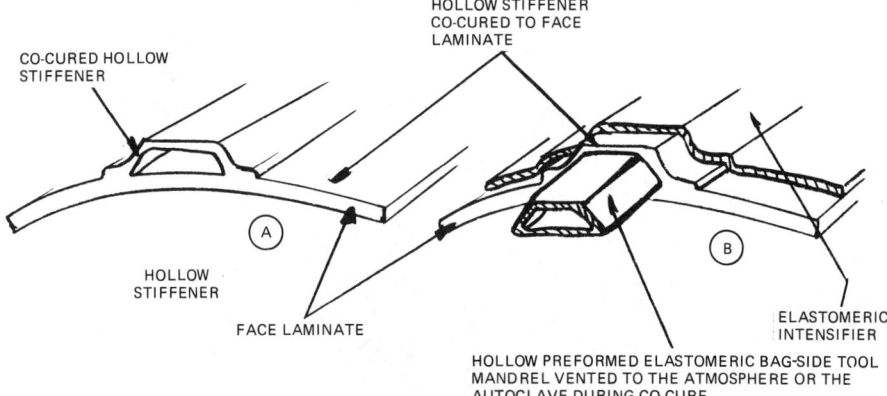

Fig. 8.83A and B. Co-cured Hollow Stiffener.

The elastomeric caul pad can be constructed so that passages within the caul pad allow the atmospheric pressure to travel to select locations as desired. This is accomplished using tubes or the manifold principle found in the embedded heating and cooling coils of a mass-cast tool or the vacuum manifold of a forming tool. An elastomeric caul pad vented in this way can be placed on both sides

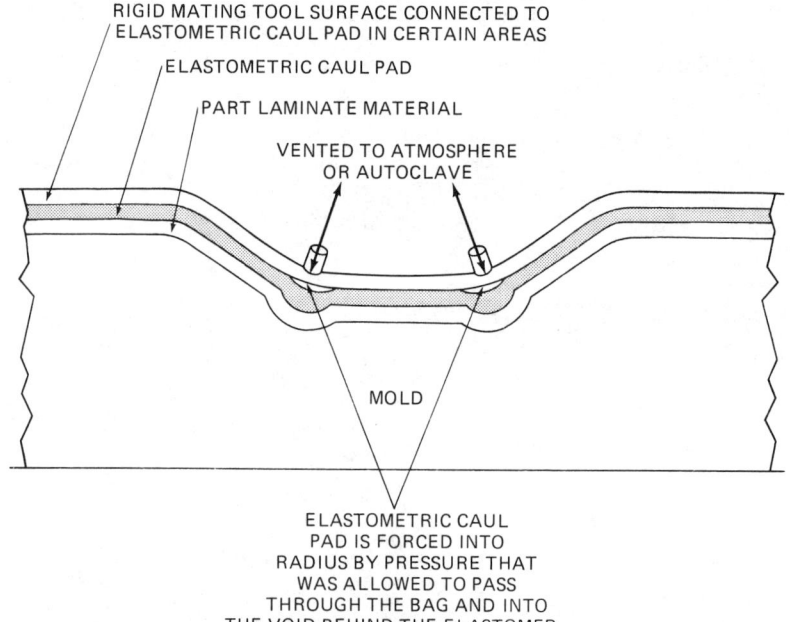

Fig. 8.84. Intensification Through an Elastomeric Caul Pad.

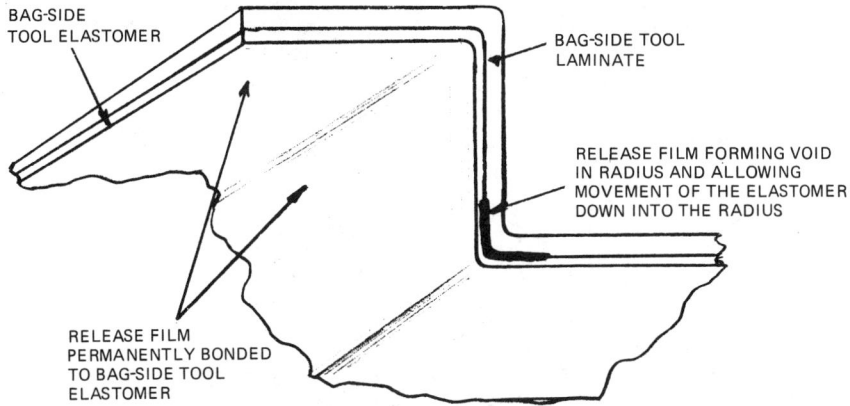

Fig. 8.85. Release Films in and on Semirigid Caul Pads.

of a part laminate being molded. Thus positioned, it will apply equal pressure in opposing directions.

It is obvious by this time that elastomeric caul pads produce highly accurate parts. The quantities from a given tool will depend upon the complexity of the part, as well as the care used in handling. A good estimate would be 100–200 parts on the average, using a 4 hr cure cycle at 350°F and 50–80 psi autoclave pressures. Of course, lower temperatures and pressures allow for the molding of a larger quantity of quality parts.

Another advantage of the elastomeric-faced caul pad laminate is that release films can be applied to the front of the elastomer and adhesion will result in a permanent release-film-faced caul pad. This is accomplished by curing the elastomer against the release film which has a bondable surface on one side.

This release film is the same type of material that can be installed between the laminate and elastomer in select areas, and is used to form a void in a radius into which the atmospheric or autoclave pressure will pass (Fig. 8.85).

The use of elastomeric-faced caul pads (laminate rear support surfaces) with selectively released internal areas, vented to the atmosphere or autoclave, allows for the formation of finished interior and exterior laminates.

An interior surface of a laminate part, such as the boat hull pictured in Fig. 8.86, is laid up simultaneously with the outside laminate; then, both surfaces are mated for co-curing and bonding. Any laminate part can be made using the approach of the internal elastomeric bag-side tool.

Once the nature and use of elastomeric bag-side tools is understood, it is time to proceed to the area of pressure intensifiers and thermal expansion molding, using foam and elastomer materials. These types of materials and process techniques, when combined with the use of caul pads or bag-side elastomeric tools, further expand the possibilities of co-cured laminate structures.

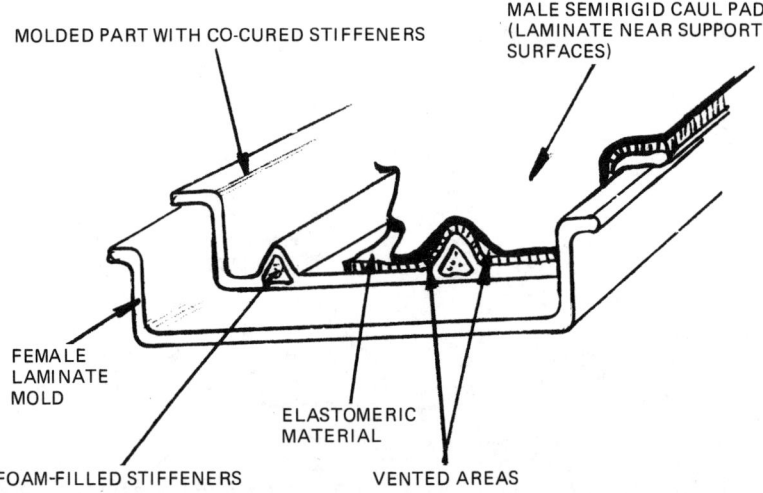

Fig. 8.86. Matched Laminate/Caul Pad Mold Set.

8.4.2. Intensifiers and Elastomeric Molding

It seems logical to combine the topics of intensifiers and elastomeric molding, since each material and process involves the expansion of a material or the enlargement and movement of a material for forming a laminate part segment and/or composite bonded assembly.

Intensifiers. These can take on many forms. Some are extrusions; some are castings; but most are elastomeric in one form or another (Fig. 8.87A and B). The cross-sections shown in Fig. 8.88A and B present an idea of some of the shapes that are available.

For instance, a silicone extrusion, intensifier cross-section configuration would include certain shapes of intensifiers.

It must be stated, however, that there is a wide variety of intensifier forms available. The forms illustrated in Fig. 8.89A–H are specifically used to push uncured laminate part material into a radius corner or recessed surface by placing the intensifier in contact with the laminate lay-up.

Intensifiers have also been applied in radius corners by installation on the outside of a bag surface. For instance, sealant tape pressed into a radius corner and acted upon by autoclave pressure will assist to apply increased pressure as shown in Fig. 8.90.

The intensifier can also be rigid and a component part of a tool set, yet still function as a forming aid beneath the bag (Fig. 8.91).

Fig. 8.87A. Extruder Manufacturing Intensifiers. *(Courtesy of the Keene Laminates Div. of Keene Corp.)*

Fig. 8.87B. Extruder Manufacturing Intensifier Material for a Radius Corner. *(Courtesy of the Keene Laminates Div. of Keene Corp.)*

324 ADVANCED COMPOSITE MOLD MAKING

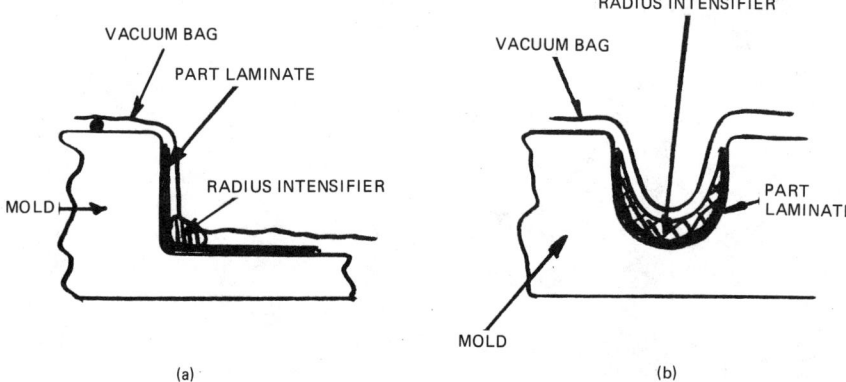

Fig. 8.88A and B. Radius or Edge Intensifiers.

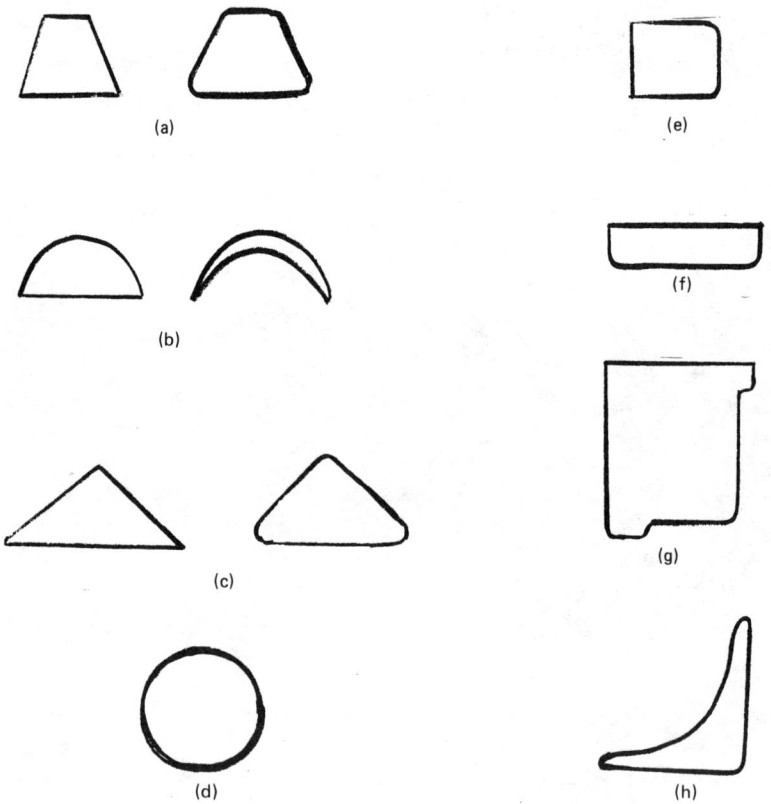

Fig. 8.89A–H. Various Intensifier Shapes.

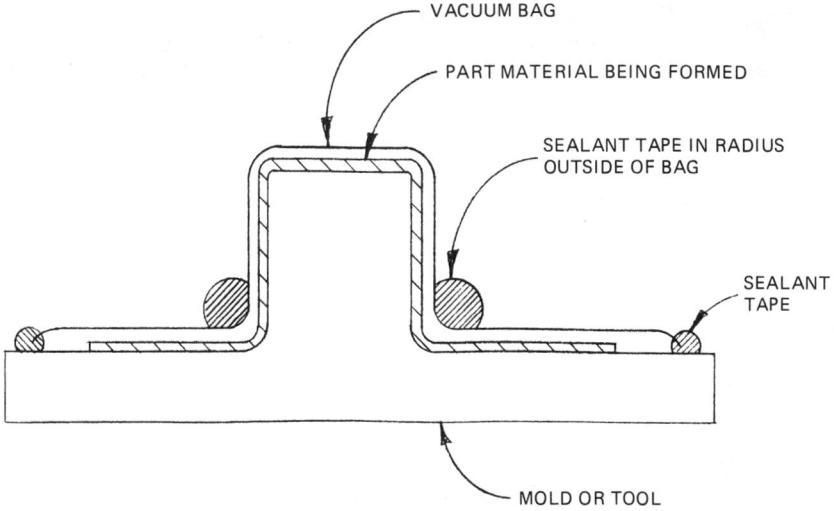

Fig. 8.90. Sealant-Tape Intensifier Outside of a Vacuum Bag.

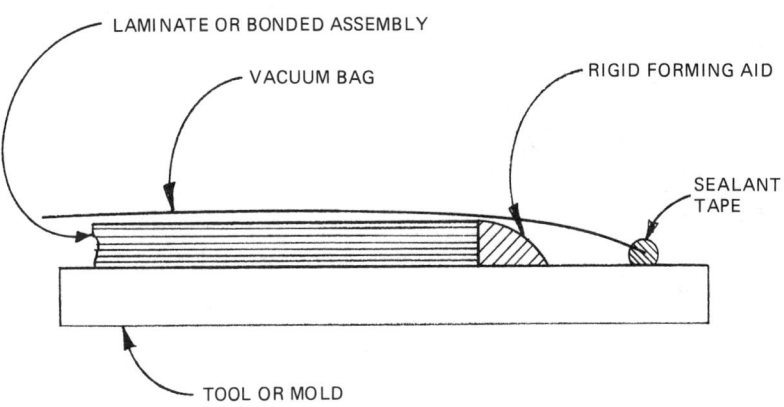

Fig. 8.91. Rigid Forming Aid.

Elastomeric Molding. Elastomeric molding is one of the more important ways to produce a composite part through the use of an expandable elastomer material, and applying pressure within the inside, confined cavity of a one-piece or split-mold container. This type of molding eliminates the need for a vacuum bag or the use of an autoclave or other pressure forming device such as a hydraulic press.

In the elastomeric molding process, an elastomer, usually silicone, is preformed by extrusion, casting, press molding, granulating, etc., and fit into a predetermined cavity. Silicones are again chosen because of their excellent reversion and chemical resistance, and high thermal and dimensional stability, as well as their high CTE. The elastomer cavity, which is within the material being formed, is filled, and all that is required to generate pressure is the application of heat. The heat causes the elastomer to expand and apply pressure to the laminate material or component parts being formed.

Two styles of this molding process exist. In one, the elastomer is totally trapped within the confines of the mold cavity, and in the other, the elastomer is allowed to expand under controlled or adjustable conditions and leave the molding cavity (Fig. 8.92).

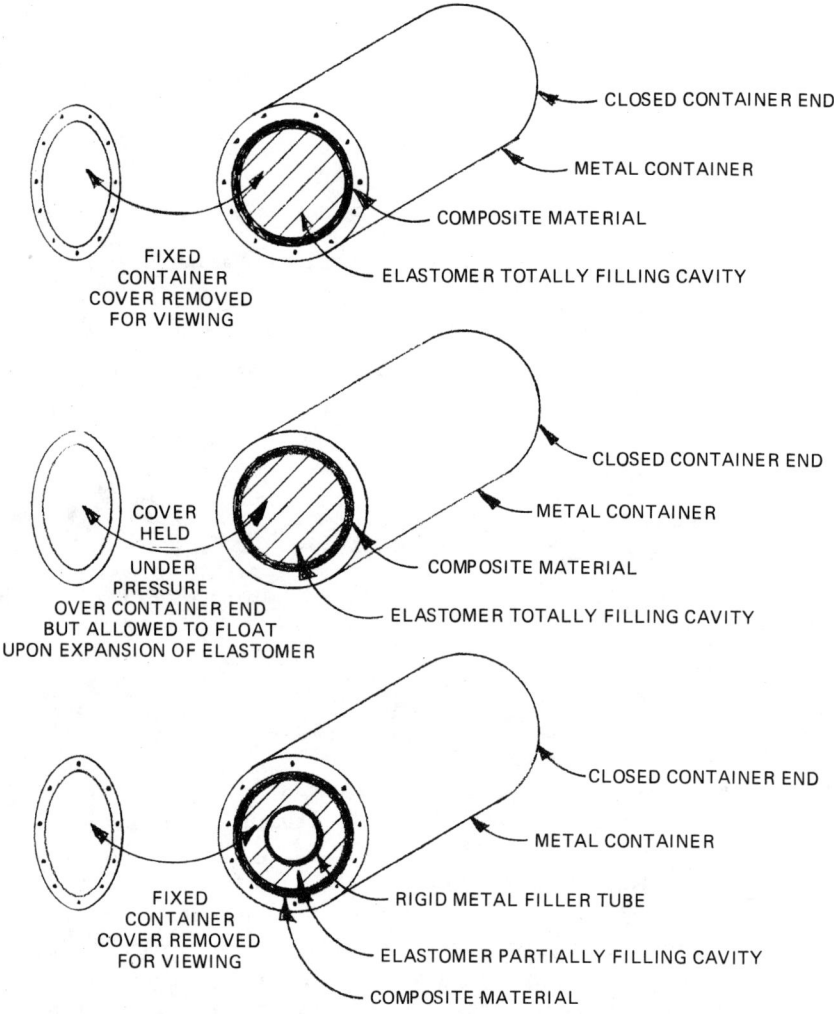

Fig. 8.92. Various Styles of Elastomeric Molding.

CONVENTIONAL OR PERMANENT REUSABLE VACUUM BAGS

Externally or internally applied heat causes the elastomer to expand and apply pressure. In the totally confined style, the pressure can be set by providing a gap around the elastomer, altering the elastomer durometer or formulation as well as changing the part configuration. Uniform pressures will also depend upon the friction developed between the elastomer and part material being formed, the mold and part material being used, and the placement of rigid filler plates within the elastomer. Experimentation is suggested with this molding method, since extreme pressures can be generated, and undesirable results may occur if the molding method is picked arbitrarily.

Thermal Expansion Foam Molding. This process is essentially the same as the elastomeric molding method, except that the elastomer is replaced with a foam, which, after expansion and forming, becomes an integral part of the formed laminate. The prefoamed or machined foam insert or preform is surrounded with the part laminate material, inserted into a split closed mold, and heated. At this point, the specially formulated foam will expand. The expansion of the foam applies pressure to the uncured laminate material being formed. Since the foam is trapped within the part material, it now becomes an integral part of the laminate upon cure.

This method, a part-forming procedure, takes place within the laminate and/or mass-cast nonmetallic mold or metallic mold (of various alloys), and has been used on occasion to intensify the radius corners within a laminate mold.

The method, however, is *not* recommended for fabrication of laminate tooling, since molds and tools are thermally cycled and the part cure temperatures usually approach the original molding temperature at which the foam was first expanded. This could cause the foam to reexpand, and bring about a laminate fracture.

The foam expansion process can lead to other forms of expansion molding techniques that are outside the scope of this text. One variation of the method described above allows for the resin to be pumped into a dry reinforcement-filled cavity around the foam. The void, once filled with the resin system, will impregnate the dry reinforcement and will compress by heating and expanding the foam. The foam expansion will then squeeze out the excess resin and the foam will co-cure with the laminate, again forming an integrally stiffened laminate.

8.4.3. Caul Plates

Caul plates, since they assist in forming laminate part sections on the surface opposite the mold surface (the bag side), belong within the family of bag-side tooling aids. Figure 8.93 suggests the location of a caul plate.

Formed from metallic or nonmetallic materials, caul plates differ from caul pads in that they are usually rigid substrates. There are, however, occasions in

328 ADVANCED COMPOSITE MOLD MAKING

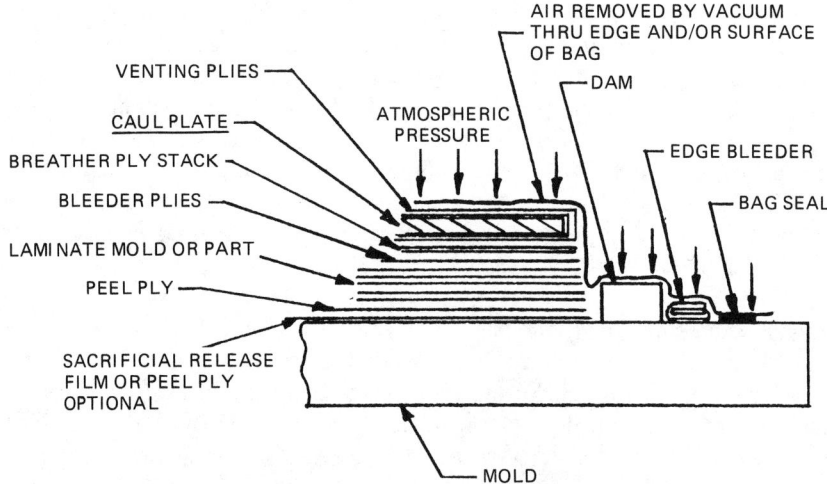

Fig. 8.93. Location of a Caul Plate.

which laminates are formed from caul plates when an elastomeric layer is attached or bonded to the part laminate side (Fig. 8.94).

The elastomeric surfaces can be applied over the caul plate in such a way that the elastomer is selectively bonded and will press into radius corners in a similar manner to the function of a caul pad.

Caul plates, as rigid surfaces, can be contoured and even molded from laminate materials to match the shape they are helping to form. They can also be formed to varying thicknesses by fabricating the plate with a honeycomb sandwich core. This results in a very light plate, especially if the caul plate is extremely large.

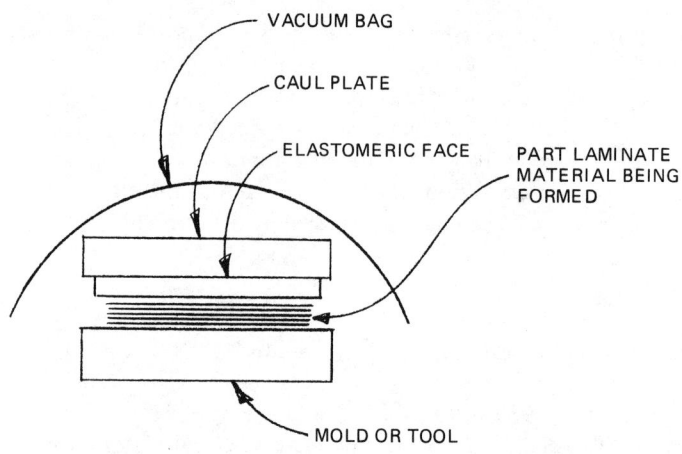

Fig. 8.94. Elastomeric-Faced Caul Plate.

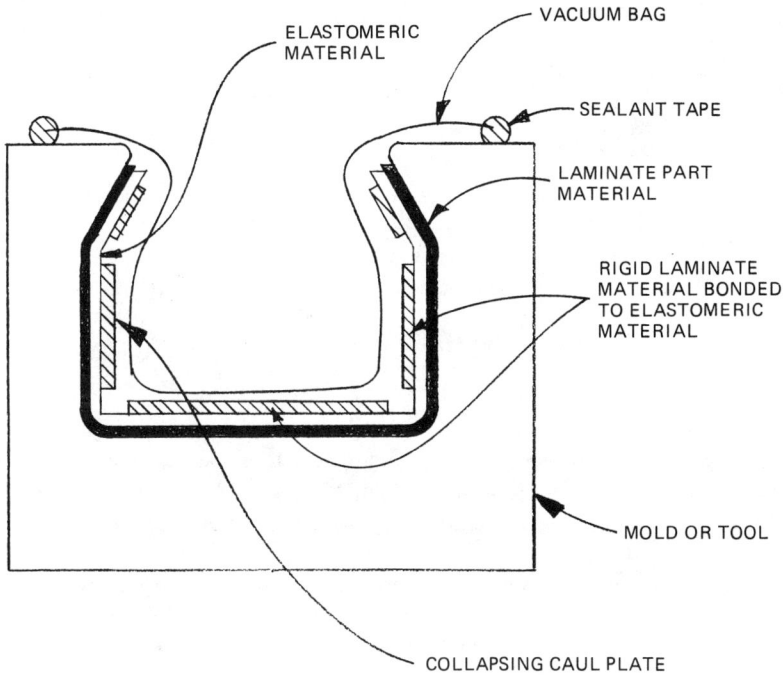

Fig. 8.95. Collapsing Caul Plate.

Part and assembly details, and even prepreg materials, can be attached to the caul plate prior to the molding cycle. This will position the details in their proper location for the final molding step. Co-curing of this type is the state of the art.

Indexing and positioning of the caul plate to the tool surface will automatically position, within the tool, whatever is attached to the plate.

Finally, caul plates can be connected to one another by an elastomeric membrane or bladder, so that if a large plate must be inserted into a deep mold cavity, it can be handled and placed into position easily (Fig. 8.95).

9
Mold Fittings, Accessories and Support Tools

Creating a mold or tool that will, in turn, create a good part is obviously a crucial consideration. But the satisfactory design, fabrication, and performance of the support tool and accessories are also important considerations which will impact upon the finished part.

For instance, if a vacuum fitting malfunctions or clogs prematurely with resin, the structural integrity of the finished mold or part will be impaired. It is for this reason that resin traps are incorporated into the mold or tool design.

Resin traps must perform as designed, without leaks or defects, to insure uniform pressure on the part or mold. Thus, it is absolutely necessary for these vacuum fittings and accessories to perform leak-free, as intended.

The finished part is of no use if it cannot be removed from the mold. Thus, part removal fixtures must be designed and constructed properly.

It is also important that, to insure a good part, the composite-reinforced plies be cut correctly and located correctly onto the mold. Furthermore, drilled holes and trimmed edges must be produced with a high degree of reliability and reproducibility to avoid any out-of-tolerance conditions.

This chapter will explain all of the above mold fittings, support tools, and accessories, as well as the complete maintenance and repair of molds and tools.

9.1. VACUUM FITTINGS

To obtain reproducible structural integrity of finished parts, a vacuum is applied after a laminate lay-up and pressure is applied during the cure. The maximum theoretical vacuum pressure is atmospheric pressure, and the maximum obtainable vacuum pressure is approximately 11 to 12 lbs/in^2. During the autoclave cure cycle, the vacuum pressure is enhanced by autoclave pressures of 25 to 200 lbs/in^2. Voids or leaks in the mold surface, or around fasteners or fittings, reduce the maximum obtainable vacuum pressure and should be avoided.

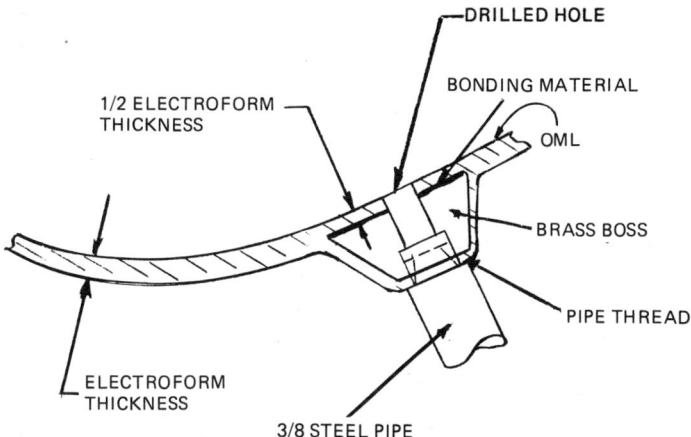

Fig. 9.1. Vacuum Port for a Nickel Electroformed Tool.

There are two major types of vacuum ports. The first is securely attached to the mold form. The second requires less tooling cost and is a part of the vacuum bag.

Although the configuration of the individual tool should be considered for the number of vacuum ports and their locations, tool surface area is the best guide for vacuum port placement. Two vacuum ports are required for a tool with a surface area of 30 ft^2 or less, one of which should be utilized for a vacuum gauge. Four ports are required for a tool of 30–100 ft^2, and six are required for a tool of 100–200 ft^2.

For metallic molds and tools, the securely attached vacuum port has three variations for three different circumstances. Electroformed metal molds must have brass vacuum bosses bonded and electroformed into the mold. They are then drilled and tapped for vacuum lines as is shown in Fig. 9.1.

Thin-skinned metallic molds have vacuum bosses welded into place, as illustrated in Fig. 9.2.

For thick metal molds, vacuum ports may be drilled directly into the mold surface (Fig. 9.3.).

To keep foreign substances out of the vacuum lines, vacuum port caps can be used (Fig. 9.4).

The through-the-bag vacuum port is quick and easy to use, and is the preferred type of vacuum port in both metallic and nonmetallic molds or tools. This type of vacuum port is also reusable, and is interchangable for employment with a multitude of mold forms (Fig. 9.5).

To insure uniformly distributed vacuum pressure, each vacuum port pot should have its own vacuum line. At the end of each vacuum line, a check valve

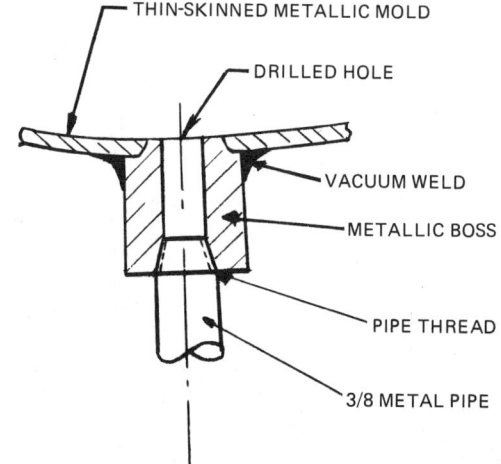

Fig. 9.2. Welded Vacuum Port.

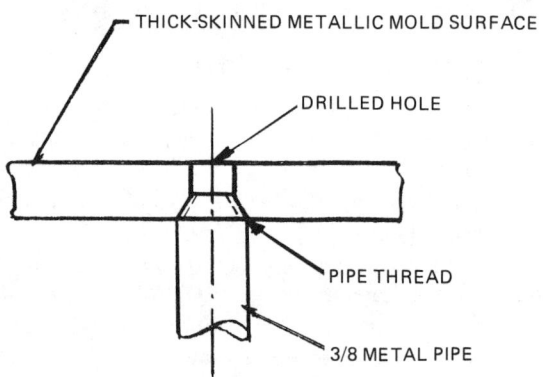

Fig. 9.3. Thick Mold Vacuum Port with Port Caps.

is installed. The check valve, which is usually attached to the mold form or the bag-type vacuum port, is the plug quick-disconnect type (Fig. 9.6). This quick-disconnect setup allows for a rapid and easy attachment of vacuum pumps and gauges.

For nonmetallic laminate molds, it is recommended that through-the-bag vacuum ports be used, as attachment of metallic vacuum ports (or any fitting through the mold or tool for that matter) is not a good idea. Any through hole in a laminate mold or tool is a location for a possible leak path at some time during its use.

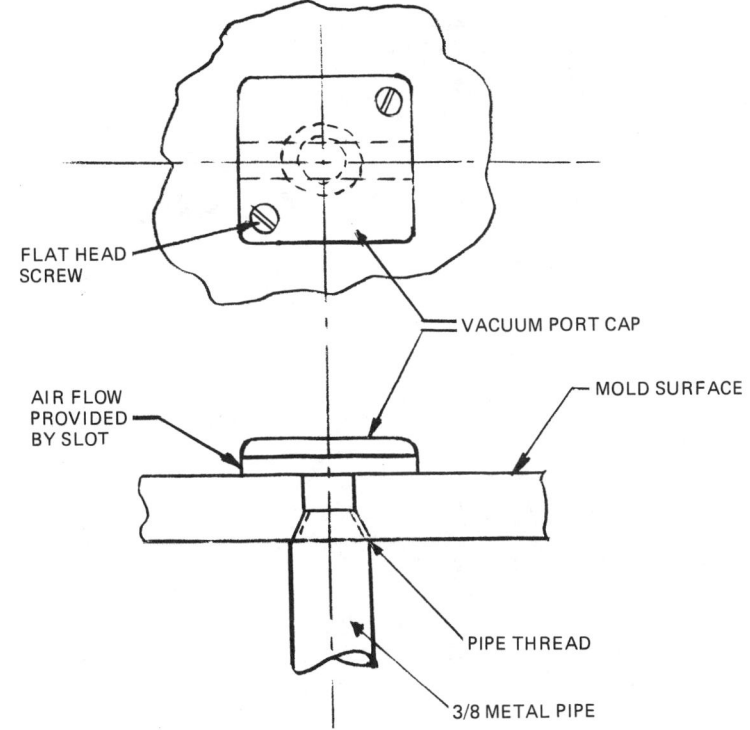

Fig. 9.4. Vacuum Port Tool Surface Cap.

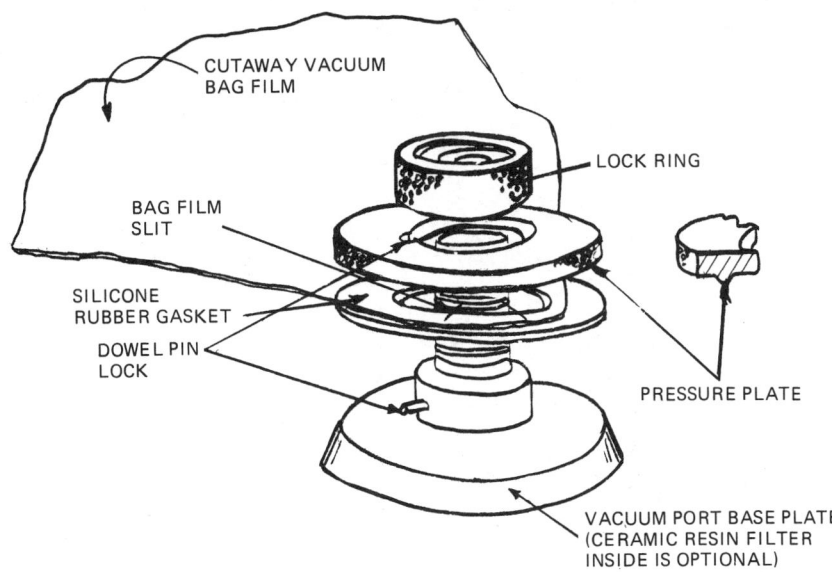

Fig. 9.5. Through-the-Bag Vacuum Port Fitting.

334 ADVANCED COMPOSITE MOLD MAKING

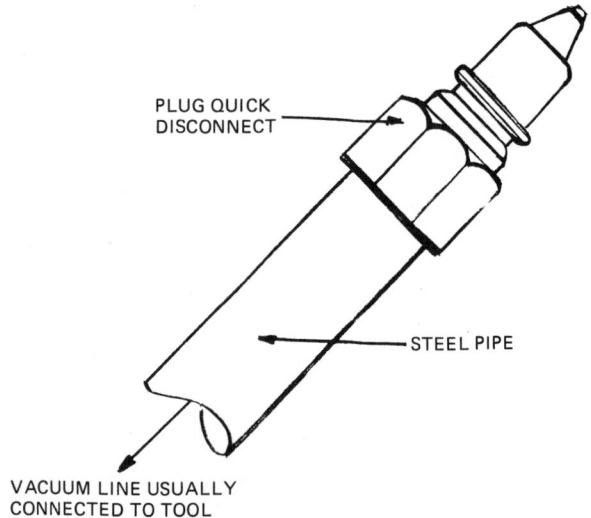

Fig. 9.6. Quick-Disconnect Fitting.

9.2. RESIN TRAPS

The resin trap's function is to provide a reservoir for excess resin collection to prevent vacuum line fouling. Excessive resin bleed can be expected when a mold has radical contours or high vertical walls. Therefore, resin traps should be incorporated into mold designs when excessive resin bleed is anticipated.

As discussed in Section 9.1, there are two basic types of vacuum ports. If a trap is added to the first type of vacuum port, then the configuration would be as illustrated in Fig. 9.7.

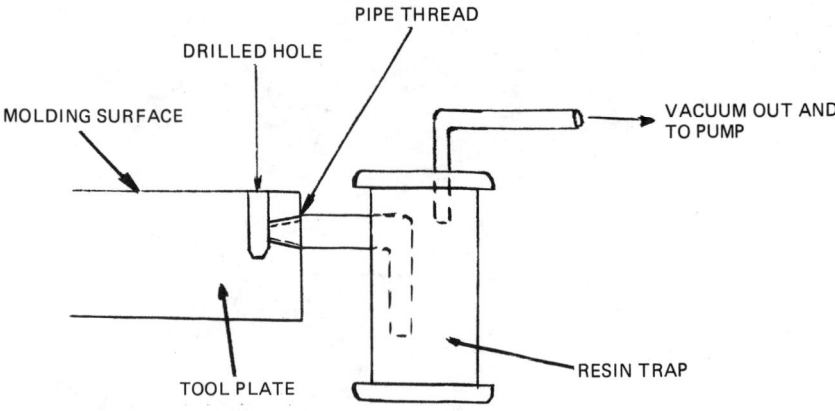

Fig. 9.7. Tool Resin Trap.

MOLD FITTINGS, ACCESSORIES, AND SUPPORT TOOLS 335

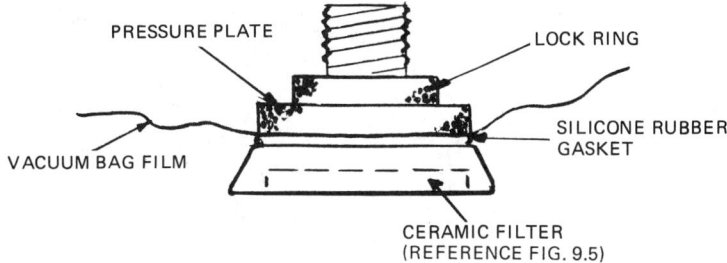

Fig. 9.8. Vacuum Bag Resin Trap.

In the second type, which is a "through-the-bag" vacuum port, a ceramic or porous filter, mounted in the port base, absorbs excess resin and clogs in order to avoid vacuum line obstruction (Fig. 9.8).

9.3. PART REMOVAL

Part removal is the equal responsibility of the mold and tool designer and the structural part designer. Parts that are not easily removed from a mold or tool are sometimes the cause of high costs in part fabrication. "Lock-on" or "lock-in" shapes can force the designer to insist upon split masters and/or molds. "Stuck" parts may cause part and/or tool damage with the possibility of the loss of both.

Therefore, careful design consideration must be given to the part and the mold with respect to part removal. Section 5.3 addresses some of the more important techniques for part releasing, since release materials are closely associated with part removal.

If a part must contain an undercut, an elastomeric intensifier or flexible caul pad can be formed to eject with the part being formed so that the part will not lock into the mold. A mold can also be formed with internal segments that will split or drop out with the formed part at the time of removal from the mold. This is necessary when the formed part shape would cause it to lock into a mold (see Fig. 9.9).

Small diameter plastic or wood dowels can be used to hold the minor mold segments in place during part forming. They can then be broken easily to remove the part and replaced before fabrication of the next bonded or laminated part. Upon completion of this, the part manufacturing cycle is repeated.

9.4. LOCATING, ROUTING, AND TRIM TOOLS

In the formation of laminated composite parts and bonded assemblies, there are four basic methods of locating subassembly components, plies, ply dropoffs, and cores.

336 ADVANCED COMPOSITE MOLD MAKING

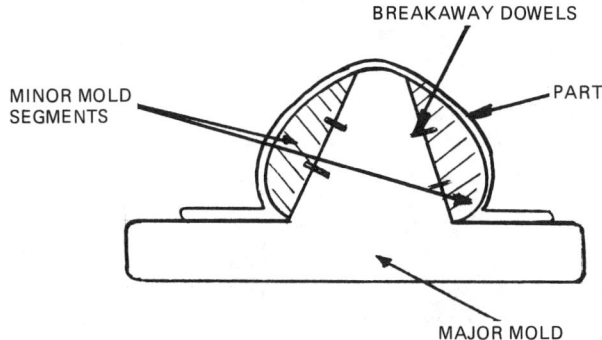

Fig. 9.9. Breakaway Minor Mold Segments.

The first method, called banking, is used to create a laminate composite part of continuous thickness with no ply dropoffs. In this method, the composite ply is banked against a fixed rail or mold wall and its location is governed by a notch in the ply and a mating protrusion on the banking rail (Fig. 9.10).

The second method of ply location utilizes a locating template (LT) which is created from a polyester film. The trade name of the polyester film is Mylar*

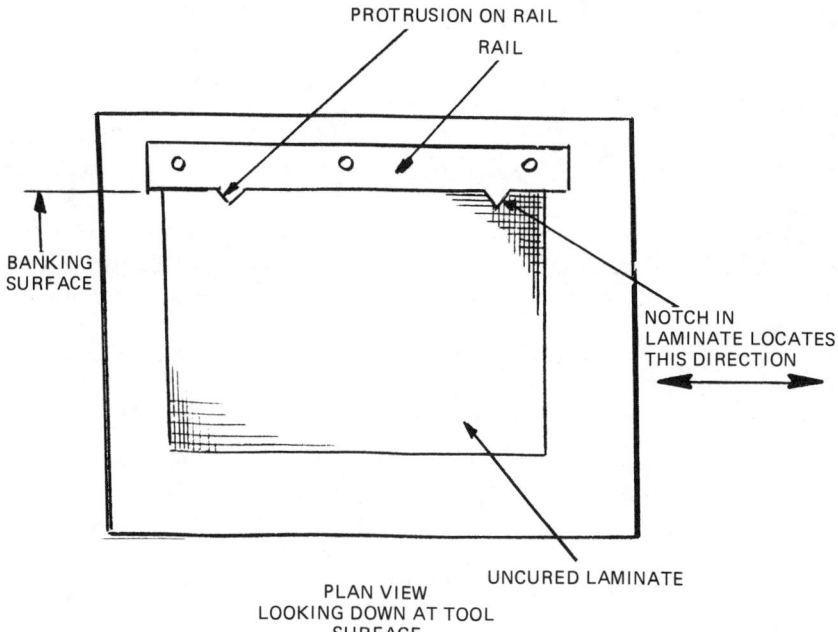

Fig. 9.10. Ply Location Template (Rail and Notch).

---
*Mylar is the registered trademark of the E.I. Dupont Co.

MOLD FITTINGS, ACCESSORIES, AND SUPPORT TOOLS   337

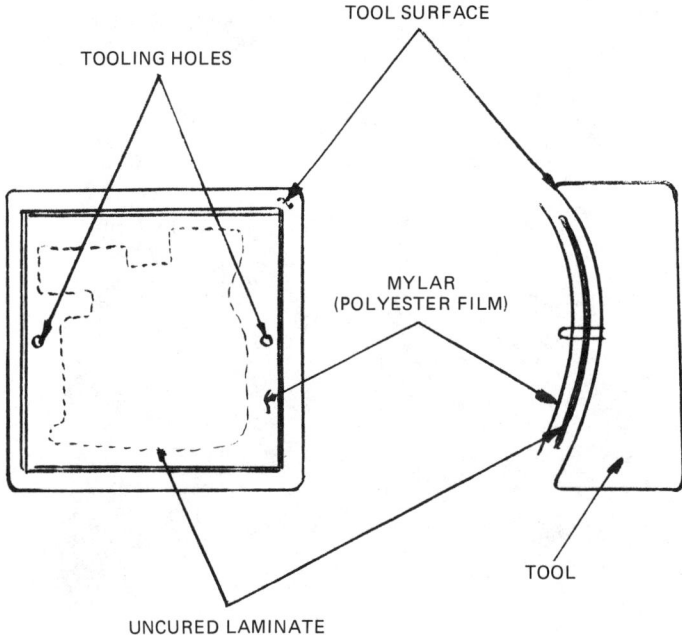

Fig. 9.11. Ply Location Template (Mylar Film).

and its intended use is for tools of little contour and short production runs. A Mylar film contains a mirror image of all data on the tool, including all trim lines and ply dropoffs. The lay-up on the Mylar film is opposite to that of a direct lay-up on the tool. For instance, the first ply down on the tool would be the last ply down on the Mylar film. Thus, when the lay-up is completed on the Mylar film, it is flipped onto the tool. Locating the relative position of the Mylar film to the tool is done with tooling holes in the tool and corresponding grommetted holes in the Mylar film (Fig. 9.11).

The third ply location method uses a polycarbonate sheet onto which a flat pattern of all ply locations, core locations, and ply dropoffs are scribed (Fig. 9.12A and B). This polycarbonate LT has location windows at every ply dropoff and core location for ply positioning. Tooling holes are incorporated into the polycarbonate to match those of the tool for relative positioning of the polycarbonate to the tool. Since the polycarbonate LT overlays the tool, the lay-up can be performed directly on the tool itself. Polycarbonates are the recommended method (Fig. 9.13).

The fourth ply location method utilizes a formed metallic sheet, fiberglass-reinforced polyester, or epoxy lay-up reproduction of the tool surface, and is called a fiberglass or rigid LT. Although the fiberglass or rigid LT has the same

338 ADVANCED COMPOSITE MOLD MAKING

A.

B.

Fig. 9.12A and B. Polycarbonate Ply Location Template Also Used to Locate Ply Direction. *(Courtesy of Grumman Aircraft Systems)*

MOLD FITTINGS, ACCESSORIES, AND SUPPORT TOOLS   339

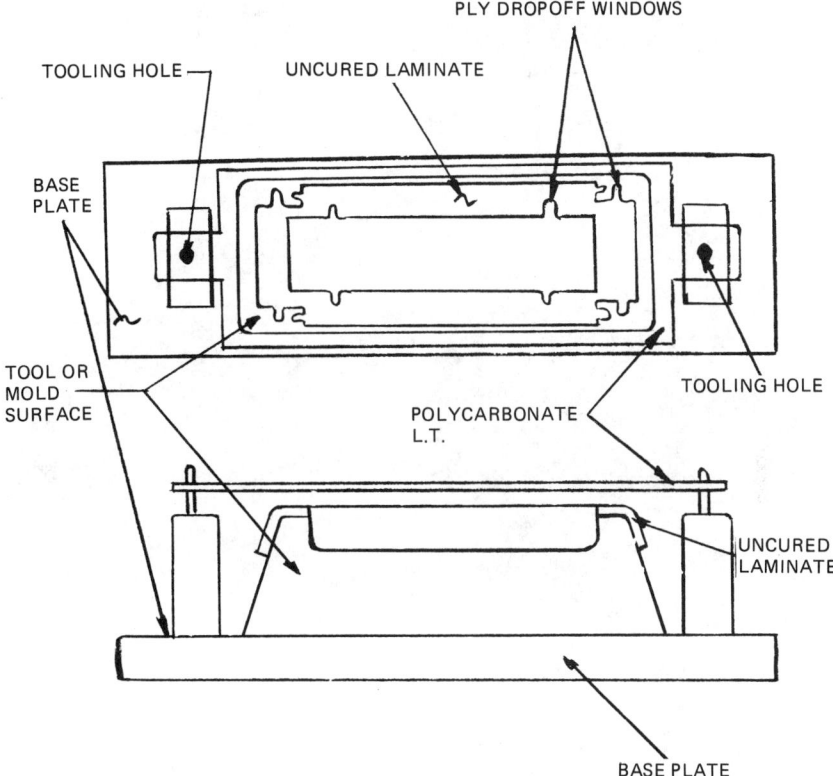

Fig. 9.13. Ply Location Template (Polycarbonate).

characteristics as the polycarbonate LT, it differs in that it has a contour that is matched to the tool. This LT is useful for tools with severe contours and long production runs (Figs. 9.14 and 9.15).

Routing and trimming of finished parts can be carried out in many different ways. A finished part can be trimmed with a conventional router. The problem with this method is that it creates large amounts of particle dust (Fig. 9.16).

Other conventional trim tools include the backsaw, sandsaw, belt sander, reciprocating sabre saw, and circular saw.

In all of these mechanical trimming and routing applications, a laminate tool can be used to guide the cutting or routing head around the periphery of the part being trimmed. This laminate guide tool is called a standard routing jig (SRJ). The edge of the SRJ laminate against which the cutting tool will ride or rub should be formed by the wet lay-up method. The laminating resin should include an abrasion-resistant filler, and this filled resin should be used to impregnate the plies along the entire edge (Fig. 9.17).

340 ADVANCED COMPOSITE MOLD MAKING

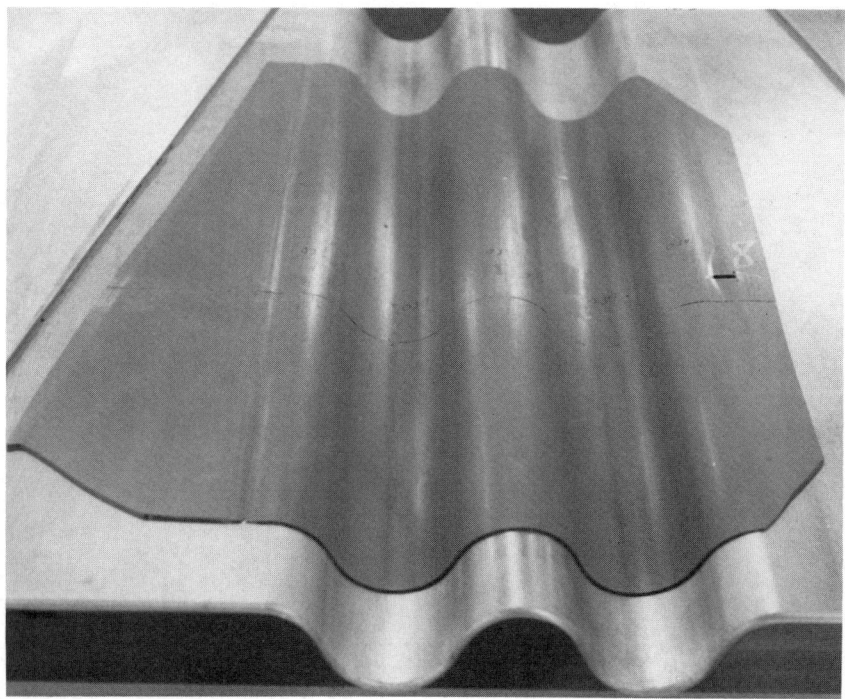

Fig. 9.14. Rigid-Contoured Metallic Ply-Locating Template. *(Courtesy of Grumman Aircraft Systems)*

All of the conventional trim tools have their own applications, but they can also be replaced with new, more accurate, and technologically advanced trim tools. Among these are diamond wire, the water-jet router, and the laser cutter.

The diamond wire has a heat-treated high-tensile-strength core with a copper-plated diamond-impregnated periphery. Since the wire is capable of cutting in both directions, the cutting action of the wire is a reciprocating motion.

The water-jet router uses water and/or cutting abrasives of various types under high pressure to locally erode the material, thus trimming or cutting the mold or part. The feed rate of the mold, part, or water jet and orifice diameter are the two controllable factors governing the quality of the cut.

There are two types of water-jet routers. The first is a hand-held router which is used on stationary molds or parts. The second is a stationary router which necessitates the mold or part being fed into the water jet. Both types produce dust-free cuts.

Lasers can be used to cut extremely accurate molds or parts — within $\pm 1/5000$ in. These close tolerances are enhanced by the high degree of reproducibility, the absence of dust development during cutting, and the low consumption of energy. Lasers coupled with computers allow for trimming or cutting molds

MOLD FITTINGS, ACCESSORIES, AND SUPPORT TOOLS 341

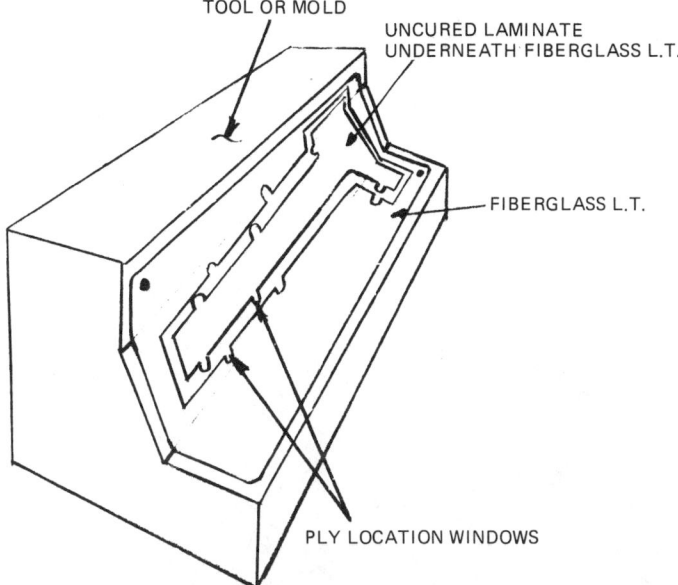

Fig. 9.15. Ply Location Template (Fiberglass LT).

Fig. 9.16. Vacuum-Assisted Composite Routing or Drilling Tool Configuration. *(Courtesy of Grumman Aircraft Systems)*

342 ADVANCED COMPOSITE MOLD MAKING

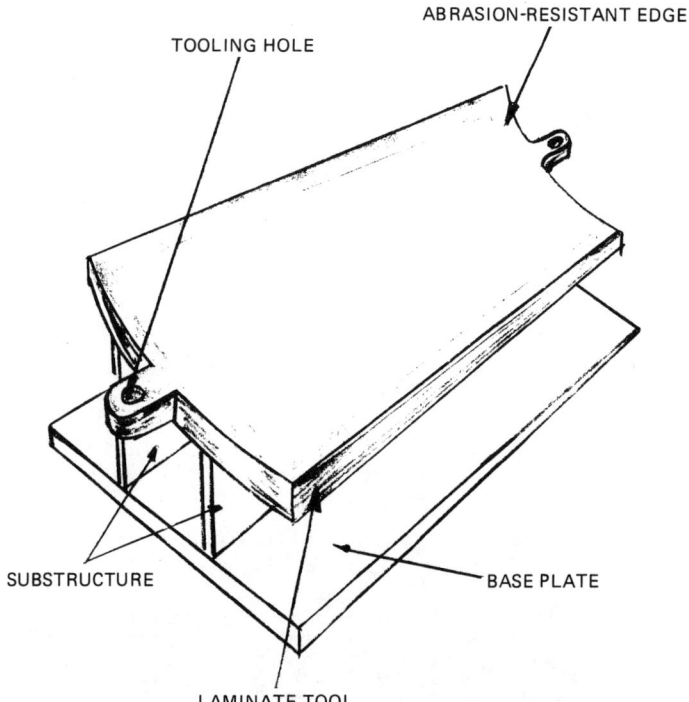

Fig. 9.17. Standard Routing Jig (SRJ) with an Abrasion-Resistant Filled Edge.

or parts of any shape while leaving a smooth, polished edge. Laser cutters can be attached to five-axis machine heads and the cutting pattern programmed to provide automated trimming or cutting of laminates up to 6 in. or even greater thicknesses.

9.5. DRILL FIXTURES AND INSERTS

Drill fixtures, or jigs, allow for consistent positioning of drilled holes in either finished parts, molds, or tools. The drill jig is usually a laminated epoxy tool into which drill bushings have been laminated during fabrication, or potted after fabrication. The locations of these bushings are determined by coordinated positions on a mandrel or master model.

9.5.1. Laminating Bushings

To laminate a bushing in place, a mandrel must have a drill bushing placed in the same location as the desired drill bushing in the drill jig. Then, potting pins are

MOLD FITTINGS, ACCESSORIES, AND SUPPORT TOOLS   343

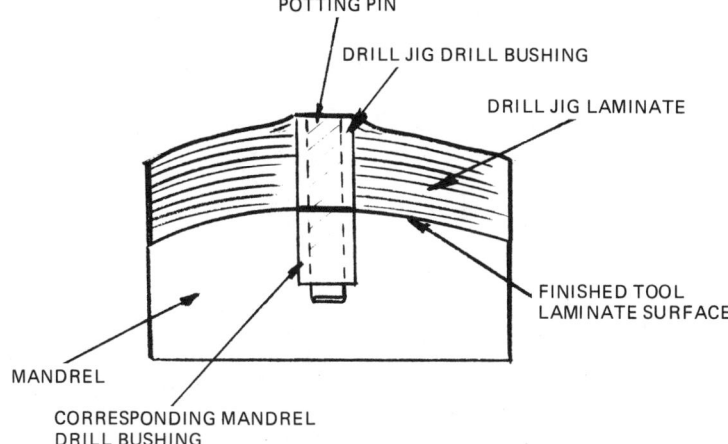

Fig. 9.18. Mandrel with a Potting Pin and Bonded Bushing.

placed into the mandrel bushings so that they do not protrude beyond the mandrel surface at a height that is more than the drill bushing height (Fig. 9.18).

Frozen adhesive films, film gel coats, paint-on gel coats, wet lay-up laminating materials, or prepreg materials are placed on the mandrel. Then, two layers of reinforced wet or prepreg material are placed on the mandrel. The knurled drill bushings of the drill jig are now placed on the potting pins, and the laminate material (reinforced wet or prepreg) is successively layered until the bushing is surrounded and the desired thickness is obtained.

Each layer of reinforced material must now be star-cut in the bushing area to accommodate the bushing (Fig. 9.19).

For blind bushings in drill jigs, a laminated square or round pad stack equal to the thickness of the bushing height and approximately 3 to 10 in. square is used to cover the rear of the laminated bushing (Fig. 9.20).

Precoating the bushing with prepreg or wet lay-up reinforcement/resin system (a couple of ply wraps) assures void-free areas adjacent to the bushing.

9.5.2. Encapsulating or Potting the Bushing

An alternate procedure is to encapsulate or pot the bushing into a drill fixture or jig. The first step is to produce and cure the reinforced laminate to the desired thickness. After the laminate has been cured, oversized holes are drilled at the locations of the desired drill bushings. Then, knurled drill bushings are placed on the potting pins of the mandrel, which are inside the oversized holes of the drill jig. The bushings are then bonded into the drill jig with a high-temperature, thermally conductive, low CTE adhesive (Fig. 9.21).

344 ADVANCED COMPOSITE MOLD MAKING

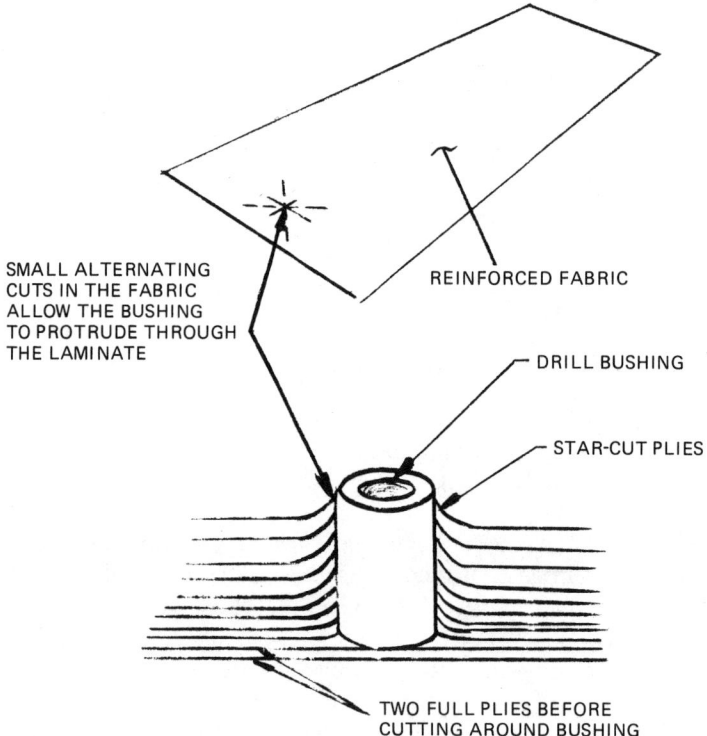

Fig. 9.19. Star-Cut Material with Bushing.

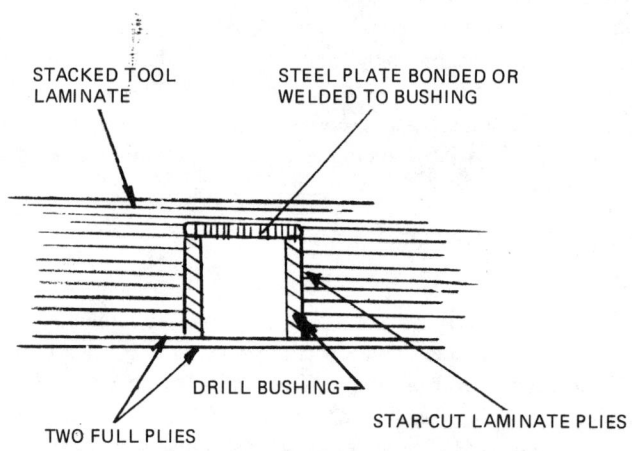

Fig. 9.20. Blind Drill Jig Bushing.

MOLD FITTINGS, ACCESSORIES, AND SUPPORT TOOLS   345

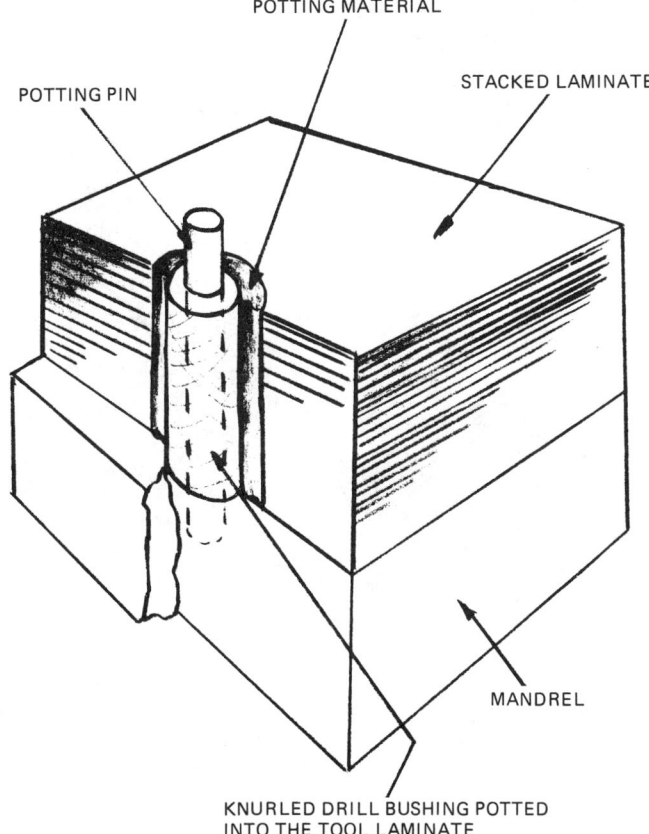

Fig. 9.21. Mandrel with Potting Pin and Potted Laminate Bushing.

When a mold requires both vacuum integrity and bushings that must be installed after the laminate is formed, the bushings must be bonded blind, into the mold. It is *always* suggested that a drill bushing be hardened and have a hardened back welded on, so that the laminate will not be pierced by a drill bit during use.

In many instances, time and money can be saved by using a combination fixture. For example, a combination drill and router fixture provides proper hole location while insuring correct part trim. Figure 9.22 illustrates this.

9.6. AUTOMATED PATTERN CUTTING

Automated pattern cutting machines possess the capability for cutting multiple composite plies at up to 600 in./min with a reciprocating knife. In addition to

346 ADVANCED COMPOSITE MOLD MAKING

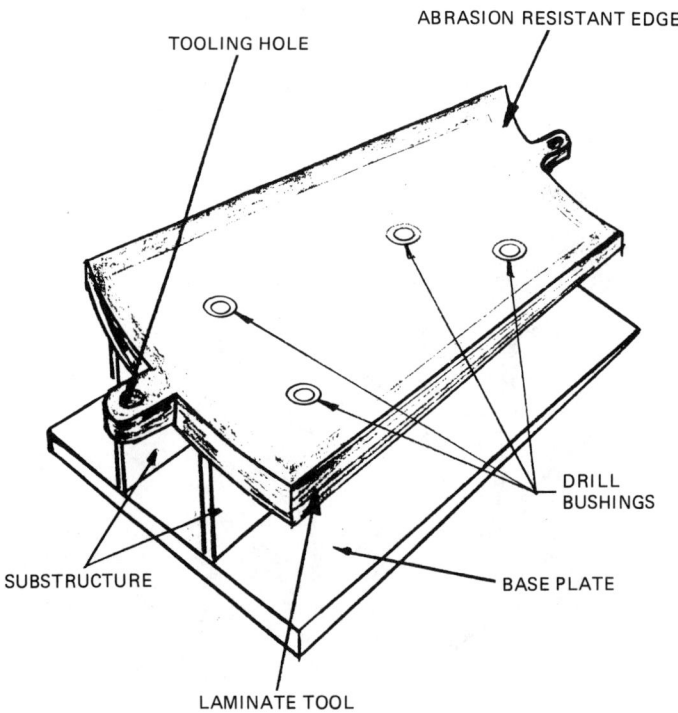

Fig. 9.22. Combination Fixture.

this, these machines can identify the pattern plies, thus eliminating misidentified parts, and ultimately reducing production delays.

Through the use of CADAM, patterns can be nested on the computer and then translated to an N/C tape. By nesting patterns, material usage can be maximized while minimizing material waste.

The patterns are cut on a bristle bed vacuum table with cutter path predetermined by the N.C. tape.

A typical nested pattern is illustrated in Fig. 9.23.

Since all cutting templates and all manual cutting labor is eliminated, substantial savings are achieved by using automated pattern cutting equipment.

The reader should also see Sections 3.4 and 3.5 for more on this topic.

9.7. MAINTENANCE AND REPAIR OF MOLDS, TOOLS, AND SUPPORT STRUCTURES

Nonmetallic molds and tools require more maintenance than metal ones, since metal surfaces are sometimes more durable than nonmetallic ones.

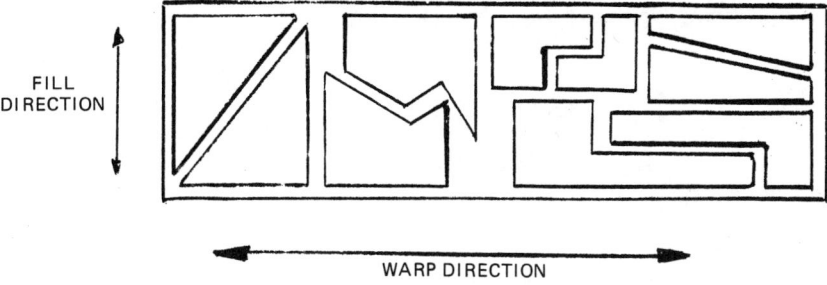

Fig. 9.23. Nested Pattern.

Still, this is no great disadvantage, since the cleaning of a mold or tool is easily accomplished, using portable blasting equipment — a method far preferable to manual and mechanical methods.

9.7.1. Cleaning Molds and Tools Before Forming

Metal and plastic molds and tools are easily cleaned of flash, mold release, and remaining part residue by blasting, using plastic beads, walnut shells, or other mild abrasives. Many forms of blasting equipment are available, but the use of a vacuum-assisted blaster is recommended (Fig. 9.24).

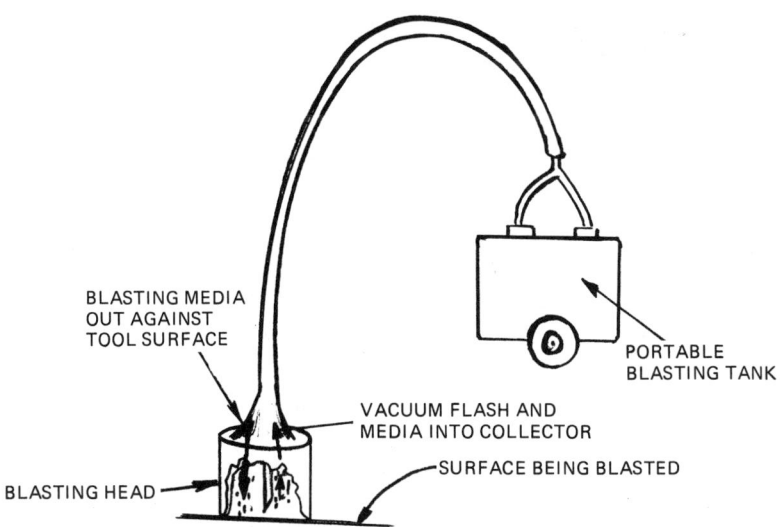

Fig. 9.24. Vacuum-Assisted Blaster.

9.7.2. Cleaning Molds and Tools After Forming

Cutting patterns, trimming gelled laminate parts, and performing other post-part-forming operations on the tool surface always damages the molding surface. The responsibility for the elimination of this damage lies with the production supervisor. If this condition is serious enough to warrant resurfacing, the method discussed below should be followed.

Resurfacing a Nonmetallic Laminate Mold or Tool. The deteriorated mold or tool surface should be abraded using portable blasting apparatus.

The surface coat, originally applied 12-15 mils thick, is all that is to be removed. The removal of this surface can be accomplished by using an abrasive blasting media such as aluminum oxide. The blasting operation is neither preceded nor followed with solvent cleaning, since this procedure might impregnate the blasted surface with a residue of mold release. All that must be removed is surface dust, and this can be done by using filtered dry compressed air.

Once the surface coat is removed, resurfacing can be accomplished by the application of frozen film surface coats or plies, or even a lightweight prepreg (2 oz/yd) or film adhesive. The frozen film, usually 12-15 mils thick, is vacuum-bag compacted and cured against the face of the tool. This is accomplished by applying a vacuum-bag sealant tape approximately 3 in. away from and around the repair area, and then applying a vacuum bag.

It should be noted that spray- or paint-on resin-type impregnation coatings exist that can be used to seal the front and/or back of a tool, but these heat-cure coatings can only be used if the molding surface of the mold or tool is just porous, not physically delaminated.

9.7.3. Repairing the Face or Backup Structure of a Nonmetallic (Laminate or Mass-Casting) or Metallic Mold or Tool

If the laminate of a mold or tool should require repair or modification, it is suggested that the procedure outlined below be used.

Before the laminate is cut into, or the substructure or backup structure removed, the laminate should be supported by installing braces (in the form of pultruded or laminated tube) to and around the area that will be removed or modified. If the *substructure or backup structure* must be modified, the new bracing members should be installed before removing the existing braces.

The purpose for supporting the laminate or the installation of bracing members before modification is to prevent the distortion, twisting, or warping that will inevitably occur should a portion of a mold or tool face or substructure be removed before it is properly braced.

Pultruded tube bracing is recommended because with its reinforcement of fibers pulled and aligned down the length of the tube, it has a much higher flexural, column, and crush strength than conventionally braided or wrapped tube.

There is an alternative. Recently, anaerobic (instant setting) adhesives have been used to bond tubular frames together, using butt seam bonds.

Another good method is the wrapping of resin-system-impregnated reinforcement tape around connections and joints, after mitering and fitting tube members together.

Mechanical fasteners are yet another connection method.

Choice of Repair Materials. It is a good rule to repair a mold or tool using the same materials from which the contour forming structure was built. The reason for this is that the CTE, other thermal and physical properties, and part molding characteristics will be the same.

Plan your modification or repair carefully. Always pretest the repair materials by producing a little sample that will allow the operator to know what the exact working characteristics of the materials are. Usually, the existing mold or tool is not only expensive, but must be returned to production as soon as possible. For this reason, the mold or tool should not be damaged, delayed, or otherwise affected by the use of untested inferior materials.

Plastic molds or tools are easily modified, since nonmetallic materials are easily machined or altered, even if reinforced. When nonmetallic laminate or mass-cast material must be modified or supplemented, it is usually wise to sandblast, grind, or otherwise abrade the surface to be modified, remove the grinding dust (do *not* solvent wash), and apply the chosen replacement material and process.

Mass-cast, reinforced laminate, fibrous-filled paste, or other forms of "buildup" material should be used to produce the necessary engineering additions, changes, or modifications. Sometimes, it is best to paint on a thin abrasion-resistant gel coat along the edge of a laminate guide tool to assist in guiding the cutting or routing head around the periphery of the part being trimmed. This laminate guide tool is called a standard routing jig (SRJ). The edge of the SRJ laminate against which the cutting tool will ride or rub should be formed by the wet lay-up method. The laminating resin should include an abrasion-resistant filler and this filled resin system is used to improve the abrasion resistance of the laminate tool edge.

It is important never to try to expedite the repair of a mold or tool by accelerating the cure through the application of heat. Heat, whether generated externally by the operator, or internally by the exothermic reaction of the thermoset material, will cause shrinkage or warpage. This is why it is best to provide heat sinks and use low-reactivity/high-temperature-cure use repair compounds.

In summation, a compound that is used to modify a mold or tool would be best chosen if it could be cured at the temperature of use. Thus, the existing tool or mold would be thermally expanded before the repair material has set, insuring that the repair material will not "pop out" of the repaired area during thermal use cycles — an extreme, but nevertheless conceivable, consequence of ill-matched thermal tolerances.

This does not mean that if a mold or tool is made to tolerate high-temperature cures, it should not be cured at a high-temperature. Room temperature materials can be chosen for ease and speed of repair, but a *matching* of temperatures is the key to safe and solid repair.

9.7.4. Replacing the Face of a Laminate or Cast Mold

Laminate. A suggested approach for laminate face replacement is to apply the new surface or gel coat to the master model. Then, apply two layers of wet lay-up epoxy or polyester fabric-reinforced material, precut to the existing tool edge size.

Locate the pre-positioned mold or tool against the wet surface, apply a vacuum by installing bag sealant tape along the rear of the laminate tool edge, and then a vacuum bag around the periphery. Atmospheric pressure will then press the existing mold being repaired intimately against the master model surface and newly applied tool surface (Fig. 9.25).

Note: Shallow 1/4 in. diameter holes may be drilled into the laminate prior to vacuum bagging it down to the replacement laminate on the master model. This will provide a mechanical lock to the existing surface.

Mass Casting. When replacing a face of a mass-cast tool, simply sandblast the surface to be altered, remove the dust from the surface to be recast, and install spacers cut to the thickness of the resurface material. Then, a PFP laminate copy, mating mold, or appropriate contour should be installed, using the shims or spacers previously installed to provide the casting gap. This mold-released contour will provide the new surface against which the mass-casting material will be formed, and the shims will insure the proper thickness of the resurfacing layer.

9.7.5. Repair of Storage-Caused Cracks

Sometimes, a mold or tool must be stored outdoors. If the mold or tool has a metallic core, the expansion and contraction of the core due to thermal shock could crack the surface. If this occurs, a repair can be effected after the crack is ground out mechanically to a gap approximately 1/4 in. wide. The grinding

MOLD FITTINGS, ACCESSORIES, AND SUPPORT TOOLS 351

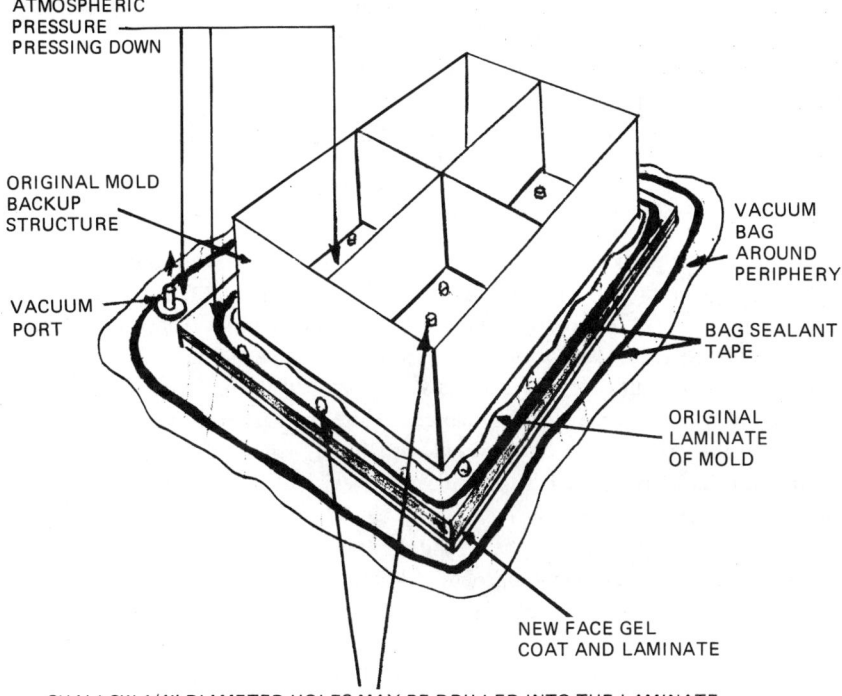

Fig. 9.25. Vacuum Bagging a New Surface to a Mold or Tool.

dust is cleared, and the crack is then filled with a material that possesses at least the same durometer (hardness) and physical and thermal properties as the original material used to form the mold.

9.7.6. Repair or Replacement of Inserts or Pins

If the mold or tool has pins or inserts in the face that must be individually repaired or replaced as part of the resurfacing procedure, the repair may be performed using the following procedure.

Remove the damaged or loose pins or bushings mechanically. Be careful not to affect the vacuum integrity of the mold or tool.

To prepare the hole for pin or bushing replacement, grind, mill, drill, or use a hole saw slightly larger than the replacement pin or bushing.

Next, prepare the master model surface with mold release, install the pins or bushings into the proper locations on the mold released surface, and finally, fill the oversized holes in the cast or laminated mold or tool, using reinforced thermoset adhesive or patching compound.

Care should be taken not to trap air into *any* repair void or bushing/pin hole, since air will expand during tool use and cause expansion of the mold surface or bonded-in pin and/or insert, thereby distorting any formed part.

9.7.7. Repair of Porous Molds or Tools

Porous molds or tools can be repaired by coating them with sealing agents. Porous surface coats or porous castings can be sealed by brushing room-temperature/ high-temperature reflow cure compounds over the entire front tool surface. This method of mold and tool repair is usually used to repair molds and tools that have lost vacuum integrity, and a very low viscosity material is required for this procedure. The sealer is then heat cured and the surface is wet sanded, using a 240 to 400 grit sanding cloth, dried, and finally, the mold is vacuum tested.

The alternate approach is to sandblast the rear and edges of the mold or tool, clean the residue from the surface, apply bag sealant and a vacuum bag to the front of the laminate tool periphery, and apply a vacuum.

Once this has been accomplished, a liberal coating of low-viscosity epoxy resin system or appropriate sealer is applied to the rear and edges of the mold or tool.

Whether the front or the rear of the mold or tool is being sealed, the procedure must be repeated until a vacuum check reveals that the vacuum integrity has been secured.

9.7.8. Repair of a Blistered, Delaminated, or Voided Mold or Tool

Sometimes, a laminate develops a void due to an existing delamination or void migration, or an imperfection occurring during tool fabrication. If this occurs, a series of holes can be drilled around the edge of the blistered area, and a high-temperature resin system injected into the holes and the holes sealed sequentially as the high-temperature resin system fills the void (Fig. 9.26).

As stated earlier, it is always best to choose a resin system that is the same as that used to fabricate the mold or tool. Matching thermal coefficients and bond compatibility is crucially important. The choice of a resin system that cures or post-cures at or above the mold use temperature is desired.

If a void or a delamination is so large that the space cannot be filled, the blister area can be removed and filled with a high-temperature, epoxy-paste mix, composed of milled reinforcement fibers and/or flock filler, usually using the same reinforcement used in the laminate.

When the blistered area is on the surface of a compound contour and the laminate cannot be surface filled, the laminate must be cut through, removed,

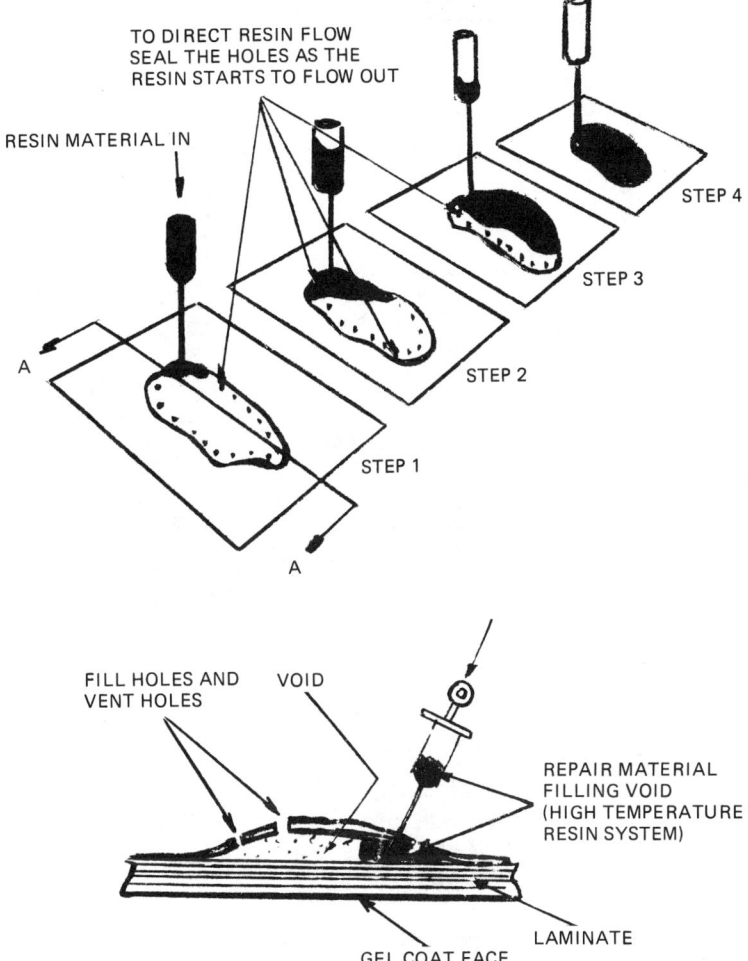

Fig. 9.26. Repair of a Blistered or Delaminated Mold or Tool.

and replaced using a PFP splash or laminate-mold contour copy to provide the original contoured surface (Fig. 9.27).

First, presand a 2 in. overlap on the rear of the laminate, to provide a better bond. Then, using the same surface coat and laminating resin system as the original tool, replace the laminate in the void, and apply 3 extra plies, to overlap the rear of the tool laminate repair area by 2 in. The PFP or contour-forming laminate is mold released next, set in place, a bag sealant is applied, and — when practical — a vacuum bag is installed, with no vacuum applied. When practical, a vacuum bag is also applied to the surface over the replaced laminate, and then a

354 ADVANCED COMPOSITE MOLD MAKING

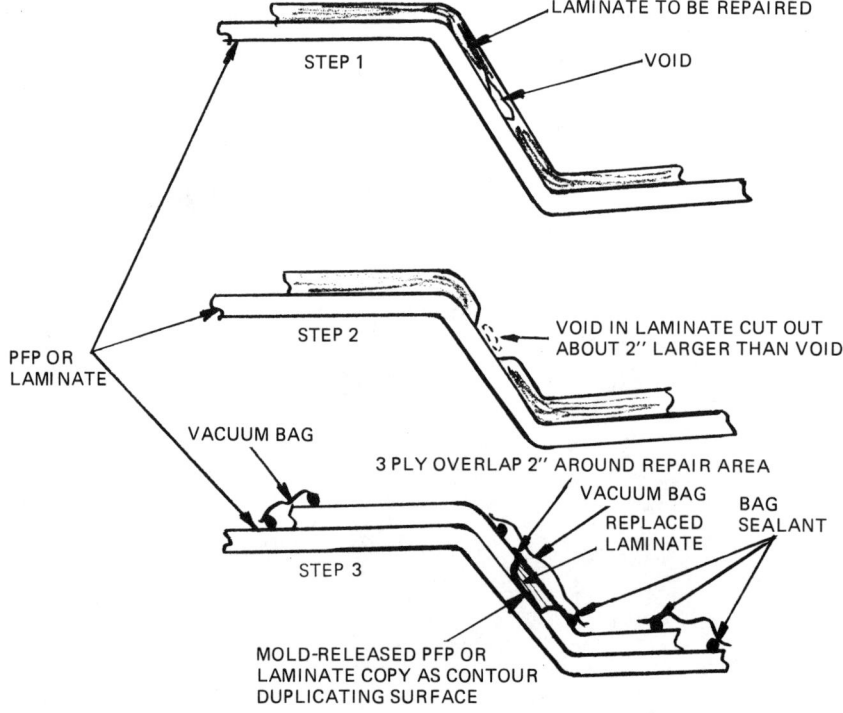

Fig. 9.27. Repairing a Void in a Tool Surface.

vacuum is applied to both surfaces, acting against each other and therefore functioning as an "envelope bag." A contact-molded and applied laminate repair can also be used, but the vacuum-bagged laminate is preferred.

9.7.9. Patching, Repairing, and Resurfacing of Metal-Faced or Machined Metal Tools

Metal-faced or machined metal tools can be resurfaced, patched, and repaired using nonmetallic materials. The size of the mold or tool and repair area will determine whether a mass-casting material, fiber-filled paste, or laminating material will be used to make the repair. Again, matching the CTEs of the original and repair materials is crucially important.

If a metal frame is to be used to replace a section of a mold or tool, the frame or structural members must be normalized. This is particularly important if the frame has been welded together. Remember, a little thought about what can happen to your mold or tool during repair, through thermal distortion, will direct your repair method and procedure.

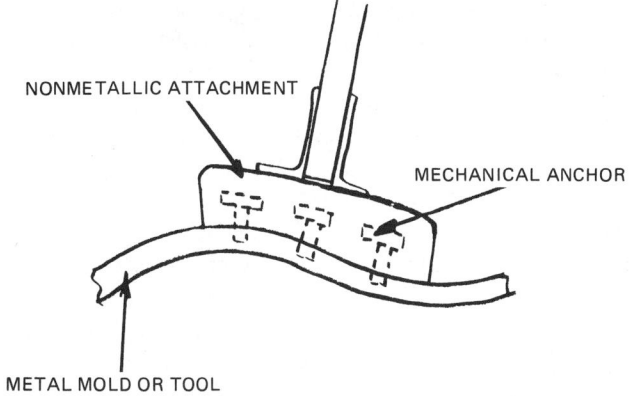

Fig. 9.28. Attaching Metallic to Nonmetallic Members.

9.7.10. Combining Metal and Nonmetallic Members

Metal frame members have been used in combination with nonmetallic members. This is allowable if the CTE of the nonmetallic structural member closely matches the CTE of the metal member, and the contact points are made using a flexible material connection. As an example, the CTE of fiberglass/epoxy at 10 to 12×10^{-6} is close to that of aluminum, which is 12.9×10^{-6}.

Whatever the mold material, whether metal or plastic, if plastic materials are attached to metal or the opposite, a mechanical anchor such as a bolt or screw should be used to secure and assure the connection (Fig. 9.28).

9.7.11. Repair to Small Areas of Nonmetallic Molds

If repairs to small areas of nonmetallic molds are required, the local application of heat of no more than 130°–150°F should be used, utilizing a heat (air) gun, blower, lamp, or other source, mounted 12–15 in. from the repair surface. At no point should the temperature exceed 150°F. This repair must be post-cured before mold or tool use.

9.7.12. Repair to Small Areas of Metallic Molds (Aluminum, Steel, Electroformed Nickel, etc.)

If repairs to small areas of metallic molds are required, a metal spray apparatus applying a kirksite-type alloy, aluminum, or nickel should be employed. This apparatus and equipment should be of a type that applies metal at a temperature of not more than 120°–150°F.

This method is so reliable that ceramic, plastic, wood, and even aircraft structural skins have been coated using the material and process, and can even be used to change the metal mold contour.

Be aware that, in all repair cases, the mold or tool must be post-cured. Post-curing should be performed with the mold or tool firmly attached to the master model, post-curing fixture, or, if the master pattern cannot tolerate the high temperature, the mold should be held in place, prior to post-cure by some other means.

In conclusion, the methods presented here are the most important and proven repair methods to consider. Others may exist and may be used to repair various defects, but if the method chosen is not a proven one, pretesting is recommended.

In all cases of repair, the matching of the thermal coefficients of the repair material to the original tool is of utmost importance.

As a final note, the testing of the repaired tool prior to use is the most significant factor, since the loss of a manufactured part or assembly due to an improper tool repair could be disastrous.

10
Facility Requirements

It is tempting, but dangerous, to address all of the requirements necessary for the total facility to produce composites, part laminates, and bonded assemblies. The temptation is toward completeness; the danger is that this chapter could very well turn into another book.

Therefore, only the following priority requirements will be addressed:

- Individual materials, molds, and tools.
- Transporters of contour-forming materials and mold forms.
- Storage of these materials in holding areas (indoor and outdoor).
- Final production use areas such as curing ovens, autoclave systems, and inspection equipment locations.
- Worker safety.

10.1. INDUSTRIAL ENGINEERING

While industrial engineering considerations focus upon plant layout and production areas in general, a focus on the mold and tool fabrication storage and use areas will suffice to assure efficient planning in the handling of materials, molds, and tools.

It is important to note that completed mold and tool handling, moving, and storage accounts for a part of the total cost of industrial engineering and facility requirements, and that material and mold/tool work flow during fabrication, as well as use, will greatly affect the end product cost.

10.1.1. Flow

Because material and mold/tool work flow during fabrication and use greatly affect the end product cost, mold and tool shops should be separated from the production facility.

The mold and tool design, mechanical engineering, and N/C programming should start the flow with automated design and fabrication teams located near each other, so that an interface with mold and tool making can effectively take place. Generally speaking, a two-hour travel time indicates the maximum reasonable distance of separation for the two facilities.

In moving molds and tools through the fabrication stages, and subsequently through production, careful planning should allow the production operators to handle the contour-forming surfaces, bonding fixtures, tooling, and accessories easily and efficiently. In some instances, it will become necessary to perform the fabrication of a mold or tool in a production area because of restrictions of material lay-up, size, weight, and worker skills. If this need should arise, the area utilized for performing the work should be planned and isolated well in advance, so that regular production is neither interrupted nor distracted by the fabrication operation.

Support equipment such as forklifts, transportation dollies, and overhead cranes should always be chosen with flexibility in mind, noting the distinct and varied needs of materials and design. Handling is an important element in facility design, since it does not add value to the parts being produced, and must therefore be minimized or performed at the lowest cost for the available equipment and labor, without sacrificing efficiency.

In summation, it is well to note that standardization of equipment, whether it be major or support, is extremely important. In order to standardize this equipment, an analysis of how the mold or tool is handled by the operator at each work station, as well as the equipment used to perform the production operation, will determine the specifications of the transportation or handling device.

To bring this down to hard specifics, let us examine the precise methods of mold and tool handling, moving, and storage.

Mold and Tool Handling and Moving. The basic considerations in mold and tool handling can be generalized to include certain basic parameters. Before choosing the production flow and method of movement, the following questions should be exhaustively studied.

What are the materials, molds, or tools?

- What is the size, weight, and shape of the material of fabrication?
- Is it metallic or nonmetallic?
- Is it new, used, or out of storage?
- Is it rugged or fragile?
- Does it consist of single or multiple parts?
- Does it come with or without accessories?

Where are the materials, molds, or tools stored?

- Are the materials and tools within the production plant, or remotely located?
- Is storage (indoor or outdoor) required, and if so, what are the conditions?

- Will the introduction of the materials and/or tools into storage be manual or automatic?
- What will the costs of transport and storage be?
- If the routing of the material, mold, or tool is interplant, what sort of access to the storage area or building is provided?

Where must the materials, molds, or tools go?

- How many stations will the molds or tools travel through in production?
- Will they be lifted, moved, and set down at each step of production?
- Will they move on transportation wagons and be stored at each station?
- Will they be stacked, lay over in a production area for a period of time, or will they continually move through the production process?
- What will the movement and storage costs be during production?
- Will the molds or tools be used (a) in plant, (b) interplant, or (c) at an outside vendor location?
- If (b) or (c) above are true, what distances must be traveled?

Once these questions have been considered, attention must turn to methods of transportation. Grouping molds and tools by quantity, configuration, and style reduces the cost of handling, when moving in-plant or interplant. Careful examination of the material, mold, or tool will dictate the method of grouping.

The routing of masters and models during intermediate mold forming and fabrication of the mold or tool as well as the routing during the use of the completed mold or tool must be carefully planned. In addition, the frequency of use of the completed mold or tool during the production of parts must be calculated and planned.

A method of doing this is to consider what must happen to the mold or tool at each station during fabrication or what the mold or tool must do at each step of the part or assembly process. Once this has been planned for, the handling pallet, cart, or carrier must be configured and placed so that it allows easy access to the working surface. Furthermore, prior planning will assure that, if support of the mold or tool is required, the materials, subassemblies, details, or molding accessories will accompany the primary contour form.

The operators in the various production areas have physical limitations in the use of a mold or tool, such as the ability to reach over and around, lift or move it. These limitations must be considered when designing the transporter or carrier, or the hoist or lift, and the associated work stations. The human engineering aspects of lifting, working environment, ability to push and pull, and finally, reach over the work area, continually come into play, and must be considered.

Figure 10.1 illustrates the physical limitations of the average operator over the work surface.

360 ADVANCED COMPOSITE MOLD MAKING

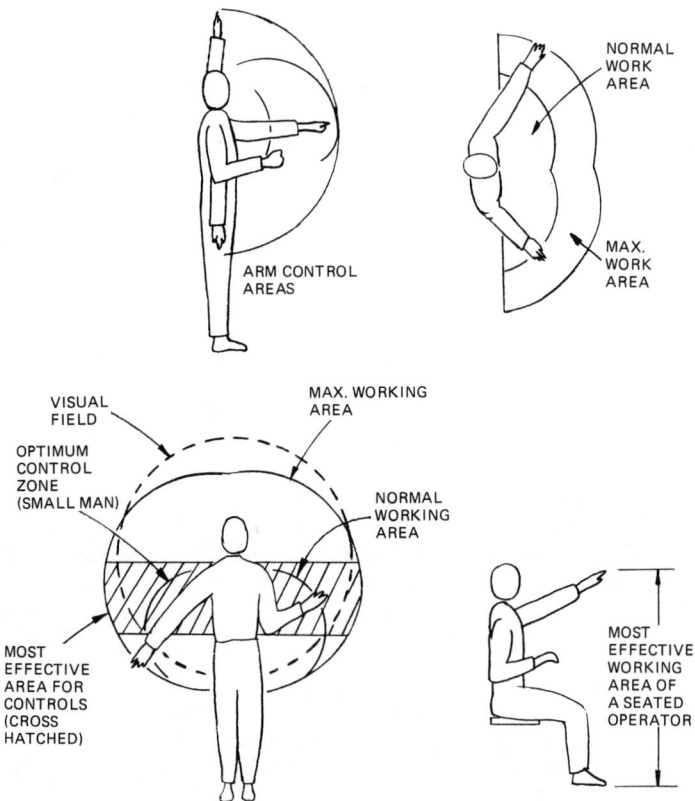

Fig. 10.1. Operator's Reach Over Work Areas.

Important aspects of the transportation vehicle to be considered are the lifting eyes, forklift provisions, and the location on each unit of its casters and wheels. As mentioned earlier, standardization, consistency, and versatility in use are the basic governing factors to apply.

Lifting eyes generally should be chosen based on tool weight and angle of the hook sling. When carts or trolleys are transported using eyes, there should be provisions to allow the supported mold or tool to be lifted straight up with cables perpendicular to the floor, and not at an angle (Fig. 10.2).

Forklift spacing and entry height should be common, and the use of all forms of lifting methods should be incorporated in each transportation vehicle.

Wheels or casters should be removable and height-adjustable so that, when adjustable work stations are not available, the transportation vehicle can be adjusted to the required height in front of the operator.

FACILITY REQUIREMENTS 361

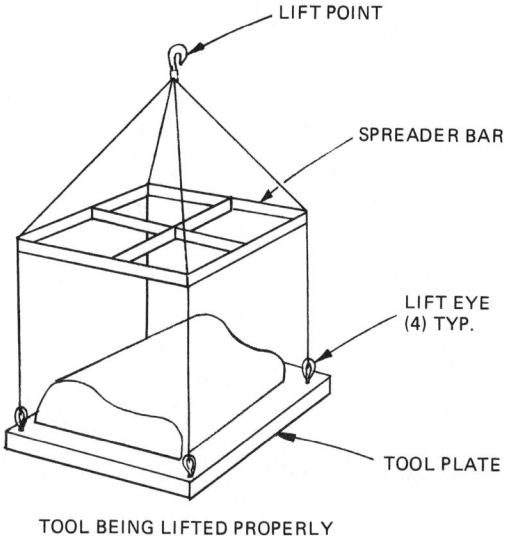

TOOL BEING LIFTED PROPERLY

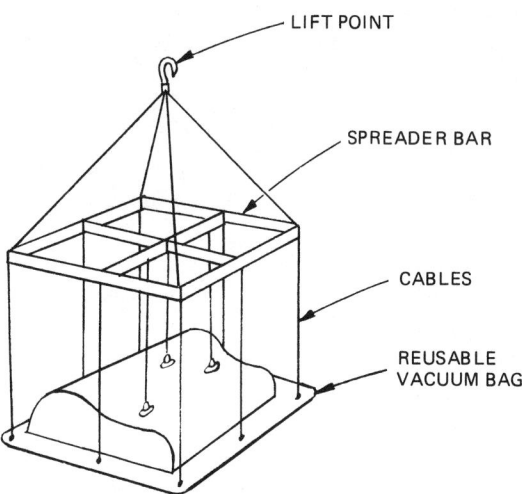

LARGE REUSABLE VACUUM BAG BEING LIFTED PROPERLY

Fig. 10.2. Proper Lifting of Work Pieces Using Spreader Bars.

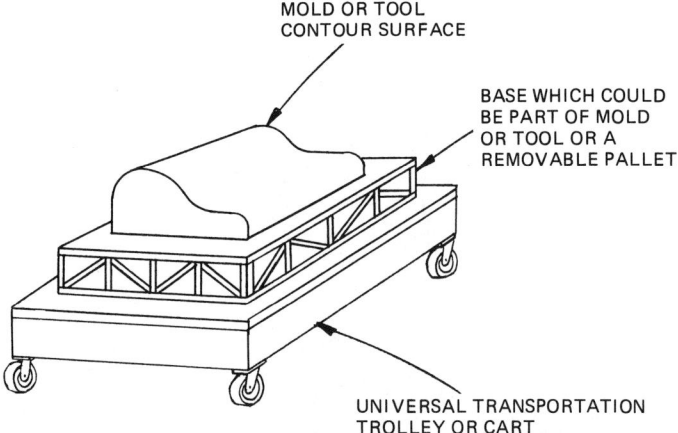

Fig. 10.3. Typical Transportation Trolley.

In the storage and production areas, as well as the ovens, pressure vessels, and autoclaves, the transportation dolly or cart should be able to move on tracks or be steered using a wire-guided system. Retractable or removable ramps (mechanically, hydraulically, manually, or automatically operated) can insert and remove production mold and tool packages from process cavities. The cart should be able to be split to allow for a universal transportation floor base and varying tool base for handling and storing on racks in an autoclave or in holding and storage areas (Fig. 10.3).

The tool, if already connected to a base, can be handled separately or set on a standard pallet which will mate to transportation dollies or pressure vessel/ autoclave/oven cavity rigging. These pallets should be of a standard and universal design.

A number of pallet sizes can be developed to suit the storage system and handling system of choice. The design should be an open grid for air circulation and access to and protection of tool vacuum ports.

For situations that occur when a number of small parts are to be processed, the pallet should contain an aluminum vacuum plate surface upon which multiple parts can be bagged. This plate should also have a vacuum port mounted in it or through the reusable vacuum bag installed over it (Fig. 10.4).

Pallets should be designed so that they may be stored on top of one another when not in use.

The system design should be carefully chosen to cover all mold and tool configurations. Mold and tool designers should be constantly aware of pallet size limitations; a unique design will dictate a unique tool base and handling system.

FACILITY REQUIREMENTS 363

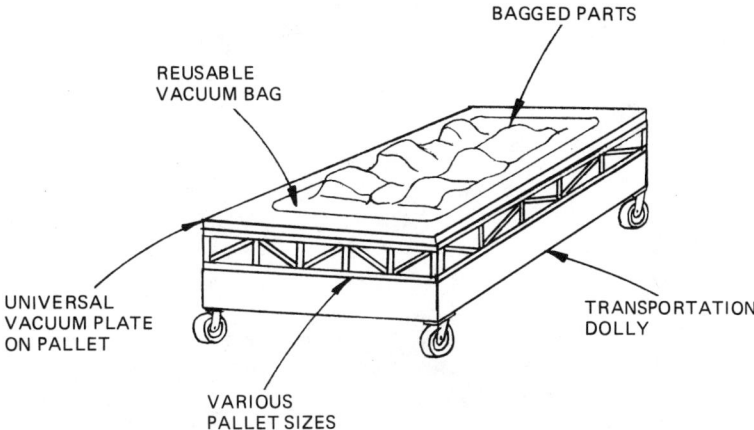

Fig. 10.4. Universal Vacuum Plate with Reusable Vacuum Bag on Transportation Trolley.

In designing the standard pallet, consider the following factors that may affect configuration:

- The transportation vehicle.
- In-process mold and tool storage and holding areas.
- Racking and rigging in autoclaves.
- Ovens and pressure vessel sizes.
- Mating handling systems for composite tape-laying or manual assembly operations.
- Final staging and storage areas.

The transportation dollies may not support the mold or tools totally throughout production because the environment in which they are used may dictate that a particular type of dolly construction material or design be used.

The route that the dolly takes and the surface over which it travels must also be considered. The dolly may require rubber or steel wheels that can tolerate low or high temperature/pressure, and it may be manually or automatically steered in the production areas.

If universal plates or multiple molds and tools are stacked to be processed simultaneously, the grouping of molds or tools should be such that materials with similar thermal coefficients are matched. A mixture of graphite epoxy tools with massive steel tools is an inadvisable idea (Fig. 10.5).

Similarly, designing multiple contours for more than one part on one mold base is not recommended. This forces all parts on the base to be processed at one time, thus confusing schedule planning. Furthermore, if spare parts are

364 ADVANCED COMPOSITE MOLD MAKING

Fig. 10.5. Condensed Organized Tool Storage Container with LTs Included. *(Courtesy of Grumman Aircraft Systems)*

required, the mold will then be tied up in the production cycle for the spares. One production operator can only lay up one part on a multiple mold surface at one time. Individual molds can be processed simultaneously by various operators. Damage to a particular mold contour on a universal mold will tie up the entire mold and all of its other contours.

The handling system, whether automated or manual, individual or group, whether it is palletized or not, should be integral. All areas such as vendors, receiving, holding, storage, shipping, production, and inspection must be integrated and included.

Gravity should be used to move molds and tools whenever practical. Automation should be considered for both production operations and storage; handling operations must always be mechanized. The "cube" of the building should be considered at all receiving, holding, storing, and shipping areas, and all available space should be utilized.

FACILITY REQUIREMENTS 365

The substantial cost savings of automating mold and tool process lines will pay for the transportation vehicles, conveyors, positioning and control equipment, overhead hoists, cranes, and elevators.

Transportation vehicles are sometimes "wire-guided" through production lay-up and work stations by the use of a wire/detection assembly in which the wire is imbedded in the floor and the detection assembly on the transportation vehicle steers the vehicle from storage and kitting through production to the process areas (Figs. 10.6 and 10.7). Computer control, scheduling, and work demand systems support the production operator, who does not then have to move from the work station.

Conveyors should be carefully chosen, based upon the material, molds, and tools to be handled. The weight of a pushed or pulled load will determine travel across the conveyor surface.

Roller conveyors are the least expensive and most popular choice for mechanized systems.

Cable conveyors or propulsion chains can mechanically move the object being processed.

Fig. 10.6. Wire-Guided Vehicle.

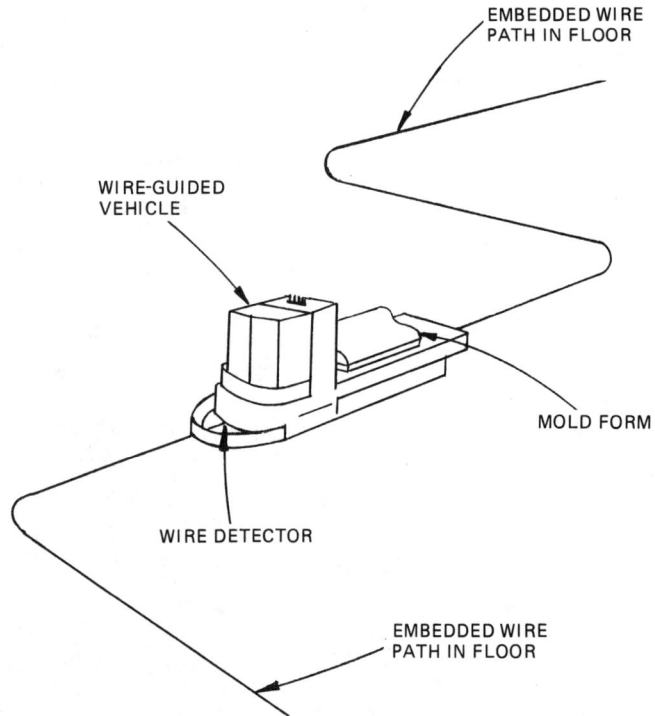

Fig. 10.7. Wire-Guided Vehicle on Embedded Wire Path.

Belt conveyors, with the elevation adjustable by the operator at the work station, should also be considered.

In fact, positioning and control equipment at each work station will assist the operator in handling the object or objects being processed. Ramps and bridges, mold and tool positioning platforms and tables, manipulators and positioning aids are just some of the support equipment to be considered at the work station and other process points (Figs. 10.8 and 10.9).

Overhead hoists, cranes, and elevators for up-down, forward-back movement should be selected for the handling of any material, mold, or tool configuration. A perfect example of this is the production process, wherein composite tape-laying machines lay laminate patterns on flat plates and the pre-cut/prelaminated stack is lifted and transferred to the mold or tool surface for continued processing.

Fixed, portable, or traveling hoists or cranes, elevators, winches, and their support equipment are part of the array of equipment used to support material, mold, and tool handling at each work and storage station. It is important to remember to practice preventive maintenance on all of the equipment and accessories.

FACILITY REQUIREMENTS 367

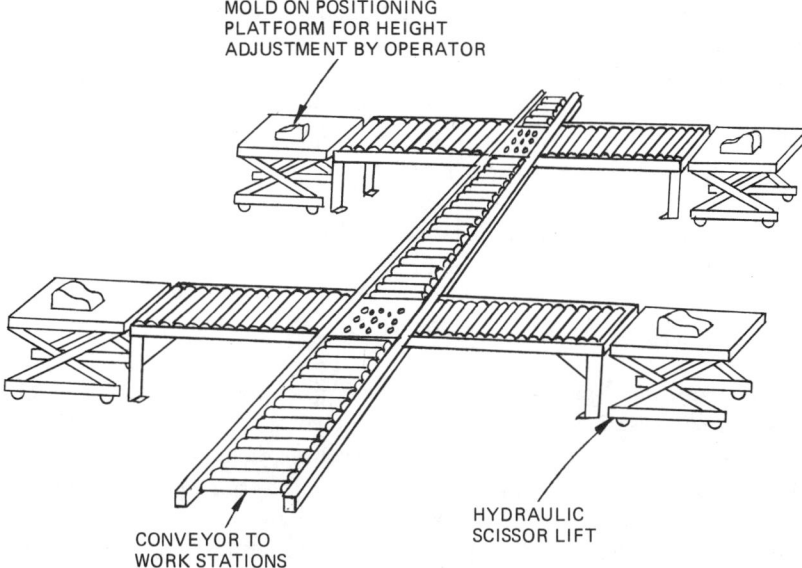

Fig. 10.8. Positioning Platform.

Fig. 10.9. Wire-Guided Positioning Work Station.

Finally, being stagnant with existing equipment for handling is a bad mistake; it must be replaced when more efficient equipment becomes available.

In summation, the handling system chosen will impact strongly on production capacity. The lack of a well-designed one will reduce capacity and capability. Therefore, it is crucially important to choose and use systems that will improve the control, capacity, safety, and performance of the operator or handler.

A final and basic axiom: Although "U"- and "L"-shaped production flow layouts are the most common, the straight flow pattern is still the simplest and cleanest layout when considering handling cost.

10.1.2. Storage

Resin systems, curing agents, chemical accelerators, reinforcements, frozen mold and tool materials, volatiles, fillers such as powders, etc., will all impact upon the design of the mold and tool materials, receiving, holding, and storage areas.

In addition, accessory materials and basic mold forms must have adequate storage areas — and sometimes individual storage areas — especially if the materials of fabrication are temperature or corrosion sensitive, or unstable when stored near one another.

Therefore, basic questions must be addressed by the industrial engineer from the very beginning. Some of these questions are:

- What materials have been received?
- Where must they go?
- How will they get there?
- How long has the material been traveling?
- How long will it be before it goes into storage?

Answers to these admittedly obvious questions will form the basis for the planning of required refrigerated, freezer, and temperature- and humidity-controlled receiving, holding, and storage areas.

Basic Considerations for Indoor and Outdoor Storage. Temperature and moisture considerations are paramount in the storage of material, nonmetallic masters and models, master molds, production molds, and nonmetallic tools. Other factors such as fragility, weight, number, exposure to outdoor elements, and transportation are important, but temperature and moisture considerations are critical.

Moisture will migrate along the reinforcement fibers of a laminate, where the resin system is bonded to the fiber pack, thereby destroying the bond and in some cases the reinforcement fibers themselves. This failure will then result in the loss of vacuum integrity and a softening of the laminate mold or tool surface.

Extreme temperature exposure or thermal shock can cause the disbonding of reinforcement fillers, resinous laminating, or binder material, which also causes softening. Once the bond failure occurs, the differences in the CTE amplify the failure until a total bond loss of the reinforcement to resinous binder occurs. Only in rare cases is this condition repairable.

The use of sealers, coating, or protective solutions will not protect all mold or tool surfaces, whether cast or laminated, from the effects of attack from the environment. The best protection is storage of the mold or tool in a temperature- and humidity-protected, controlled area.

As a general rule, it is always cheaper and more efficient to store up than out, so this must be considered when choosing or building material, mold, or tool storage areas.

Prior to defining the storage area, one must categorize the materials, molds, or tools. By doing this, patterns will emerge and the material and objects to be stored will logically fall into groups, in the following way:

Shape: Common categories are: flat, curved, irregular, nestable, compact.
Weight: This can be measured per unit or per volume or by the density of the material, mold, or tool. Certain racking systems contain weight limitations. Thus, this is a critical consideration.
Size: An obvious consideration that should not be overlooked.

In addition to the above priority considerations, certain ancillary ones should be used in handling the materials used in the fabrication of molds and tools. *Classification* of materials is important, and should be limited to a maximum of from seven to ten classes. In *subclassifications* under this, one would categorize materials into three classes: resins, curing agents, and fillers.

Examples should be given to the individual who will be categorizing materials in each case, and those not falling under a major or subclassification can be grouped under the category "other." Also, *first in, first out (F.I.F.O.)* procedures for date-identified materials are a must.

If large quantities of materials are stored, further breakdowns may be necessary. Suggested categories for this would be: Physical characteristics; use frequency; liquids; solids; and gases.

The storage and holding areas are directly associated with the layout of the total material and work flow. When the layout is already in existence, the distance for each move seems to be already established. However, redesign of existing areas to more firmly link material and work flow is suggested to generate cost savings. Indoor and outdoor storage remote to material, mold, and tool handling is obviously unwise and antithetical to certain fundamental prerequisites already mentioned.

Movement both within and out of these areas should be charted carefully, analyzed, and redesigned to allow movement of the shortest distances to exist between points both within and extending outside of the storage area.

Distance, incidentally, should not merely be measured in terms of flat planes. Up/down distances on one floor and between floors must form an important consideration in measurements of flow. The condition of flow with relation to intensity per hour is, then, as important in the storage area as it is on the production floor.

One way to look at these moves is to define the start and finish points and then calculate the average lapsed time at each point for all average items moving through production.

Another way to analyze cost, related to moves, is to define the function of the material, mold, or tool in the storage area and then to load as many support materials and accessories with it as possible before its travel to each stop in the process. This will allow analysis of the route of a specific item, rather than a general path for all materials, molds, or tools.

The design of storage areas and production areas depends on fundamentals. The materials, molds, or tools, the methods used to move them, and the number of times they must move are all related to each other. The available space where the product is being produced, the manufacturing process being used, and the location — fixed or moving — of the manufacturing point are some of the beginning fundamentals.

The storage area layout of large materials, molds, or tools usually relates to small quantity levels of finished parts, with the process to produce them being a simple one. When the layout of the storage area is based upon a particular product, the quantity produced can be large if a simple manufacturing procedure, coupled to a standardized storage plan, is utilized. On the other hand, when the layout of the storage area is based upon a layout dictated by the manufacturing process, not only is the process expensive but the product is usually complicated, the quantity produced is small, and the storage area involves much more consideration of layout and expense problems (Fig. 10.10).

Finally, an analysis of the characteristics of the total production flow will help with the design of the receiving, holding, and storage areas that support part material, mold, and tool flow.

The questions to be answered by this analysis are these:

- Where does the material, mold, or tool begin its trip (start) and where does its journey end (set down)?
- What is the starting space requirement in the storage, staging, and holding areas during each phase of the flow?
- How does this impact on the structural integrity of the building through which the material, mold, or tool must travel?

Fig. 10.10. Complicated and Versatile Mold and Tool Storage Area. *(Courtesy of Grumman Aircraft Systems)*

- What sort of room or space must the product travel through?
- What is the strength of the substructure beneath the space, and will the substructure support the product's weight?
- Is the area heated or unheated?
- Is it indoors or outdoors?
- Are there particular environmental requirements, such as freedom from dust, that the product demands?
- Are there multiple storage buildings?

If the frequency of put/pull/put away is great, automation in the storage room, whether partial or total, should be considered.

A final storage area consideration is movement within that area. A "wire-guided" system, similar to the one that steers transportation vehicles through production, is also used for storage areas, and is either fully automated or consists of a "man-aboard" to start, stop, and pick up objects (Fig. 10.11).

A last, but not unimportant, point is that when designing and installing a revised or new storage system, whether it be into a rebuilt or new area, it is suggested that operating personnel should be briefed and directed through the process flow by the use of models, prints, and photographs (see Fig. 10.12). This will allow complete and common understanding, from the individual worker to the department head. This will not only increase efficiency in installation and use, but maintain morale during the installation of new storage and

372 ADVANCED COMPOSITE MOLD MAKING

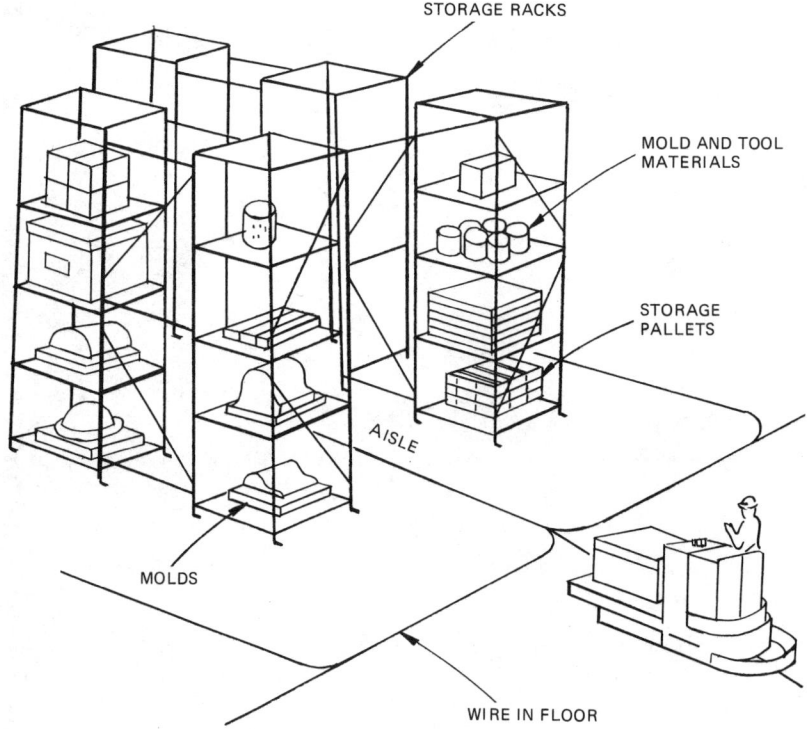

Fig. 10.11. "Man-Aboard" Wire-Guided System.

handling equipment, and forestall future complaints of confusion and blame on the system for failure, when in actuality the fault lies in a proper explanation of, and instruction in, the system's use.

10.1.3. Final Production Use Areas

The most basic method of processing parts is through a convection (air circulating) oven. Ovens or heating and curing cavities assume various forms, but the most universal part-forming and mold- and tool-making cavity is the autoclave.

Thus, it might be well to explore its requirements in some detail at this point.

Within the controlled environment of an autoclave, a pressure at a specific temperature is developed. Autoclaves must all be fabricated, because of their functionality, to an **ASME** code. In order to refabricate, add to, or modify the equipment in any way, it must be recertified.

FACILITY REQUIREMENTS 373

Fig. 10.12. Indoor Storage Area with Increased Height, Set Up for Wire-Guided Aisles. *(Courtesy of Grumman Aircraft Systems)*

Care must be exercised when considering a used autoclave or pressure vessel. *Always* investigate, through a consulting engineer, the original registered fabrication data to see if the vessel has been altered.

The reason for this is that if certain qualifications are not met, a complete disassembly of the modifications, and their reinstallation, might be dictated by the approval inspector. A complete pressure vessel or autoclave is extremely complicated internally, and the modification might be beneath the completed internal accessories and apparatus.

Safety requirements also dictate the presence of certain details, such as semiautomatic and automatic control systems. These will greatly enhance the safety of this system and the quality and quantity of the units produced.

Besides laminating various forms of reinforced thermoset and thermoplastic composite materials, the autoclave has also been used in the aerospace, aircraft, medical, marine transportation, and construction industries to deaerate, bond, cure, and dry, through the introduction — into the equipment chambers as the pressure transfer media — of inert gases, air, steam, and other substances. Usually, but not always, the parts, assemblies, or surfaces to which pressure must be applied are covered with conventional plastic film or reusable rubber vacuum bags. A vacuum is applied beneath the sealed bag, atmospheric pressure presses the laminate or bonded assembly, and the mold is inserted into the autoclave for pressure curing or bonding.

Recent advancements in rubber technology have aided in the formation of reversion-resistant rubber vacuum bags, allowing for the use of the autoclave to laminate, bond, and even vulcanize the bag rubber itself.

When choosing this type of equipment, it is necessary to carefully examine the mold and tool size to be processed, since the autoclave cost increases substantially as the diameter increases. This is becoming more prevalent because of the increased use of thermoplastic composites which require pressures in excess of 200 psi at temperatures up to approximately 750°F.

Of course, although autoclaves are the ideal method of compaction, it must be remembered that pressure vessels without internal heat have been used successfully for years in the initial compaction of mold and tool laminates prior to curing in an oven. If no autoclave is available, this is the method to employ.

Some general principles to remember when investigating equipment are listed here.

First, if a vacuum system is required, the style of this system, heating systems, and chamber insulation configuration should be inspected. Also, keep in mind that some parts are processed in an autoclave without the use of vacuum bags.

Next, the vacuum system should be carefully planned to match the requirements of the transportation dollies, racks, or other interior rigging that exists or that is used to transport the objects into and out of the pressure chamber. It is best to select automated control systems so that a bag failure might be detected and compensated for by zone isolation or operator warning signals. The selection of vacuum pumps will depend upon the anticipated surface area and projected number of objects loaded into the equipment at any one time.

The heating system should be carefully controlled and the historical operation recorded onto a hard copy that will be preserved for future use by the engineering, manufacturing, and quality control departments.

The thermocouple locations on the molds, tools, or objects will define the location for installation of the mating jacks on the internal and external walls of the chamber. The controls to operate the system must be such that the critical time/temperature cycles of mold and tool heat-up rates, as well as the duration at temperature, along with pressurization and depressurization, happen exactly at the required times.

Computer control of these cycles can greatly enhance the versatility, as well as the accuracy, of the system. Furthermore, existing computer apparatus can be altered to interface with, and operate, more than one of a group of autoclaves in an installation.

If computer processing equipment does not exist, or its addition is prohibitive in terms of cost, a bank of interconnecting timers will provide adequate control.

So much for control. What of the process itself?

Steam, gas, or electricity is used to supply heat, individually or in combination, to provide either a dry or wet working climate to the atmosphere within the cavity chamber. Heat exchangers provide the dry heat required, with limitations when steam or gas is the basic heating source.

Preheating with these sources of heat also provides a reduced cost in overall heating when the heat sources can be mixed within the same system. Not all autoclaves have air circulation provisions, because some applications involve the use of oil and inert gases, besides the usual air and steam working substances. In those that do, circulation of the heated air or gas is provided by either internal or external blowers.

When a used autoclave is being considered for purchase, the chamber insulation configuration can be considered critical. Used autoclaves may have been designed for wet use internally, and as such would be very costly to operate if altered to a dry interior.

New designs can be produced with either an internal or external insulation approach. (Mold and tool use in autoclaves for aerospace, aircraft, and composite molding, bonding, and shaping is almost always of the internal insulation type.) However, autoclaves with external insulation force the user to heat the entire vessel, which results in poor energy efficiency.

Lastly, it is important to examine the pressure chamber as basic facility equipment. Unlike part-making materials and processes, mold- and tool-making materials and processes can require the use of an unheated pressure chamber to compact the prelaid plies of reinforced epoxy mold prepreg or tooling laminate elastomers used as flexible pads or vacuum bags.

Whatever equipment is required to process the mold and tool materials, it is necessary to measure the maximum width, length, and height of the mold or tool before ordering. Material, mold, and tool storage areas, and lay-up and preparation areas, should all be environmentally controlled. Access into and out of the building and enclosed department should be carefully planned so that any configuration of model or master or completed structure can pass through.

Machining operations, whether manual or automated, must be planned to handle the size of the object being machined. Model stock to be N/C machined must be prepared, or a mass-cast master poured, near the finish machining point. An "in-line" flow, or "L" shaped or "U" shaped manufacturing sequence, will reduce the cost of producing molds and tools.

In order to assure the maintenance of dimensional accuracy, the master or model should be handled as little as possible coming out of the model machining step directly into the reproduction of contour, whether it be a plaster splash, PFP, lay-up, or mass-casting.

The final inspection area for molds and tools can be located near the holding and prestorage area so that completed molds not required for in-house production can be proved-out and shipped from that location. Others that will be used in-house can, from the same location, be staged for the installation of support accessories such as blocks, rails, permanent reusable elastomeric caul pads, and reusable vacuum bags, and shipped to production for prove-out.

Finally, the safety of the worker, operator, or associate employee — working around, near, or next to the mold or tool maker and user — is of prime importance. Certain constituents and components of nonmetallic mold- and tool-making systems can harm humans after continuous exposure.

10.2. SAFETY, HYGIENE, HEALTH, AND WASTE DISPOSAL

It has been proven that good housekeeping, protection devices around moving machinery, and safety devices for the machinery operator are not all that is required to prevent accidents in the work place. Statistics show that accidents, in more than 75% of cases, are directly attributable to human factors.

Thus, the elimination of, or reduction in, accidents lies with both the operator and management. It is up to the operator to follow training rules. It is the responsibility of management to properly train the operator or worker in the use of materials, equipment, and machinery.

In fact, a planned, repetitious program of materials and process safety instruction is a daily task of management. The management team must organize, through its supervisors and foreman, a plan that will motivate the worker to maintain a daily program of keeping a clean, organized, and efficient work area and production system.

The careful planning and scheduling of both work and flow will also assist in promoting safe working conditions. A worker who has an adequate amount of time to complete a task will not take unnecessary short cuts causing unsafe conditions. An aware management team will frequently observe the workers in the daily performance of their work, assuring that they are safety-conscious and interested in generating good working habits. It is for this reason that the management team must be knowledgeable of the materials and safe-handling procedures for them, in order to translate this information through strong leadership to the worker.

10.2.1. Preplanning for the Handling of Hazardous Material

All materials should be reviewed by the designated safety group, supervisor, or management personnel, to assure that there are no constituents present in the formulation of the materials that could harm the user. The existence of suspected carcinogens or other hazardous components in a material selected for the production of a mold or tool is not an uncommon occurrence. Just because the material has been purchased in a cured form, such as a slab, block, or other preformed shape, does not (in some cases) eliminate the presence of the hazardous ingredient. Post-working operations such as machining, cutting, drilling, etc. could decompose or alter a hazardous ingredient, causing the material to become dangerous in that state.

Care should also be taken when processing materials, especially new in form, near or around other known or unknown materials. Toxicity and the awareness of its presence, for instance, has been a material and process consideration in the use of epoxy and related materials for more than 40 years. NIOSH, OSHA, and the EPA (as well as other federal and state government agencies) have defined, redefined, and are still defining various materials, new and old, as toxicity problems.

Resins, curing agents, catalysts, fillers, reinforcements, and other material additives should be studied carefully before they are used as prototype or production materials. Chemicals like methylene dianiline (MDA) and vinylcyclohexane dioxide (VCHD) are just two of the common materials used in the past that have been cited as potentially hazardous.

Given the proof of the past, management should thus plan for the use of both cured or uncured version of mold- or tool-making materials prior to their use. If there is a need for the collection of process waste such as dust, gas, etc. that contains carcinogenic materials, then professional and local governmental agencies should be contacted prior to choosing such a system.

Some considerations for the suggested handling of precured material that must be machined, drilled, cut, etc. would include provisions for the proper removal of dust, cuttings, or chips. The material compounder or formulator usually plans for the processing of the material so that it will result in low amounts of dust formation. Monitoring of machining and cutting speeds, and proper cutter or machinery selection will increase safety in this area, since, in some cases, dust can be caused by the improper speeds, feeds, and cutters chosen for a particular machining operation. In addition, collectors at the workplace, as well as intakes on the bench level and floor level, will prevent dust entry into the air conditioning or heating systems.

When installing new equipment, it is also of prime importance to consider whether dust collection will be necessary in order to maintain proper safety and health protection. Permanent installation of dust collection equipment then becomes critical to the production operation, because a complete system will restrict floor planning. The duct work of this equipment, whether above or below the machine surface, is fixed. This means that if a piece of equipment is removed and replaced by one generating chips instead of dust, a portable-type collection device must be installed in addition to the fixed one. It is for this reason that portable units are sometimes more practical than permanent ones.

When handling uncured materials, the prevention of rashes, dermatitis, and other irritations is a necessary precaution. Rubber gloves should be worn. Other exposed parts of the body should be covered by long-sleeve shirts, laboratory coats, etc. If exposed skin is mandatory, then protective approved skin creams should be considered.

Good ventilation, respirators, and goggles are required in almost all cases. Solvents should not be used to remove resinous materials from the skin, since the solvent will dilute the compound, and possible skin penetration can occur.

With new safety cans required by the federal government, all material containers will have instructions for use. The respective data sheets for a particular material being used will have a set of use instructions, and they should be rigorously read and followed.

Polyurethane materials should be treated with particular care. Some of the elastomers contain reactive toluene diisocyanate (TDI), and are classified as toxic. A system should be chosen with a lower-vapor-pressure curing agent system or one that does not contain TDI. In cases such as this, well-ventilated work areas that pull heavy fumes away and down, instead of up, are preferred. In addition, bench-top ventilators, instead of overhead units, have also been suggested and are preferred.

Epoxies as urethanes require the same careful treatment. Fumes are usually given off during the reaction of the resin, hardener, and diluent. Hardeners usually affect the operator's skin. Thus, gloves, protective clothing, and frequent washing of exposed areas, as well as respirators and eye protection, are a must. Plastic-coated canvas gloves have worked well in the past when handling fillers. Forced-draft bench or downward exhaust has been recommended to remove fumes (Fig. 10.13).

10.2.2. Autoclave Safety

Autoclave or pressure vessel safety is a special and important case. All doors on existing equipment should be equipped with a safety device which secures the door from opening during use. A fail-safe device of this type is usually standard, but it is important to be sure that it is there. A combination of electrical, pneumatic, and mechanical safety devices is suggested.

In addition, audible as well as visual alarms should be in place to inform operators of a lack of cooling water, motor failures, excessive autoclave pressure or temperature, the presence of personnel within the work cavity, or any other danger that must be identified.

Inert pressurization gases such as nitrogen should be used, to eliminate the possibility of autoclave fires, which will cause damage to molds, tools, or chamber interiors.

FACILITY REQUIREMENTS 379

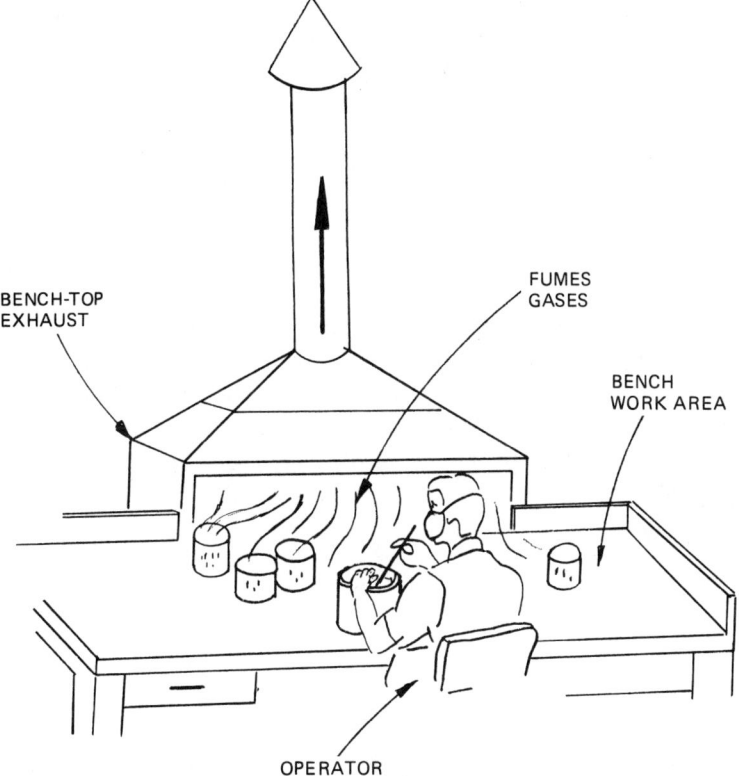

Fig. 10.13. Bench-top Exhaust.

10.2.3. Precautions for the Handling, Transportation, and Storing of Equipment

Safety is of the utmost importance in these areas. The movement of something is what usually causes an industrial accident. When purchasing handling equipment, an insistence upon a demonstration of use will pinpoint the safety features available, and deficiencies that can be cured by additions will become apparent.

To have this equipment installed and operating and then be shutdown due to safety reasons could be a disaster. Therefore, safety features should not be taken for granted. They should – in all cases – be demonstrated before purchase.

In use, authorized personnel should be the only individuals, who, after extensive training and supervision, should move anything. Careful determination of the handling methods and the kind of moving system, plus careful operation of that system, all result in safe handling.

In summation, intense safety awareness will allow for close participation of the many people who are required to be involved with, interested in, or affected by the handling of materials, molds, and tools.

10.2.4. Waste Disposal

The matter of waste disposal has become so complicated that careful attention should be given to the subject. For fear of a total production halt, management should employ the advice of on-staff facility personnel, local government, and federal agency groups to advise and direct the disposal of various materials and waste. Nothing should be discarded, whether it be a resin, filler, curing agent, unique reinforcement, solvent, chemical, or other substance, until it is identified and the disposal procedure documented in writing and approved by the required authority. Above all, it is important that the decision to dispose or discard be made by an informed group rather than one person.

11
Inspection and Quality Control

Much emphasis has been placed upon the automated and complicated methods developed for the inspection of composite parts and assemblies. But until recently, there has been very little emphasis upon the area of mold and tool inspection and quality control.

Quality control and the inspection of advanced composite molds and tools should start with the originator of the structural design, as well as the mold and tool designer.

The demands that are placed upon the mold or tool maker are based upon requirements to produce a part or assembly, using a special material or process. It is for this reason that sometimes only two, perhaps even only one method exists to produce the complete part or assembly. Therefore, the molding or forming method might dictate the mold making method and thereby define the inspection criteria (Fig. 11.1).

An understanding of what happens to the mold and tool materials during use is a good way to understand what the inspection and quality control requirements must be.

A basic example is the choice of dissimilar materials within a mold or tool, and the expectation that these materials will stay dimensionally stable during thermal cure cycles. The coefficient of thermal expansion (CTE) is only one governing factor directly linked to mold and tool thermal stability. A careful study of the chapter on materials, as well as of the chapters dealing with design, design related areas, and fabrication, will inform the responsible inspector and/or quality control individual of the criteria that should be applied to define the inspection criteria, identify a discrepency, and also be aware of the reasons for the discrepency (Fig. 11.2).

Some fairly exotic equipment designed to inspect materials, parts, and assemblies has been developed in recent years. This equipment can also be used to check the materials, or the mold or tool, that will form the part or assembly.

382 ADVANCED COMPOSITE MOLD MAKING

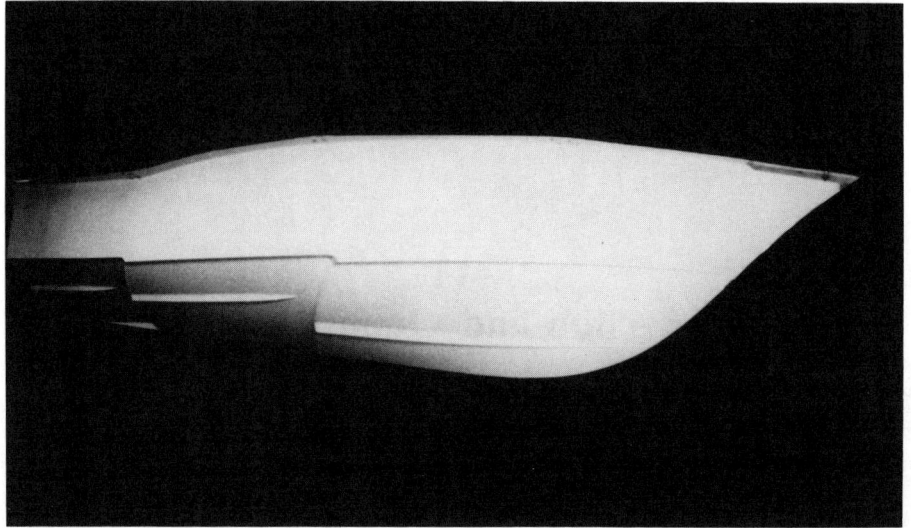

Fig. 11.1. A Large Composite Boat Hull Formed by a Unique Molding Method in a Female Mold Using Integrally Molded Foam Stiffeners, Unique Planing Steps, and Outer and Inner Skins, All Molded Simultaneously and Co-cured Together During the Curing Cycle. The Mold Techniques Involved Use of Collapsing Caul Pads and Reusable Vacuum Bags. *(Courtesy of Connell Engineering and A.C.E., Inc.)*

Some of these nondestructive inspection (NDI), nondestructive testing (NDT), and destructive testing (DT) methods of inspection are:

- Radiographic methods
- Optical systems and equipment
- Electrical measuring
- Sound or sonics in air, such as acoustic emission
- Sound in other transportation media, such as water
- Mechanical testing
- Thermal techniques, such as survey, infrared, and thermographic reading
- Diffuse light, lasers, and liquids as penetrants
- Physical testing through various tensile, compressive, flexural, and other related methods
- Computer testing

Following the examination of inspection and quality control, will be some of the predictions and directions that are evolving in the advanced state of the art of composite mold and part making.

INSPECTION AND QUALITY CONTROL   383

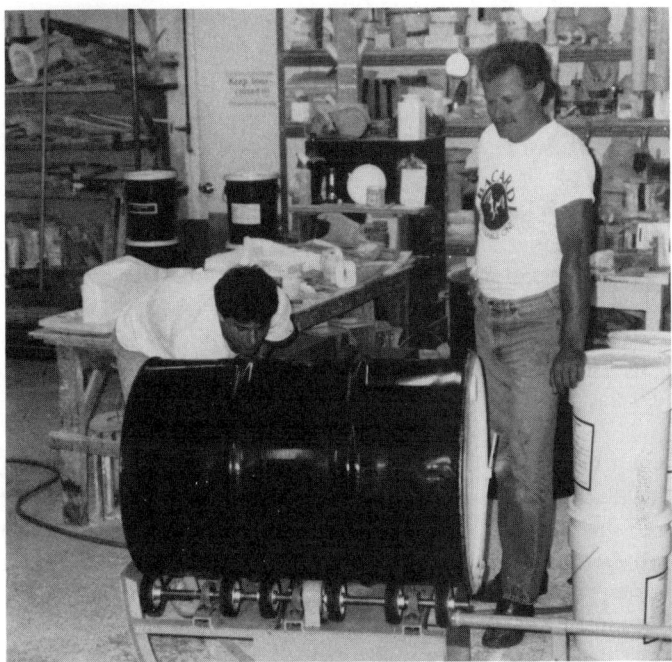

Fig. 11.2. Low CTE (.4 x 10^{-6}), Lightweight, Room-Temperature-Pourable Ceramic Syntactic Foam Being Mixed Using Inexpensive Mixing Equipment. *(Courtesy of A.C.E., Inc.)*

11.1. NONDESTRUCTIVE INSPECTION (NDI), NONDESTRUCTIVE TESTING (NDT), AND DESTRUCTIVE TESTING (DT)

The following methods form some of the procedures available to test the materials, and even the processes, used in the fabrication of both metallic and nonmetallic molds and tools. Although these are only a few of the available methods, they are considered to be the most important and frequently used ones at present.

11.1.1. Nondestructive Inspection (NDI)

The quality assurance and inspection methods for molds and tools, like composite parts, are based predominately upon nondestructive inspection and test methods. On the other hand, destructive testing, such as certain physical and mechanical test procedures, is performed on coupons or extensions of the metallic or nonmetallic mold or tool. These extensions are usually prepared during the formation of the mold or tool. They are removed from the exterior of the trim line of the finished article, and then used for testing.

Like "Human Engineering," one of the most rewarding NDI methods or techniques for assuring quality processing during fabrication of molds and tools is the assigning of a process engineer who has worked with the mold and tool designer to the task of following the item during each production phase. This could almost be considered "Human Quality Control."

Ultimately, the mold or tool designer himself should really be the individual to follow the process step by step, through mold, tool, and part production, assuring that the materials and processes chosen are applied properly.

It is possible to say that a master, model, mold, tool, or accessory cannot be over-inspected, due to the fact that a quality part or assembly depends directly and totally upon the surfaces from which it is created.

Radiography. Various forms of radiative energy (X-rays, gamma radiation, etc.) can reveal inclusions, voids, and delaminations. The radiometric absorption will take place in mold materials, but only to certain depth. Therefore, one unique and innovative approach to using radiographic detection is to take advantage of available radiopaque tracers, and use them to locate laminate mold or tool plies, and laminated imbedments, and to even identify plies, their orientation, and their existence beneath the mold surface. The tracers, then, can be used as locators, markers, and/or quality aids.

Some manufacturers have even used selectively placed tracers within the materials used to form the structural parts. This has been accomplished by dispersing the tracer in the base resin system or precoating the reinforcement of the structural part with the tracer.

Optical Measurement and Alignment. Optical measurement and alignment instruments provide a means for checking and/or positioning the components of, or entire surfaces of, extremely small and large master patterns, models, molds, tools, or finished parts. Among these instruments are the optical comparator, transit square, and theodolite.

The *optical comparator* is a device used to inspect small tools or finished parts with complex contours or minute radii. This instrument employs a bright light and a condensing lens to obtain a shadow image of the contour in question. Then, the image is enlarged through a projection lens and, in turn, reflected off a mirror onto a translucent visual screen. The screen usually contains a series of circular diameters to which radii can be checked.

Furthermore, complex contours can be checked against accurately enlarged line drawings which are placed on the screen. Thus, extremely small radii and complex contours can be inspected to within 0.0001 in. with the aid of this instrument (Fig. 11.3).

INSPECTION AND QUALITY CONTROL 385

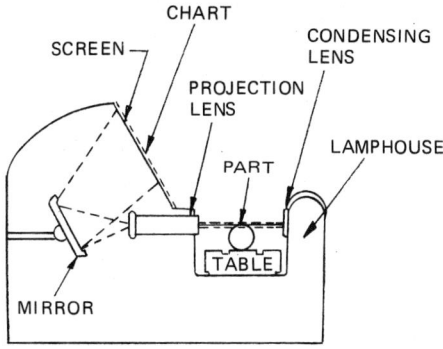

Fig. 11.3. Optical Comparator.

A *telescopic sight* is used in conjunction with a transit square, sometimes referred to as a transit jig.

The telescopic sight contains an objective lens at the front end of a tube, while a pattern with cross hair lines is positioned at the rear. Behind the pattern is an eyepiece. Although an object, when viewed through the telescopic sight, appears to be inverted, the telescopic sight is valuable in determining a straight line, or line of sight (Fig. 11.4).

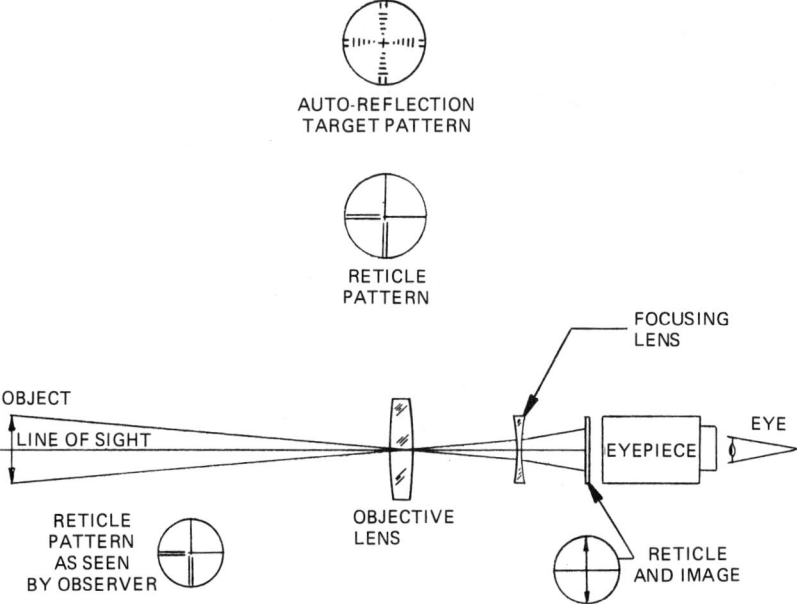

Fig. 11.4. Telescopic Sight with Line-of-Sight Viewing.

386 ADVANCED COMPOSITE MOLD MAKING

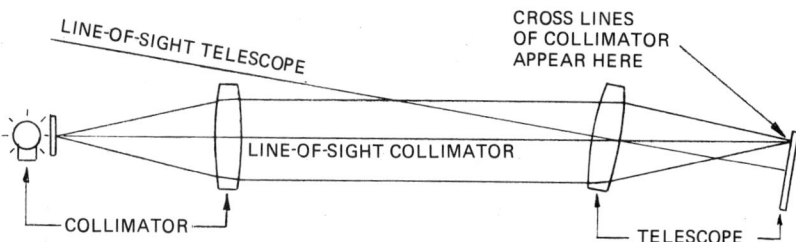

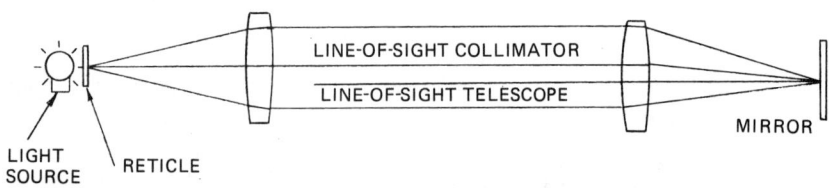

Fig. 11.5. Line-of-Sight Autocollimator Needing Adjustment.

An *Autocollimator* is similar to a telescopic sight, but its functions are slightly different. The autocollimator's primary function is to utilize an internal lens pattern to align a mirror so that it is perpendicular to the line of sight.

In the autocollimator, a light source illuminates a semitransparent glass mirror. The reticle and the returning image of the reticle which is deflected off of a mirror, are viewed through the eyepiece. Both are compared. If the cross hair patterns do not coincide, adjustments are necessary, since the mirror has been proven to be not perpendicular to the line of sight (Fig. 11.5).

While not as accurate as the autocollimator, *autoreflection* may be substituted when an autocollimator eyepiece is unavailable. Either an autoreflection target mounted on the front end of a telescope, or an imprinted target on the objective lens, is necessary to adjust the system. Once the telescope is focused, the mirror may be adjusted to make the reticle patterns coincide (Fig. 11.6).

Autocollimation or autoreflection, when used in conjunction with a jig transit, can create an optical square. The jig transit has a telescopic line of sight which is utilized to take measurements. An autocollimator telescope is used with a jig transit that contains a hollow-ground elevation axis, but an autoreflection target requires a jig transit containing a mirror concident upon its elevation axis (Fig. 11.7A and B).

Once the system is set up as shown in either part of the previous figure, the jig transit may be elevated or lowered to check a station-plane or to perform a vertical alignment of a mold or tool component part or assembly (Fig. 11.8).

INSPECTION AND QUALITY CONTROL   387

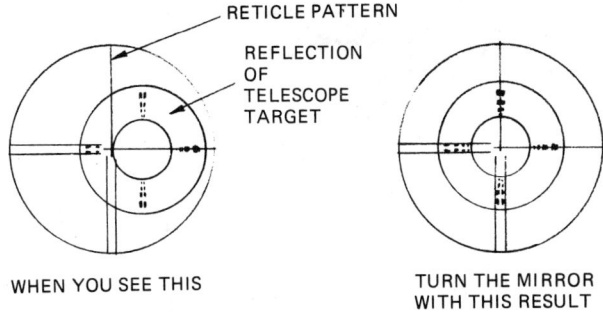

Fig. 11.6. Autoreflection Mirror Adjustment.

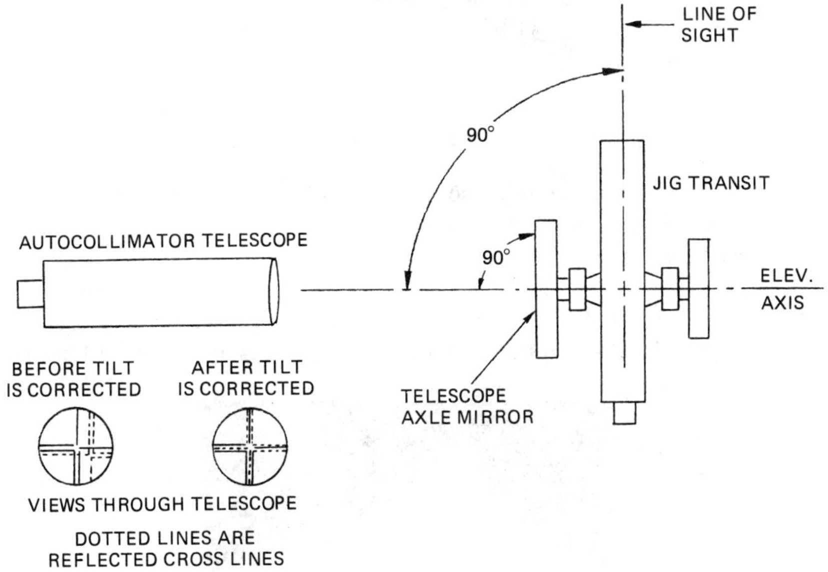

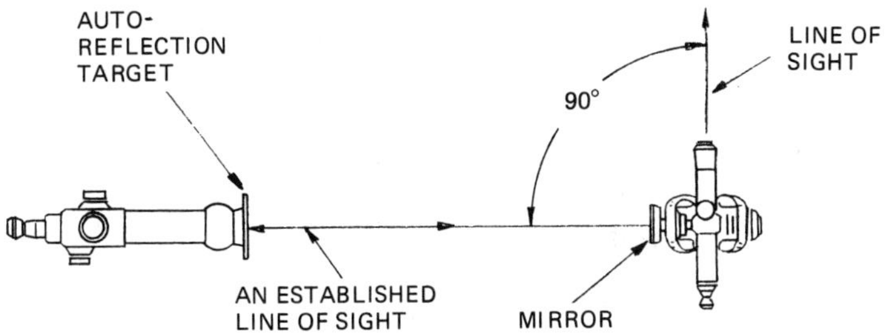

Fig. 11.7A and B. (A) Autocollimator Telescope and Jig Transit; (B) Autoreflection Apparatus and Jig Transit.

388 ADVANCED COMPOSITE MOLD MAKING

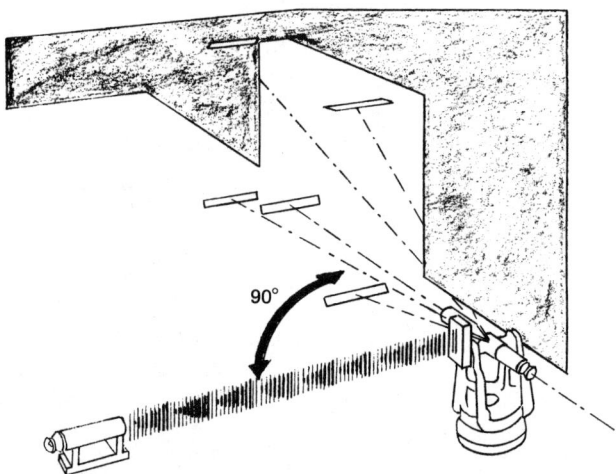

Fig. 11.8. Jig Transit Checking Station-Plane.

The jig transit telescope may also be locked and rotated horizontally to take a flatness measurement, as shown in Fig. 11.9.

Theodolites are precision angle-measuring instruments. They are capable of accurately measuring pitch and yaw (vertical and horizontal angles). These angles can then be used for triangulation, which will determine the coordinates of a point in a three-axis system. Once the coordinates are obtained for all of

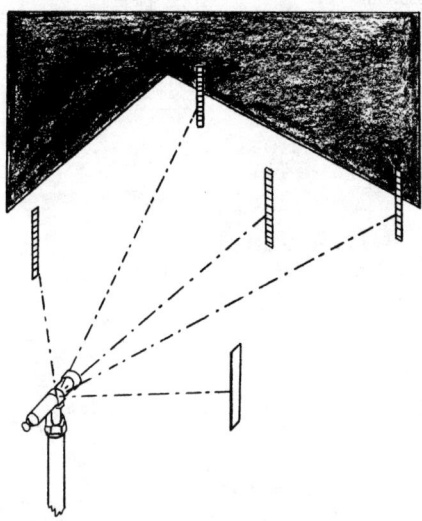

Fig. 11.9. Flatness Measurement.

the necessary points, their values can be compared to nominal numbers which will determine the error in the surfaces or contours of a pattern, mold, or tool.

Theodolites are available in two model types: a conventional scale-reading type and a digital computer-controlled type. The conventional model is accurate to within 5 seconds of arc, and triangulation calculations must be performed separately to obtain point coordinates.

In contrast to the conventional model, the digital version requires no separate calculations. Here, the computer performs all of the necessary data manipulation, including error deviations with respect to nominal values. Furthermore, the digital model is accurate to within 2 seconds of arc.

Robotic optical inspection utilizes specific support equipment.

The accuracy, reliability, and reproducibility of the results of optical inspection can be improved through the use of support equipment. A robot and computer coupled with optical inspection instruments is a highly efficient means of inspection in that it reduces costs and can operate unattended.

Computer-controlled robotic equipment is available with reproducible measuring accuracy within the range of .001 in.

Articles being inspected need not be aligned in accurate attitudes with respect to the inspection equipment, since the measurements are performed in a coordinate measurement system. This system compares the three-dimensional relative distances of predetermined inspection points to those preprogrammed into the computer. Both sets of data are then compared to each other to determine the tolerance conditions.

The twin advantages of speed and accuracy, then, make computer-assisted robotic optical inspection ideal for a quick determination of the acceptance or rejection of a mold or part.

Process Inspection. This is the most critical type of nondestructive visual inspection, and thus it is the best method that can be used to assure that there are no foreign inclusions, such as release materials, hardware, or even hand tools, included in a mass-cast tool or mold laminate.

For metallic molds and tools, monitoring the process will always reveal a defect, if such is present when the mold or tool is being fabricated. In-process inspection of trim tools, drill fixtures, routing tools, mold forms, bonding tools, etc., can also be performed using templates and checking tools.

The fabrication process for laminate molds and tools should be monitored to assure that the age or shelf life of the materials used is within allowable storage limits. To assure that the orientation of the precut plies (warp and fill direction) of a laminate is correct, the information should be noted in the design drawing or the manufacturing procedure. The gaps between ply ends, the weights of fabric materials, compaction, debulking, and final vacuum bagging are usually planned by the mold and tool designer and process engineer.

For nonmetallic molds and tools, the laminate or mass-casting *must* cure to a certain level. The chemical reaction must occur and molecular crossinking will allow the contour-forming surface to be used for duplication. In order to determine that this has happened, and in order to assure that this level of cure has occurred, a number of methods besides mechanical and visual ones must be used.

One such method is based upon *electrical testing.* This system can be used for all thermoset materials, including various types of elastomers and rubber. The system monitors cure by recording the dielectric property changes.

Thermoset laminate and mass-casting resin systems undergo a chemical reactive change in viscosity, from the low to gel point. The gel point, usually described as from 100 to 200,000 centipoises of viscosity, indicates a significant cure point level. But beyond that, it is no longer an indicator, and conventional DT and NDT methods must be used to assure full cure.

Dielectric monitoring, measuring electrical conductivity, can reveal the extent of cure. Dipoles in the resin system transform from the unhindered to hindered state during a cure, and ion movement reduces as the cure progresses. Measuring the dielectric property of loss factor and permittivity reveals the extent of cure.

Loss factor is the indicator of energy expended in dipole orientation and ionic movement in an AC field and *permittivity* is the extent of dipole orientation and ionic movement in an AC field. Loss factor will be a measure of viscosity in the beginning of the reactive cycle and will then be later used to measure the extent of the cure.

Sonics, another NDI method, is available to test molds and tools in many forms. In this method, various frequencies, transmitted through air, water, or other media, are recorded by reflecting the source off or through the surface of a mold or tool in order to detect various kinds of defects such as cracks, voids, delaminations, etc. The sound frequency response, usually displayed on an oscilloscope, can be recorded and a printed scan of the surface will map out a pattern which reveals possible defects.

Mechanical Inspection. Mechanical inspection is still another method of control. Prior to inspecting reinforced composite molds, tools, or finished parts by mechanical means, the structures in question are usually subjected to careful visual examination. In a tool laminate, cast mold, or part, nondestructive visual inspection can reveal gross defects or errors, such as scratches, blisters, wrinkles, voids, delaminations, and resin starved and resin rich areas.

Also, the types of errors or defects can indicate the possible cause and determine the corrective solution. For instance, dark discoloration of a fiber-reinforced composite laminate may be attributed to overheating. Thus, the cure temperature and/or the cure cycle may require compensation.

INSPECTION AND QUALITY CONTROL

Mechanically inspecting a mold, tool, or part can be carried out with a multitude of measuring instruments designed to inspect or compare specific characteristics against a certain standard. Among these gauges and inspecting instruments are: the micrometer, the vernier caliper, height gauge, sine bar, gauge blocks, coordinate measuring machine (CMM), and durometer gauge.

A *micrometer,* or "mike," measures outside dimensions, and is often referred to as an end measuring instrument. Although use of the "mike" is relatively simple, its accuracy is dependent upon the proper amount of applied thimble torque. Too much torque causes the frame to elastically deform or spring, and thus error is created.

A *vernier caliper* typifies the vernier principle of measurement and is also an end measuring instrument. The main beam or scale on a typical vernier caliper is numbered in increments of 0.025 in., with the smallest scale division being 0.001 in. This instrument is also available in a digital version which provides LCD readouts, in which the smallest scale increment is 0.0005 in. Since this instrument is capable of measuring outside dimensions, inside dimensions, and depth, it becomes a versatile inspecting tool.

Height gauges are available in either vernier or digital-type readouts. These instruments are similar to a vernier or digital caliper, except that the fixed jaw has been replaced by a fixed base. The simple versions of these gauges usually measure perpendicularity, flatness, straightness, and center distances.

A *sine bar* is an accurately ground flat steel straight edge used for precision measuring and checking of angles. It also contains identical circular cylinders attached at each end with a definite separation distance.

Gauge blocks are practical length standards of industry. They are fabricated from either steel or carbide with two faces flat, level and parallel within 0.000001 to 0.00006 in., depending upon length and grade of accuracy.

The *coordinate measuring machine* (CMM) measures points on a surface or holes which are dimensioned from coordinate axes and from each other. The object to be examined is securely fastened to an inspection table, while an indicating probe travels along the surface to be checked. Then, the instantaneous prove position, in a two- or three-axis coordinate system, is accounted for and displayed as a digital readout.

The *durometer gauge* operates in a similar fashion to that of a Rockwell tester. Both have a penetrator or stylus which pierces the surface to be tested, and the depth of penetration determines the material hardness. Since the penetrator is preloaded with a specific force or weight, various hardness ranges are obtainable by changing the preload weight. Two ranges of durometer gauges are used on rubber and plastic type materials, "A" for soft, and "D" for harder materials. The Rockwell tester is reserved for metals or very hard, dense metallic or nonmetallic materials.

Quality Assurance Programs. The programs for quality assurance, and the people who administer them, are directly responsible for the successful fabrication of both metallic and nonmetallic molds and tools. Although structural part and tool designers, manufacturing engineers, and others monitor the materials and processes in-house, it is suggested that quality assurance personnel be employed to constantly observe the production of materials at the raw material manufacturing site as well.

The quality control procedures observed at the manufacturing site of the raw material suppliers, compounders, and material processors is an exact indication of how consistent material formulations will be, and what the reliability level of the processed material will be.

Let us examine one such quality control procedure that is used to enforce the quality of the vendor process manufacturing of prepreg laminate mold and materials.

In this procedure, prepreg reinforcements are continuously impregnated on processing equipment. From yard to yard along the broad goods, the prepreg machinery operator should know exactly what the curing characteristics of the resin systems are. It is very easy, for instance, for variations in resin system viscosity to occur due to changes in the amount of solvent used to dilute the resin system required to impregnate the reinforcement.

Through inspection and monitoring, these variations are easily and quickly detected as the reinforcement material is processed by sampling 4 in. X 4 in. pieces of material as it is being impregnated and then running resin gel times and resin to reinforcement content ratios as the impregnation equipment is processing.

The results are then quickly transmitted back to the equipment operator, and if a problem exists or an adjustment must be made, an immediate response is effected.

Depending upon the processing steps used in the preparation of various mold and tool material resin system formulations, the vendor surveillance (quality control) personnel can either monitor, suggest material formulation changes, or even alter the check points during raw material preparation, compounding, or processing.

Whereas metal materials used in the fabrication of molds and tools only require the use of mechanical operations for transformation to a mold form, composite molds and tools, fabricated as mass-castings, laminates, and by other methods, require inspection operations before, during, and after their manufacturing process.

These forms of nonmetallic materials change state during processing. The inclusion of any foreign matter, variations in cure temperatures, pressures, or times, might be cause for a mold or tool failure during fabrication. Deviations from weight, or volumetric mix ratios, or other variations, could affect the integrity of the mold, tool, or even finished product.

When a review of the inspection steps and process methods of the quality assurance program takes place, the raw material inspection step and the mold or tool design review usually start the cycle. After the design has been reviewed and the incoming materials are checked for conformity to a material specification, in-process inspection for conformity to process specifications takes place. In other words, first material inspection takes place, then process inspection. One or more of the following nondestructive raw material inspection or testing methods are then used to check for defects of various kinds:

- Checking constituent materials' properties by: chemical composition; infrared spectroscopy; differential scanning calorimetry (DSC); rheometrics; or liquid chromatography (Fig. 11.10).
- Mold or tool design review, involving: thermal conductivity and tool heat-up rates; and compatability of different CTEs.
- Automated Equipment such as computer-aided curing equipment, as well as automated controls to monitor other functions, are most desirable (manual adjustment of pressures, temperatures, etc., in presses, ovens, autoclaves, and other process equipment is costly and requires the presence

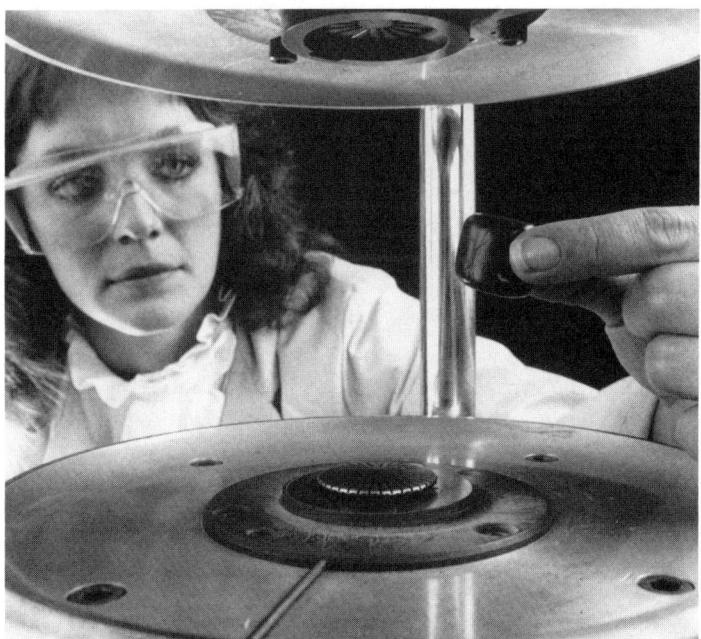

Fig. 11.10. Preparing Elastomeric Compound Test Samples. *(Courtesy of the Keene Laminates Div. of Keene Corp.)*

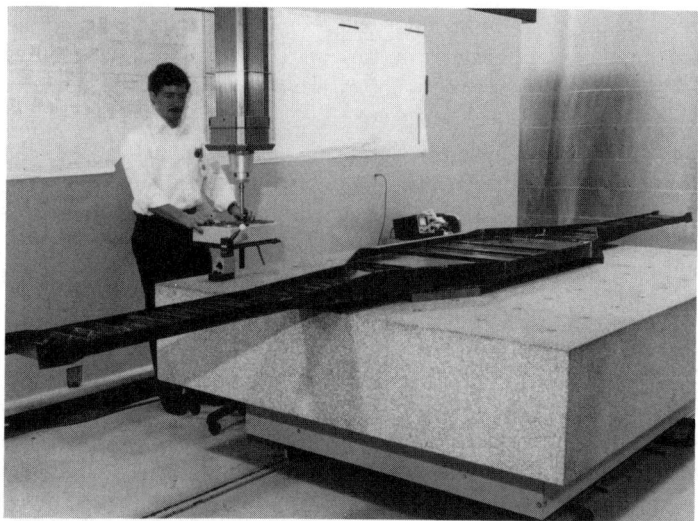

Fig. 11.11. Automated Nonmetallic Laminate Inspection and Contour Verification. *(Courtesy of Grumman Aircraft Systems)*

of both a manufacturing engineer and an inspector (see Fig. 11.11). Digital contour measuring is also included in this category.
- "Proofing" of a mold or tool.

The method of proofing a mold or tool deserves some explanation: A mold or tool should be subjected to a cure cycle (temperature, pressure, etc.) that will prove that the designed heat-up rate exists, no leaks in pressure or vacuum exist, and that the part or assembly formed on the surface will be as structurally designed.

To provide this information, a sacrificial part, called a "greenie," is formed on the mold or tool surface. If the part materials are so expensive that a complete part cannot be formed, a fiberglass-reinforced copy is made. On occasion, in fact, the advanced composite reinforcements are used to produce only a portion of the "greenie."

A small portion of the part is instrumented or monitored during cure, and this segment is physically tested after curing to assure that the part materials have been processed properly.

During tool proofing, the structural designer, mold or tool designer and builder, and quality personnel must adjust design allowances, should all the parameters originally chosen not be met.

Thermal survey and *pressure checking* of mold forms and bonding fixtures takes place after fabrication of new, and even altered, tooling. The purpose of the thermal survey is to reveal areas of the mold or tool, whether metallic or nonmetallic, that might heat up at a rate not consistent with the rest of the mold or tool surface, since consistency in the heat-up rate of the surface is critical to a uniform curing of the film adhesive on a bonded assembly or resin system of a laminate lay-up (Fig. 11.12).

Thermal surveys are conducted when curing parts or assemblies on a mold or tool in an autoclave or oven, and preparation of the mold or tool includes placement of thermocouples on the front and rear of the mold contour. The thermocouples are placed over the maximum and minimum masses of the mold or tool. Bonded assemblies or laminate lay-ups simulating the part are installed, and vacuum bags are applied. Then the mold or tool, with the laminate or assembly, is set up in an oven or autoclave after the vacuum pressure has been assured and no leaks exist.

The heat-up rates usually vary from $1°-15°F/min$ of rise. If the heat-up after the thermal survey is not uniform, the masses of the mold or tool must be adjusted, heat sinks added, thermal insulating blankets applied or removed, or some other thermal adjustment made to make the contour surface more thermally uniform.

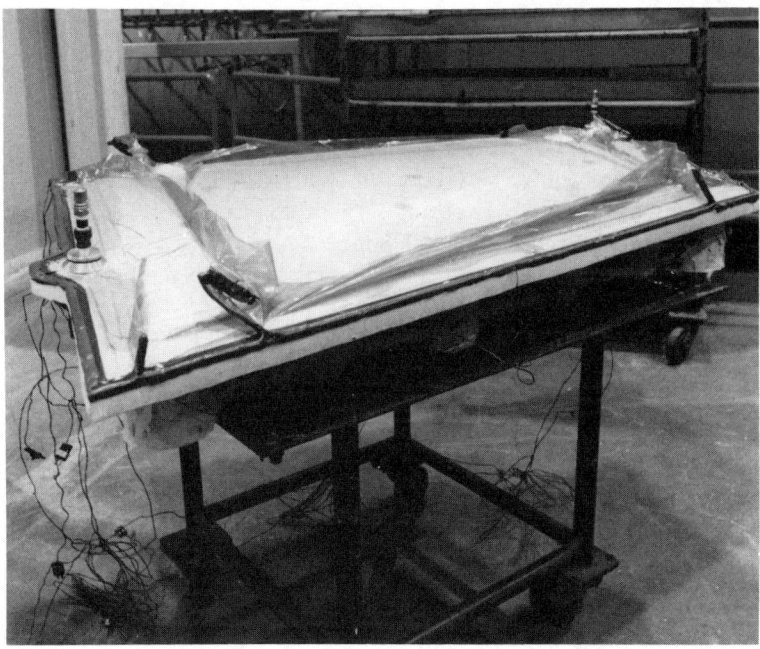

Fig. 11.12. Thermocouples are Used to Perform Thermal or Heat Surveying of the Tool Surface. *(Courtesy of Grumman Aircraft Systems)*

Pressure checking, usually performed under vacuum bag (atmospheric) or autoclave (usually up to 200 psi) pressure, takes place before, during, and after the thermal survey. Vacuum gauges indicate the presence of leaks in the mold or tool surface. Two gauges, one on the intake and one on a blind extension across the mold or tool from the inlet, should be installed and monitored.

Leak rates (dropoff) before, during, and after the thermal survey should be in the range of from 1 to 5 in. Hg/5 min. This is a guide, and some companies, because of a specific application, require more stringent requirements such as 1-5 in. Hg/10 min.

Another form of molded replica is the *"white master"* which is an "engineering laminate" showing the exterior trim lines, hardware locations, exact contour, cut-outs, etc. This laminate then becomes the master from which the contour and all dimensions are preserved, verified, and referenced. Sometimes taken from the master pattern or master mold, this laminate becomes both the quality assurance and engineering reference. These types of laminates are sometimes made even if the coordinated data for design is available as CAD or CAM.

The use of *computer-aided controls, systems, equipment, and processes* in the manufacture of various kinds of advanced composite molds and tools varies from basic digitizing and coordinate-measuring equipment, used to check contours, through verification of machining cutter paths, proofing of machining cutter paths, proofing of machining programs, design tolerance compensation, and even the control of the actual curing of the mold or tool when a laminate or casting is employed.

The computer-assisted cure controls aid the capability of storing many different process profiles for various types of materials through the use of laser-based dimensional-measuring apparatus, with tolerances of accuracy up to .00016 in. at resolutions of up to .00008 in.

These systems are based upon spot illumination, with the light source generated by a modulated laser or LED infrared light source.

11.1.2. Nondestructive Testing (NDT)

Sometimes, it is difficult to separate NDI and NDT methods. Thus, for the purposes of easy identification, some of the methods may be found mixed within the same category.

In either case, the method chosen should result in an inspection process that will assure the highest level of quality, even if a portion of the mold or tool material must be tested to destruction.

Holography is a form of presenting recorded or viewed images. It is important, when using this method, to use a viewer to eliminate any possibility of laser injury to the eye.

In this method, coatings and penetrants, applied to the contour surface, transmit the holographic image to computer-controlled recording equipment. Thus, the images become a reflection of the interior stresses that are or could be present during the use of the mold or tool. And thus, the technology allows for nondestructive inspection based upon the application of a thermal or mechanical stress load to a rigidly mounted mold or tool, and then measuring the projected surface or structural deviation as referenced to an original unstressed mold or tool surface.

Acoustical holography, on the other hand, can be used to provide a three-dimensional view inside a section of a mold or tool, and can even detect weld flaws or defects on the surface of a metal tool by measuring and analyzing ultrasonic reflections sent and received by a single transducer.

Thermographics, another NDT (nondestructive testing) method, can reveal surface defects such as voids, inclusions, etc., by observing temperature variations on the surface of the mold or tool. The theory of thermographic imaging or photochromatic paints is based upon the thermal conductivity of the mold or tool material, and especially the variations of it in the area of a void or inclusion.

In this method, the mold is heated and cooled, while the surface temperature is observed during the thermal cycle. Sensitive infrared measuring equipment detects variations as small as $0.1°F$. on the mold or tool surface. Automatic scanning equipment can then produce a thermal profile of either color or black and white images, which can pinpoint a defect by measuring the thermal pattern variations.

Color inspection reveals flaws and voids. These are detected by the change of a color pattern in a coating. Photochromatic, strippable paint coatings applied to a contour surface will reveal any defects after exposing this coating to an ultraviolet light source.

Heating the mold or tool will cause the paint coating, under ultraviolet light, to transmit a color pattern that will show voids and inclusions.

11.1.3. Destructive Testing (DT)

As mentioned earlier, destructive testing is usually only necessary when qualifying the mold or tool material being considered for use. DT usually allows for determination of physical properties not provided by any other test methods. For this reason, as with NDT, the definition and placement of a particular test within a category may be difficult. Again, there may be an overlap of categories and definitions within categories.

Using DT methods, whether for metallic or nonmetallic materials, will allow for the establishment of raw material manufacturing parameters.

A good example of DT, relating to nonmetallic material methods is shown in the following partial listing of test methods.

The *physical properties* of laminate mold and tool materials are measured using various methods, but the basic requirements to be established before and after cure are usually the same from material to material. Some of these can be verified by certification from the manufacturer, but the following must be verified by the user:

- The ultimate tensile strength and modulus of elasticity of the reinforcement.
- The specific gravity.
- The weight/unit length of yarn or material.
- The finish on the reinforcement fiber.
- The resin content, as a percentage by weight.
- Flow, as a percentage by weight.
- Gel time in minutes.
- Tack time in minutes.
- The impregnated fabric or material weight.
- Visual uniformity of fabric or material appearance. For example, defects in excess of 5–10% of total roll length could be cause for rejection of material roll.
- Alignment of warp and fill of impregnated fabric so as to be perpendicular to each other, with the warp parallel to roll edge.
- The roll widths, size, and roll core configuration to be defined by the user.
- Storage life to be as per manufacturer's data.
- Physical and mechanical properties of the cured laminate: ply thickness and weight of reinforcement to resin content; void content; and longitudinal flexural strength, modulus, and horizontal shear.

In summation, it is interesting to note that quality and control, usually occurring at the end of materials and process manufacturing and techniques, should also happen at the very beginning of a structural, conceptual, engineering design. And this beginning means even before the mold and tool design process is chosen, since the parameters of what is designed and produced must be chosen before design can proceed.

Now, let us take a look into what the future holds in composite mold and tool making and molding processes.

12
The Future of Composite Mold Making

The future may be hard to predict, in general, but the future of the art of advanced composite mold making (and we must refer to it as an art), is fairly easy to forecast (Fig. 12.1).

The direction for the future will be towards mold and tool materials capable of higher temperature processing, and, in some instances, processes that will allow for the fabrication of parts and assemblies through shorter and less expensive cycles (Fig. 12.2).

Reduced cost has always been a direction setter, but in the multi-industry processing of various metallic and nonmetallic mold materials for the forming of composite parts, one can easily predict what the mold materials must do and what they must be (Figs. 12.3 and 12.4).

It is rare that new mold and tool materials are developed for the formation of composite parts and assemblies. What usually occurs is that existing materials are reformulated and altered to perform new process functions in the part-making process (Fig. 12.5).

Thus, these could not be considered materials and process innovations, but merely improvements to allow existing mold materials to be used differently and at a reduced cost. A good example of this is the formulation of mold materials to produce room-temperature-cured prepreg molds. These materials can be processed using atmospheric pressure and vacuum bagging only. Conventional mold and tool laminate prepregs, on the other hand, require higher processing temperatures and pressures (Fig. 12.6).

Still, these mold and tool materials can only be used to temperatures of around 400°F, since the formulations are based upon organic nonmetallic thermoset compounds (Figs. 12.7 and 12.8).

One processing advantage within the limited-use temperature range of up to 400°F is the fact that the laminate mold materials, when reinforced by various types of graphite fabric, can subsequently be machined and finished to exacting dimensions after molding. This additional machining step allows for the production of highly accurate graphite-reinforced laminate copies, if the parts and assemblies are fabricated using graphite-reinforced laminate materials.

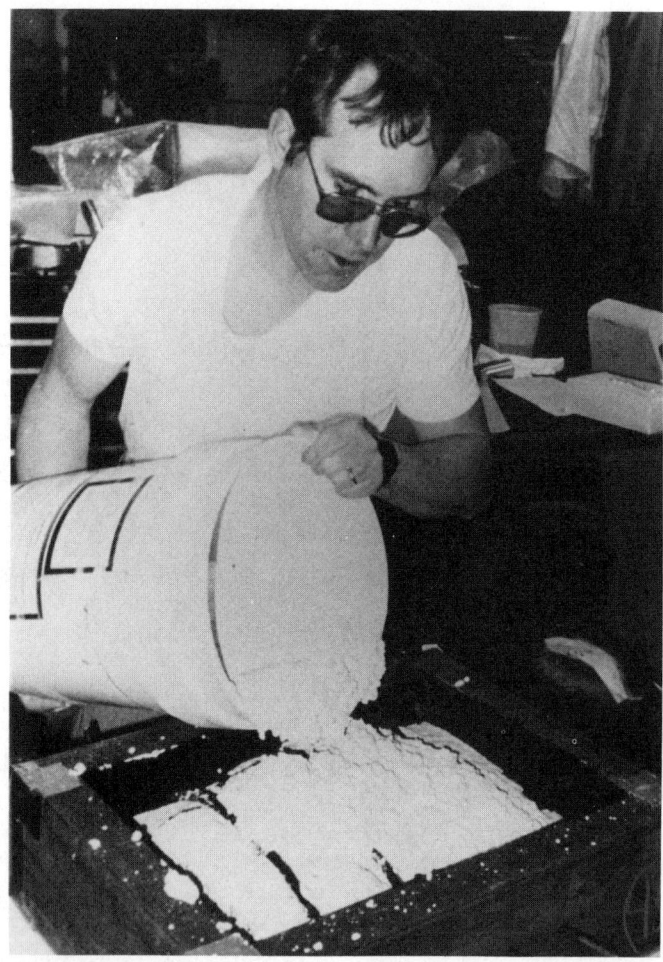

Fig. 12.1. Pouring an Integrally Heated Ceramic High Temperature Mold for the Forming of Structural RIM Parts. *(Courtesy of A.C.E., Inc.)*

THE FUTURE OF COMPOSITE MOLD MAKING 401

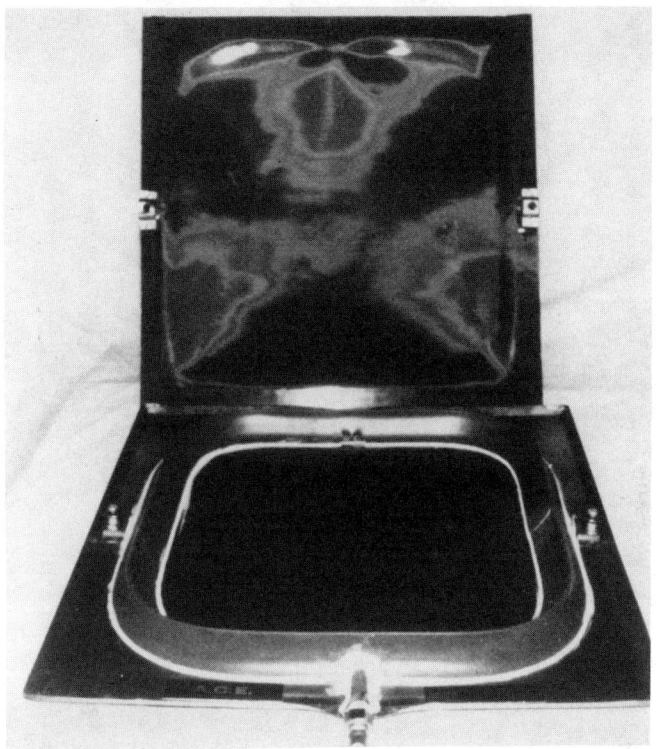

Fig. 12.2. High-Temperature, Low-Profile, Reusable Vacuum Bag, High-Production Laminating System. *(Courtesy of A.C.E., Inc.)*

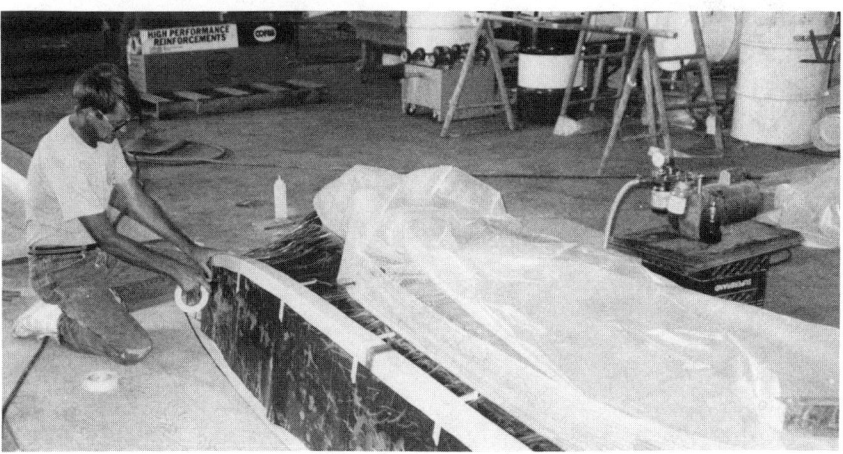

Fig. 12.3. Application of Seams to High-Production, High-Temperature, Reusable Vacuum Bag and Caul Pad System Used to Form Composite Co-cured Boat Hulls. *(Courtesy of Connell Engineering and A.C.E., Inc.)*

Fig. 12.4. N/C-Controlled 5-Axis Laser Cutter for Trimming and Cutting Kevlar and Other Composite Laminate Materials. *(Courtesy of Russell Plastics)*

Fig. 12.5. New Elastomers and Reusable Vacuum Bag Systems to Mold High-Temperature Advanced Thermoplastic Laminate Stacks in an Autoclave. *(Courtesy of A.C.E., Inc.)*

THE FUTURE OF COMPOSITE MOLD MAKING 403

Fig. 12.6. Fabrication of a Large Reusable Vacuum Bag to Form Composite Co-cured Boat Hulls. *(Courtesy of Connell Engineering and A.C.E., Inc.)*

What must be considered is the quick processing of new materials being developed, redeveloped, and to be developed. These will be molded or formed at temperatures up to 1000°, 2000° or even 3000°F. These part-forming materials, some of which do not even exist at this point, must be processed to the finished contour. Advances in the technology for mold and tool materials, instead of structural part materials, will finally allow this demand to be met (Fig. 12.9).

404 ADVANCED COMPOSITE MOLD MAKING

Fig. 12.7. Integrally Stiffened Laminate Tool Using Laminated Foam Core Stiffeners Cured to the Rear of the Laminate Shell. Aluminum Frame Support is Used as a Lay-up Tooling Aid. *(Courtesy of Grumman Aircraft Systems)*

Fig. 12.8. Integrally Stiffened Laminate Tool Inside of Aluminum Frame Support. *(Courtesy of Grumman Aircraft Systems)*

THE FUTURE OF COMPOSITE MOLD MAKING   405

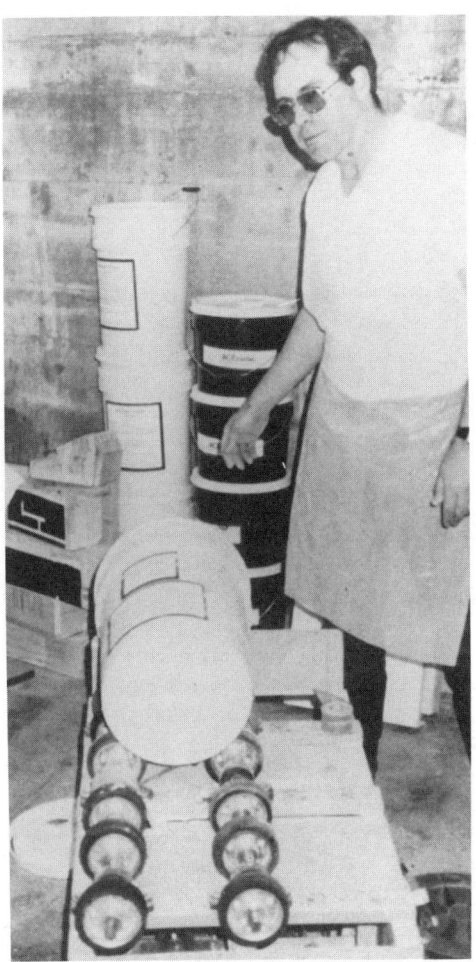

Fig. 12.9. Mixing Room-Temperature Castable Ceramic High-Temperature Mold Material on an Inexpensive Mixer. *(Courtesy of A.C.E., Inc.)*

As far as the present thermoset laminate lay-up of parts and/or assemblies is concerned, nonmetallic reinforced laminates will, in the future, be produced in large quantities; using less expensive molding techniques. The parts will be made in less expensive molds using inexpensive equipment and apparatus. Mold temperatures, pressures, molding times, etc., will be altered by being increased or reduced using more reactive polymer systems. The higher reactivity will allow for quicker setting and curing of part materials at new temperatures and pressures (Figs. 12.10 and 12.11).

The experimental composite front crossmember shown in Fig. 12.10 performs the same functions and is essentially the same size as the steel part that it

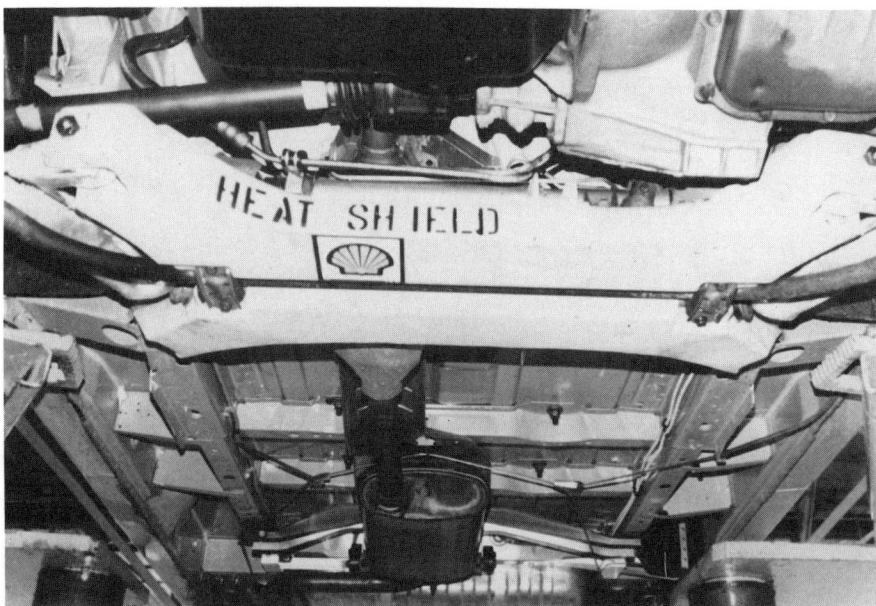

Fig. 12.10. RTM Composite Auto/Van Part Formed in a Mass-Cast and Laminate Tool. This Experimental Part was Designed and Manufactured by the Shell Development Co. of Houston, TX According to Chrysler Engineering Department Specifications, and 6 Prototype Parts were Manufactured by Ardyne, Inc., a Shell Chemical Subsidiary. *(Courtesy of Shell Chemical Co.)*

Fig. 12.11. High-Temperature Ceramic Mold Forms for Forming Very Accurate RTM and Composite Parts. *(Courtesy of Aero Patterns and A.C.E., Inc.)*

replaces. The crossmember supplies the reaction to all load input forces from the attached components — primarily consisting of high forces from wheel loads.

The crossmember was fabricated by resin injection into a dry preform (RTM) from epoxy resin (due to the close proximity of the catalytic converter) and a variety of woven and stitched E-glass fabrics. The part has a urethane closed-cell foam core, and contains several steel attachment plates and bearing surfaces. The installed crossmember weighs approximately 10% less than the steel part that it replaces.

The pictured part was tested statically and dynamically on Chrysler's road simulator, in the testing laboratory, on public roads in hot and cold climates, and at the Chrysler proving grounds in Michigan and Arizona. After approximately two and one-half years of testing, the composite crossmembers had experienced relatively minor cosmetic damage and some cracking — none of which affected their structural integrity. Chrysler's overall evaluation is one of qualified success, the qualifications due to the cosmetic damage, some loosening of bolt torques (fairly well-controlled in the later tests), and because longer-term testing may reveal other problems.

Significant contributions to this program were made by the following people: Edward J. Lesniak, Jr. and David T. Langstone, Chrysler Motors; Robert D. Farris and Dr. King Him Lo, Shell Development Company; and John A. Neate and Robert J. Cavalier, Ardyne, Inc.

In the area of thermoplastic composite part molding and processing, only the higher-temperature nonmetallic materials will be capable of being used to form and shape required contours. The reason for this is the fact that advanced thermoplastic composite materials are and will be processed at temperatures in excess of 600°F. The time necessary for forming these composite materials will be decreased, and the temperature of use of the finished article will be increased, through the introduction of higher-temperature mold-making materials. It is obvious that high-temperature mold materials such as ceramics will allow these part materials to be processed easily.

Unique processing techniques involving RF (radio frequency) or microwave energy to heat, cure, or shape composite part materials will be considered in the future. Ceramic mold materials are invisible to these heating sources and these methods are less expensive than conventional convection methods. Mass-cast molds will be merely loaded with a reinforcement, the cavities filled using low-pressure injection, and the preclamped mold fed through the RF or microwave heating source. Thermoplastic composite materials will be processed using similar methods (Fig. 12.12).

Self-reinforcing polymers will eventually replace some of the reinforced materials for part making. It will be much easier to process these more thermally, chemically, and mechanically stable materials than present ones.

On the other hand, reinforcements (when used) will be preformed dry to various shapes and, in some instances, will be placed automatically and robotically into mold cavities. The placement will allow for three-dimensional stacking,

408 ADVANCED COMPOSITE MOLD MAKING

Fig. 12.12. Room-Temperature-Cast, High-Temperature-Use Ceramic Composite Mold Forms. *(Courtesy of Aero Patterns and A.C.E., Inc.)*

weaving, stitching, and formation in the x-, y-, and z-axes. The three-dimensional reinforcement technique provides properties in the cured composites that are unattainable by other methods. Isotropic to anistropic properties are possible through combinations of various types of fibers in all three axes.. These newer reinforcement types — fiberglass, kevlar, graphite, and others — are stronger than woven forms, since a woven material might have crimps where the rovings meet and touch each other.

The reinforcements, because of versatile manufacturing methods, are tailored in form for use in such processes as RTM, reinforced RIM, structural RIM, and many other molding methods. The formation of the preforms are, in some instances, almost fully automated, thereby assuring consistency and quality reproducibility from preform to preform. The multiple layouts of the fiber preforms can be preoriented to any angular direction over each other. Selectively placed material within the preform can provide thin or thick sections easily, throughout a part.

The cost of the finished composite part will be reduced by the formation of fiber preforms, since this process is highly automated and can even be computer-

THE FUTURE OF COMPOSITE MOLD MAKING   409

Fig. 12.13. Large Ceramic High-Temperature Mold Being Removed from a Very Accurate and Precise Pattern. *(Courtesy of Russell Plastics, Aero Patterns, and A.C.E., Inc.)*

Fig. 12.14. Large, Very Accurate, Cast Ceramic High-Temperature Mold with Ceramic Syntactic Light Weight Foam Back-Fill. *(Courtesy of Russell Plastics, Aero Patterns, and A.C.E., Inc.)*

controlled. Conventional prepregs cannot be formed or shaped into the configuration that a dry preform, impregnated in a mold, can be.

And finally, there will be no limitation on the size of the thermoset nonmetallic parts that will be formed in the future. Ten, fifty, even one hundred foot, preformed impregnated reinforcements of excellent quality will be produced through resin system injection, during cycles as short as one minute. These composite parts will be integrally stiffened, stronger, and much less expensive than their "hand-laid-up" counterparts (Fig. 12.13).

The days of the thermoset vacuum-bagged or autoclave-molded part will come to an end (Fig. 12.14).

Bibliography

The briefness of the following bibliography is due to the fact that most of the information contained in this text is based upon state-of-the-art and original materials and processes developed through recent and continuing experimentation, research, and development.

A.C.E., Inc., Plainview, NY, Technical Literature.
Advanced Composites Design Guide, Third Edition, Second Revision, Air Force Flight Dynamics Lab, Wright-Patterson Air Force Base, Dayton, OH, 1976.
Advanced Composites Design Guide, Aircraft Systems Division, Grumman Corp., Bethpage, NY, 1986.
Airtech International, Inc., Anaheim, CA, technical literature.
Boeing Plastics Tooling Manual, The Boeing Co., Seattle, WA, 1966.
Burnham Products, Inc., Wichita, KS, technical literature.
Composites Manufacturing Engineering Design Guide, Grumman Aerospace Corp., Bethpage, NY, 1983.
Control Technology, Inc., technical literature.
Crandall, S. H., N. C. Dahl, and T. J. Lardner, *An Introduction to the Mechanics of Solids,* New York: McGraw-Hill, 1978.
D Aircraft Products Co., technical literature.
Deutschman, W. J., W. J. Michels, and C. E. Wilson, *Machine Design Theory and Practice,* New York: Macmillan Publishing Co., Inc., 1975.
Doyle, L. E., L. A. Keyser, J. L. Leach, G. F. Schrader, and M. B. Singer, *Manufacturing Processes and Materials for Engineers,* Englewood Cliffs, NJ: Prentice-Hall, 1985.
Eshbach, O.W., and M. Souders, *Handbook of Engineering Fundamentals,* New York: John Wiley and Sons, 1975.
Frekote Corp., Boca Raton, FL, technical literature.
Fiberite Div., ICI, Orange, CA, technical literature.
Hawkeye Enterprises, Inc., Inglewood, CA, technical literature.
Hexcel Corp., Chatsworth, CA, technical literature.
Hoiman, J. P., *Heat Transfer,* New York: McGraw-Hill, 1981.
Keene Laminates Div., Keene Corp., Bear, DE, technical literature.
Kovach, G., *Encyclopedia of Polymer Science and Technology, Release Agents,* New York: John Wiley and Sons, 1970.

Lee, H. and K. Neville, *Handbook of Epoxy Resins,* New York: McGraw-Hill, 1967.
Lipton, A., *Autoclave Systems for Building,* Lipton Steels and Metals Products, Inc., 1985.
Lubin, G., ed., *Handbook of Composites,* New York: Van Nostrand Reinhold, 1982.
Lubin, G., ed., *Handbook of Fiberglass and Advanced Plastics Composites,* New York: Van Nostrand Reinhold, 1969.
Optical Metrology, Grumman Aerospace Corp., Bethpage, NY, 1983.
Parker, E. E., and J. R. Peffer, *Polymerization Processes,* New York: Wiley–Interscience, 1977.
Ren Plastics Div., Ciba Geigy Corp., East Lansing, MI, technical literature.
Roark, R. J., *Formulas for Stress and Strain,* New York: McGraw-Hill, 1975.
Systematic Approach to Plastics Material Selection and Design, Plastics Encyclopedia, 1984–1985.

Index

Index

Abrasion resistant fillers, 40
Accelerator, 96, 102
Accessories, mold, 330
Accidents, 376
Accounting, 37
Acoustical, holography, 397
Acrylate, elastomer, 317
Adhesives, 122, 127
    film, surface coat, 124
    flexible, 122
    rigid, 122
    thermosetting, 123
Adjustable,
    molds, tools, 154, 219
    structure, 216
    tools, nickel, 252
Advanced, contouring, 60
Advantages of plastic tooling, 28
Aging, 40
Aliphatic, amines, 100
All-plastic mass-cast molds, 172
Alpha gypsum cement, 75
Alphanumerics, 45
    data base, 47
Aluminum
    heatup rates, 25
    molds, 1, 61, 69, 231
    tool repair, 355
Amine, 96, 99, 100
    curing agents, 101
Anhydride curing agent, 102
Anisentropic
    laminate, 12
    material, 11
Anisotropic properties, 112
Application of bag sealant tape, 268
Applied load, 17
Approximate, life, 38
APT
    AC, advanced contouring, 60

automatically programmed tool, 48, 51, 59
Area storage, 369–372
Aromatic curing agents, 96
Autocollimator, 386
Autoclave, 357, 373–375, 378
    ASME code, 372
    fabrication of, 372
    heating, 25
    heatup rate, equation, 25
    modify, 372
    vented, 258
Automated
    design and drafting, 43, 58
    design and manufacturing, 40, 55
    pattern cutting, 345
    plotter, 42
Autoreflection, 385–386
Average life, 38
Axial loads, 10
Axially loaded member, equation, 12
Axis, two, five, 44

Back fill
    ceramic syntactic, 409
    syntactic, 117, 409
Backup structure, 38, 159, 199, 202
    attachment, 214, 287
    boards, 126, 283, 287
    configuration, 208
    floating, 216
    frames, PFP, 161
    metal-faced molds, 225
    modification, 348
    nickel tools, 251
    spacing, 205, 207
    templates, 204–205
    tubular, 215
Bag
    bladder, 273

414　INDEX

Bag *(continued)*
    bleed, vertical, 282
    boat, vacuum bag, 401, 403
    bridge elimination, 263
    cast, 305
    double seal, 257
    lift support, 292
    no-bleed, 282, 286
    pattern, 293
    reusable frame, 306
    sealant, 253
    sealing, 274
    seam, 294
    side tools, 154, 196, 314, 315, 321
        reinforcement of, 319
    sizing, 267
    skin, 262
    thermocouples, 306
    vacuum port, 292
    vacuum, reusable, 253, 254, 259, 290
Bagging
    ear, 265, 269
    envelope, 258, 259, 272, 273
    face, 289
    internal, 273, 275, 276
    peripheral, 204, 214, 282, 283, 286, 289
    prepreg, 277–278
    pressure, 259–260
    surface, 258–259, 262, 272
    tool face, 192
    uncured rubber, 300
    vented, 258
    wet lay-up, 278
Banking plies, 336
Bat, hemp, 80
Beams, 17, 21
Belt conveyor, 366
Bending moment, 18, 21
Bezier patches, 54
Biaxial fabric, 108
Bisphenol acetone, 93
Bladder
    bag molding, 259
    envelope, bag, 276
Blanket, heat, 25
Blaster, grit, 149
Blasting, vacuum assisted, 347
Bleeder
    edge, 261, 279, 281
    resin, 254, 260, 279

    surface, 281
    vertical, 261, 282
Blistered
    mold, 352
    repair, 353
    tool, 352
Blocks, gauge, 341
Board
    back-up, 283
    contour, 206
    header, 206
Boat hull, 382
    vacuum bag, 401
Bonding
    bushings, 133
    fixture, 180
        adjustable, 217
    life, 38
    model stock, 137
Boron, trifluoride, 102
Boss, vacuum, 239, 331
Braces, metal, 72
Brake, forming, 234
Breakaway mold parts, 70, 179, 335, 336
Breather
    edge, vacuum, 285
    vacuum, 254, 261, 284–285
Bridge bag, 262
Brush, wet lay-up, 197
Burden, labor, 34
Bushing
    blind, 344
    bonded, 344
    drill, 342–343
    encapsulation, 343
    installation, 282, 344
    knurled, 345
    laminating, 342, 344
    locating, 145
    potting, 343–345
    replacement, 72
    serrated, 72
Butt seams, 198
Butyl elastomer, 317

Cable conveyors, 365
CAD, 396
    computer-assisted design, 29, 40, 42, 49, 240
    work station, 44

CADAM
    computer-aided design and manufacturing, 43, 46–47, 49–50, 51, 55, 346
CAD–CAM
    computer-aided design and manufacturing, 29, 43, 45, 55
    future, 48
    system selection, 46
Calcination, 75
Calculation, cost, 36
Calendered fabric, 110
CAM, 396
    computer-assisted manufacturing, 29, 40, 43–45, 240
Carbon
    cloth, 109
    graphite, 167
    molds, 30, 167
Carcinogen, 1, 100, 376
Cart, mold, tool, 360, 362
Cast vacuum bags, 303, 305
Castable ceramic, 81, 164
Casters, mold, tool, 360
Casting, 92
    face, 175
    foam, 114
    mass, repair, 350
    surface, 117
Catastrophic failure, 12
Cathode ray tubes, 47
CATIA
    computer graphics aided three-dimensional interactive application, 46–47, 50–51, 55
Caul
    elastomeric faced, 328
    pads, 255, 314–315, 320–322
    plates, 255, 314–315, 327–328
    collapsing, 329
Caution, curing agent, 100
CCM, 391
Cement, contact, 123
Cementitious,
    compounds, 81, 166
    properties, 6
Ceramic
    compounds, 81
    heated, 164
    hollow microspheres, 223
    integrally heated mold, 400

    machinable, 165–166
    mass-cast mold, 383, 408
    metal-faced molds, 223
    microspheres, 116
    mixing machine, 405
    mold, 164, 201, 406
    base, 158–159
    repair, 356
    releases, 150
    syntactic foam, 409
    tools, 2, 30, 406
Checking
    engineering programs, 44
    machine models, 44
    pressure, 394, 396
Chemical
    analysis, 95
    foam, 119
Choosing mold and tool materials, 41
Chopped strand mat, 106
Cleaning molds and tools, 148, 347
Closure, reusable vacuum bag, 291, 306
Cloth, fibrous, 107
Co-curing, 318, 320
Coefficient of thermal expansion, 26–27, 37, 41, 68, 215, 232
Cold flow, 85
Collapsing caul pad, plate, 329
Combination fixture, 346
Compact, 257, 283
    laminates, 374
Composite
    forming hollow, 179, 199
    low-cost, light-weight, 410
    member, 12
    moldmaking future, 399
    thermoplastic, 73, 256, 407
Compound
    cementitious, 166
    complex curved surface, 47
    fairing, 139
Compression molds, 119
Computer, 44
    aided design and manufacturing (CADAM) 43, 51
    assisted design (CAD), 29, 43, 49
    assisted manufacturing (CAM), 20, 40, 45
    controlled preform manufacturing, 408
    hardware, 46
    network, 46

416 INDEX

Computer *(continued)*
    plating mandrel, 70
    software, 46
Condensation, polyimides, 98
Conduction, 21–22, 24
    equation, 22, 25
    heat transfer, 22
Cones, 45
Connections, bolted, riveted, 14
Consolidate, 257, 283
Contact
    cement, 123
    molding, 199
Continuous path N/C machining, 58
Continuous strand mat, 106
Console, 56
Contour
    board, 206
    change, 71
    headers, 206
    verification, 354
Convection, 21, 24
    fluid motion, 24
Conveyor belt, 366
Conveyors, 365
Coordinate measuring machine, 391
Core, 335
    sandwich, 117
    splicing, 117
Cork dam, 284
Correct tool design, 16
Cost
    accounting, 37
    class of labor, 33
    machinery, 37
    material, 370
    mold and tool, 2, 29, 31
    prepreg molds, 181
    reduction, 187
    tools, 37
Covers, molds and tools, 40
CPU, 47
Cracking
    gel coat, 198
    repair, 350
Creep, rupture data, 17
Cross linking, 98
    anhydride, 102
Critical loading, 17
CTE, 27, 37, 41, 68, 181, 189, 191, 215, 232, 343, 349, 354–355, 381, 383, 393,
Cure
    monitoring, 103
    test, 393
Curing
    agents, 98–99, 101
    accelerator, 96
    aliphatic, 96
    aromatic, 96
    caution, 100
    epoxy resins, 94, 100
    microwave, 407
    monitor methods, 95, 393
    R.F., 407
Curls, machining, 61
Cursor, 42–43
Cusp, 61
    removal of, 177
Cutters, 61–62
    feeds, 44, 49
    laser, 340
    path, 44, 49
        verification, 136
    speeds, 44, 49
    water jet, 340
Cutting
    aid, 40
    patterns automatically, 131, 345
    prepreg, 189
Cycle, life, 38

Dam, cork, 284
Dead, piling wood, 91
Debulking, 193, 257
    prepreg, 284
Decisions to design, 41
Defects
    mold, 356
    tool, 356
Deflection
    of beams, 17
    maximum, 21
Degradation
    thermo-oxidative, 98
Delaminated mold, tool, 352–353
Densify, laminate, 257
Design
    automatic, 40
    considerations, 21
    drafting, 43
    manually, 129

INDEX

parameters, 2
preparation, 41
productivity, 43
responsibilities, 41
stress, usable, 11
Designer, mold and tool, 130
Destructive testing, 382, 397–398
Dielectric
   measuring, 95
   monitoring, 390
Diethylenetriamine, 99
Digitizer, 47
   cursor, 43
Digitizing, 29, 42, 45, 136
Diluents, 103
   elastomers, 305
Dimensional
   stability, 1
   tolerance, 163
Disconnect, quick, fitting, 334
Display terminal, 47
Disposable, vacuum bags, 253
Disposal, 376, 380
Distortion of patterns, 81
Dollies, transportation, 358
Double vacuum bag seal, 257, 270
Doubler, vacuum bag seam, 294
Drafting, automated, 58
Drawing
   frames, 47
   functions, 47
   stretch mold, 236
Drill fixture, 342
   jig, 342
Drophammer, mass-cast tool, 179
D.T., destructive testing, 382, 390, 397–398
Durometer, 391

Ear
   formation of, 268–270
   vacuum bag, 264–266
Edge
   breather, 285
   rolled mold, 221
Educational aid, 4
Eggcrate
   attachment, 214, 283, 287
   backup boards, 126, 206, 287
   backup structure, 20, 38, 109, 199
   elimination, 128
   fasteners, 212

solid board, 127
templates, 205
tie-in, 207
Elastic limit, 10, 12
Elastic region, 10
Elasticity, theory of, 5
Elastomer, 83
   caul pad, 316
   reversion resistant materials, 3, 255
   vacuum bag, 304
Elastomeric molding, 325–326
Electroformed
   formation of molds, 242
   tools, 242
   mold vacuum port, 331
   nickel tools, 2, 57, 65, 70, 159
   plating mandrels, 30, 45, 70
Electromagnetic waves, 24
Electroplated
   formation molds and tools, 242
   plating mandrels, 30, 45, 70
Electroplating process, 247
Emissivity, 24
Encapsulation bushing, 343
Energy
   absorbtion equation, 24
   radiation equation, 24
   region, low, high, 22
Engineering
   changes, 65
   considerations, 29
   cost, 37
   guidelines, 30, 31
   human, 384
   industrial, 357
   mechanics, 7
   parameters, 29–30
Envelope
   inside vacuum bag, 275–276
   vacuum bag, 258, 273
Environmental exposure, 369
EPA, 377
Epoxy
   adhesives, 124
   applications, 99
   characteristics, 95
   diluents, 103
   epichlorohydrin, 93
   esters, 102
   flexible, 85
   hydrophobic hardener cured, 160
   resins, 92, 103

418 INDEX

Epoxy *(continued)*
  resin, curing, 94
    molecular weight, 93
  stripper, 143
  surface film, 110
Equipment
  honeycomb, 203
  mixing, 165
  support, 366
Error, detecting logic, 47
Estimating
  allowances, 35
  costs, 31
  factors, 32
  forms, 33
  maintenance, 35
  process, 33-34, 36
Estimator, 31
Ethylene, propylene, 83
Exhaust, bench, 379
Exotherm, 199
Expansion
  rubber, 326
  thermal, foam molding, 327
Expected life, mold, tool, 37
Exposure, environmental, 369
Extending mold and tool life, 38
Extrude, intensifier, 324
Eyes
  mold and tool, 178
  tube, 203
  wood, 203

Fabric
  biaxial, 108
  calendered, 110
  graphite, 181
  knitted, 108
  mold, styles, 277
  patterns, 196
  preimpregnated, 103
  release, 147, 199, 255, 302
  satin, 108
  triaxial, 108
  unidirectional, 108
  woven, 107
Fabrication
  feasibility, 24
  master pattern, 134
  metallic molds, 231

Face
  bagging, 192, 289
  casting, 172, 175-176
Facility requirements, 3, 357
Factor, safety, 9, 13
Failure modes, bolted and riveted
    connections, 14
Fairing
  compound, 66, 139
  process, 123
Fasteners, eggcrate, 212
Fiber
  ply orientation ($0°$, $90°$, etc.), 12
  preformed, 408
Fiberglass epoxy board, 104, 112
Fibrous
  cloth, 107
  fillers, 104
  reinforcement, 105
Filament wound molds, 182, 201
File management, 47
Files, N/C, 47
Fillers, 104
  spherical, 114-115
  surface coats, 40
Film
  epoxy, 110
  gel coat, 124, 191, 193
  locating templates, 337
  release, 150, 285, 321
  surface coat, 124
Final estimate, 36
Finishes, sealer, 143
Finishing
  cut, 61
  molds, tools, 243
Fitting
  mold, 330
  quick disconnect, 334
Five axis machining, 44
Fixture
  adjustable bonding, 217
  checking, 127
  combination, 346
  drill, 342
Flat bed plotting, 47
Flat pattern, mold, tool, parts, 42
Flexible, caul plates, 83
Flexcore*, 209
Flexural, strength, hot, 183

*Flexcore, Reg. TM Hexcel Corp.

Flow
  material, 357
  work, 357
Fluid motion, 24
Flush, mold edges, 38–39
Foam, 104, 117
  chemical, 119
  core stiffeners, 219
  isocyanurate, 121
  master models, 139
  preform, insert, 327
  polyimide, 121
  polyurethane, 121
  syntactic casting, 114
  thermal expansion molding, 327
  trimer, 121
Fold, vacuum bag, 269
Force, negative, tensile and compressive, 6, 7
Forklift, mold and tool, 360
Formed metal mold, 233
Forming
  brake, 234
  patterns, 298
  reusable bag seams, 294
  vacuum bag ear, 268
Frame
  metal, 233
  plaster, 137
  repair, 354–355
  reusable vacuum bag, 303, 306–307
Future, composite mold making, 399

Gauge, blocks, 391
Gel coat
  cracking, 198
  fillers, 40
  film, 124, 191, 193
  liquid, 189
General accounting, 37
Glass cloth, 104
Graphic input tool, 42
Graphite
  block, 39, 111
  cloth, 104
  epoxy board, 112
    mold, 1
  fabric, 109, 110
  heatup rate, 25
  solid, 167

Graphitization, 112
Graphitized carbon, 167
Greenie laminate, 38–39, 394
Grit, blasting, 125
Guided-wire vehicle, 366
Gypsum
  cement, 74
  pattern storage, 80
  pouring, 77
  screed, splash, 74
  storage, mix, 76

Hammer, mass-cast, tool, 179
Hand blending, 61
Handling practices, 38
  molds, tools, 358
Hands-on mold design, 5
Hard
  maple, 89
  plaster, 75
  wood, 88
Hardener, hydrophobic, 71
Harmful constituents, 1, 377
Hat, sections, 181
Hazardous
  ingredients, 1, 377
  materials, 376
Headers, 42, 56
Health, 376
Heat
  blanket, 25
  debulking, 193
  emission equation, 24
  microwave, 165
  resistant, resins, 97
  RF, 25, 202
  sinks, 350
  transfer, 5, 21
    modes, 22
  up rate, 30, 395
    equation, 25
    table, 30
  time, 68
Hemp, 80
Hetrocyclic polymers, 97
Hexagonal core, 211
  assembly attachment, 211
  corners, 211–212
High energy region, 22
High production molds, tools, 67

420 INDEX

High temperature
  molds, 407
  relesses, 147
Hooke's law equation, 11
Holes in structural members, 16
Holography, 396-397
Hollow
  composites, 180, 199
  microspheres, 116
Hoist, reusable bar brackets, 313
Homogeneous material, 10
Honeycomb
  aluminum-faced, 127
  back up structure, 203
  cutting, 127
  flexible, 209
  panels, 125
  plastic-faced, 126
Hot melt
  adhesives, 123
  prepreg, 110
Hourly rates, 33
Human engineering, 389
Hydrophobic hardener, 71
Hydrotel machining, 239
Hygiene, 376

IBM, system 360, 47, 59
  370, 59
  3250, 50
Idaho pine, 88
Imbedment bushing, 72
Impreg, 85, 153, 194
Impregnated wood, 85
  fibers, 97
Impregnation, machine, 194
Impact strength, 104
Improper storage, 40
Industrial engineering, 3, 357
Inelastic action, 12
Inorganic acids, 101
Inspection, 3, 103, 381-383
  area, 376
  mechanical, 390
  process, 389
Insert
  foam, 327
  replacement, 351
Installation, thermocouple wires, 271
Integral stiffeners, 109, 128, 159, 180,
  202, 319

Integrally
  heated ceramic mold, 400
  stiffened S-RTM and S-RIM mold,
    404, 410
Intensifier, 196, 255, 320, 322
  radius, 320, 324
Interactive
  computer programs, 44
  graphics, 48-49
  numerical control, 44
  phases, 54
Intermediate molds, 153
Internal release, 147
Isentropic material, 11
Isotropic materials, 115

Jig
  drill, 342
  routing, 339, 342
  transit, 387-388
Joint failure, 16
Juxtaposed models, 47

Keller pattern, 239
Keyboard, 44, 45
Kinematics, 50-51
Kneading foam, 161
Knife coated fabric, 110
Knitted fabrics, 108
Knockouts, 179

Labor, 33
  class, cost, incentive, 33-34
Laminate, 73
  board, 127
  compaction, 257, 283
  consolidation, 257, 283
  debulk, 257
  densify, 257
  drill bushing, 343
  eggcrate, 127
  face replacement, 350
  greenie, 39
  integrally, stiffened, 404
  low
    energy region, 22
    cost mass-casting, 142
  mold
    repair, 348
    room temperature prepreg, 162
    shield, 39

INDEX 421

phenolic, 114
radius, 198, 279
splash, 38, 394
spring back, 217
thickness guide, 197
tube structure, 216
vacuum bagged, 198
wet lay-up molds, 182
Lap joint strength, 5, 14
Large prepreg molds, 188
Laser
cutter, 340
N/C controlled cutter, 402
Latex, 83
elastomer, 305
Lay-up, 92
radius, 279
tools, 198
Leak rates, 396
Lewis
acid, 99
bases, 101
Life
cycle, 38
mold, tool, 37–38, 67
Lift
attachments, 370
mold and tool, 360–361
vacuum bag, 292
Lignin, 85
Light pen, 44, 47
Limitations, design, 40
Lines, trim, 144
Loading, cyclic, impact, 9
Loads, tolerance, 6
Locating
bushings, 145
plies, 335
template, 336–337, 339
tool, 335
Lofting, 42, 56
templates, 130
Low
cost light-weight composites, 410
production rate molds and tools, 64, 66
Loss factor, 390
L.T. (locating template), 336–337, 339
Lumber tallying, 91

Machinable, foam, wax, 44
Machine, impregnation, 194

Machined
checking models, 44
mandrels, 70
metal molds, 240
repair, 354
tools, 240
Machining
graphite block, 112
mandrels, 70
masters, 114
N/C, 240
nickel tools, 70
speed, 61
Macrospheres, ceramic, 116
Mahogany, 88
Maintenance, molds and tools, 346
Man aboard storage vehicle, 372
Mandrel
drill jig, 343
filament winding, 202
flexible, 181
mass-cast, 246, 249
plating, 45, 70, 112, 114, 136, 245, 247
removable, 200
salt, 199
Manual design, 40
masters, 129
Manual nesting, 42
Manufacturing methods, 29
Marine syntactic foam, 119
Mass, 24–25
cast
alteration, 173
ceramic mold, 30, 65, 227, 408–409
S-RTM mold, 406, 408
face, 175
flexible, 178
mandrel, 246, 249
masters, 136–137, 141, 170
metal-faced, 227
models, 155, 165
molds and tools, 65, 142, 172, 174–180, 348
reinforced mold, 173
repair, 350
spray metal master, 224
surface, 177
centroid, 27
Master, 1–3, 111
digitizing, 136
fabrication sequence, 134–135

Master *(continued)*
    flexible, 178
    frame, plaster, 131
    inexpensive, 64
    mass-cast, 137, 141, 169, 175
    model, 41, 143
    mold, 153–154, 160
        economics, 155
        foam backed, 161
        wet lay-up, 158
    pattern, 130, 155
        bushing, 133
        hole locations, 133
        scribe lines, 133
    preparation, 129
    storage, 359
    white, 396
Mat, 106
    chopped, 106
    continuous strand, 106
    surfacing, 106
    veil, 191, 193
Matching curing agents and resins, 100
Material
    anisentropic, 11
    brittle, 9
    hazardous, 376
    isentropic, 11
    mass, 25
    master, molds, 29
    pricing, 35
    routing, 359
    storage, 357
    work flow, 357
Mathematical model, 47
MDA, 100, 377
MDI, 121
Mechanical inspection, 390
Mechanics, 7
Member, mold, tool, 8
Metal
    aluminum molds, 1, 61, 69, 231
    base frame, 233
    braces, 72
    composite, contrasts, 12
    faced
        back-up, 225
        integrally heated, 227, 229
        molds, 222–224
        spray application, 225

mold
    advantages/disadvantages, 232
    deformation, 14
    fabrication, 231
    finishing, 243
    repair, 354
    powder, 104
    rolled, mold, 238
    sheet locating templet, 337
    sheet molds, 237
Microballoons, 115
Microspheres, 115–116
Milling machine, 45
Mixing
    ceramic molds, 165, 405
    plaster, gypsum, 76
Model, 1, 2, 3, 45, 61, 111, 127, 137
    changing, 65
    materials, 61
    preparation, 129
    stock, 112, 135
    storage, 359
Modify
    mold substructure, 348
    mold, tool, 71
Modulus
    elasticity, 10
    highest, 12
    longitudinal, traverse, 12
Molds
    accessories, 330
    accuracy, 37
    adjustable, 216, 219
        nickel, 252
    aluminum, 67, 231
    automated, 3
    bleed air, 281
    breakaway, 335–336
    bumper, 39
    cart, 362
    ceramic, 164, 201, 405–407, 408–409
    classification, 369
    cleaning, 148, 347–348
    compression, 119
    cost, 30
    delamination, 352
    design, 2, 5
    edges, 38–39
    elastomeric, 322, 325
    electroformed, 242

INDEX   423

engineering, 3, 28
fabrication, 3
fabric styles, 277
face, cast, 175
    repair, 348
filament wound, 182, 201–202
fittings, 330
graphite epoxy, 181
handling, 358, 377
hazardous material, 376
high temperature, 164, 406–407
inexpensive, 66
inspection, 388, 398
integrally stiffened, 404
intensifier, 83
laminate, N/C machined, 156
laminated, 180
life, 37–38, 41, 67
lifting, 360–361
making, future, 3, 399
mass-cast, 3, 65, 172–174
master, 45, 178
materials, 1–2, 29, 41, 69, 73
metal, advantages/disadvantages, 232
    faced, 222–223
moving, 358, 370
nickel, 67, 69
pallet, 362
picture frame, 194
porous, 352
positioning platform, 367
post-cure, 192, 290
prepreg, 182, 185–187
proofing, 394
prototype, 62, 181
quality, 31
refurbishment, 40, 70
releases, 143
repair, 40, 70
resin trap, 334
rolled edge, 222
roll formed, 233, 238
routing, 359
spring back, 217
S-RIM, S-RTM, 170, 223
steel, 231
storage, 40, 357, 363–364, 368, 371
stretch draw, 236
support tools, 330
testing, 398

treatment, 143
thermoform, 119
undercuts, 179
vacuum form, 119
weight, 45
wet lay-up laminate, 182–183, 195
workflow, 357
Molding
    autoclave, 259
    contact, 259
    elastomeric, 326
    pressure bag, 259
Molecular action, 22
Moisture, molds and tools, 71
Moment, 7
Monitor, 45
Monolithic graphite tools, 2, 30, 111–112,
    167–168
Moving molds and tools, 358
Multiple tooling, 33

N/C. *See* Numerical control
NDI. *See* Non-destructive inspection
NDT. *See* Non-destructive test
Near-shape machined laminate, 155
Necking, 10
Nest, 42
    pattern, 42, 347
Newton's law of cooling, equation, 22
Nickel tool, 65, 67, 70, 242, 250
    adjustable, 252
    back-up tool structure, 251
    electroplating, 30, 159
    handling, 62
    models, 60, 62, 75
    proofing, 139
    repair, 355
NIOSH, 100, 377
No-bleed vacuum bag, 281, 286
Nondestructive
    inspection, 382, 396
    testing, 382, 390, 396–397
Nonmetallic
    molds, tools, 29, 153, 164
    tube frame, 215
Normal stress, 13
Northern white pine, 87
Novalac resins, 93
Numerical control (N/C)
    continuous path, 58

424   INDEX

Numerical control (N/C) *(continued)*
  equipment, 50, 57
  machined, 29, 44–45, 48, 55–57, 75
    graphite, 111
    handling tools, 62
    laminate, 157
    mandrel, 70
    masters, 112, 133, 135, 156, 177
    metal, 240
    rolled metal, 238
    thermoform tools, 58
  models, 60, 62, 75
  point to point, 56–57
  processors, 60
  programming, 51
  surface machining, 50, 53, 60

Ocean
  submersibles, 119
  vehicles, 119
OML (outside mold line), 20
One step master pattern, 134
Operator reach, 360
Optical
  alignment, 384
  comparator, 384
  inspection, robotic, 389
  measurement, 384
Organic
  acids, 101
  microspheres, 116
Orientation of plies, 199
OSHA, 100, 377
Outdoor mold and tool, 40
Oven, 357

Pad
  caul, 255, 314–316, 320–322
    collapsing, 329
Paint-on
  elastomer bags, 305
  surface coat, 191
  vacuum bag, 292, 303
Pallet, mold and tool, 362
Panels, honeycomb, 125
Paper
  based phenolic, 114
  pattern, 298
Part
  design, 29

  material and tolerances, 41
  removal, 335
Parting
  agent spray metal, 224
  film, 66
Patching
  compound, 97
  molds and tools, 354
  process, 123
Pattern, 3, 111, 130
  automated cutting, 131, 345
  foam, 298
  keller, 239
  lumber, 85
  N/C machined, 133
  paper/plastic, 196
  preparation of masters, 129
    metalfaced masters, 224
  prepreg, 196
  reusable vacuum bag, 298–300
  storage, 80
Pay back, automation, 42
Peel ply, 125, 147, 199, 254, 283–285
  location of, 261
Period of elasticity, plaster, 79
Peripheral, bagging, 71, 204, 214, 284, 286–288
PFP. *See* Plastic faced plaster
Physical
  properties, 2
  limitations of worker, 359
Phenol formaldehyde, 96
Phenolic, 96
  bonding agents, 97
  laminated, 114
Picture
  framing, 194
  prepregs, 283–284
Pin, replacement, 351
Plant layout cost, 37
Plaster
  equipment, 75
  gypsum compounds, 74
  master, 131–132
  mixing, 75–76
  reinforcing, 80
  screeding, 79
  splash, 133

INDEX

Plastic
    deformation, 10, 12, 14
    dies, 28
    faced plaster, 133, 153–154, 185, 245
        frames, 160–161
    fixtures, 28
    machinable, 130
    model stock, 137
    molds tools, 28
Plate
    caul, 255, 314–315, 327
    collapsing caul, 329
    elastomeric, 328
    tear-through, 15
Plated,
    nickel tools, 70, 242
        backup structure, 251
Plating
    forward/reverse, 250
    mandrel, 45, 70, 112, 245–246
Platform working position, 366
Pleat, vacuum bag, 263–264, 313
Plotter, 42, 44
    hidden lines, elements, 53
Ply
    banking, 336
    build-up, 278
    drop off, 335
    location, 199, 335
    template, 339, 341
    rigid, 340
    sequence, 279
PMDI, 121
Point to point, 56
Polyamide hardener, 94
Polycarbonate ply locating template, 336
Polyester
    film, 336
    locating template, 336–337, 339
    resins, 97
Polyether, 101
Polyhedral solid geometry, 50, 53
Polyimides, 98, 101, 121
Polysulfide, 85, 179, 305
Polyurethane, 179
    foam, 121
Polyvinyl alcohol film, 256
Porosity test, 252
Port, vacuum, 239, 292, 332, 334
    through bag, 333

Post-cure, 192, 290, 356
Post-processor, 59
Pour
    gypsum cement, 77
    in place, mold back fill, 161
    mold and tool, 352
    vacuum bag, 305
Precast mandrel, 245
    model, 169
        stock, 57, 136–137
Precured, postcured vacuum bag, 293
Predicting cost, 36
Preimpregnated fabric, 74
Prepreg, 108–109, 186, 188–189
    debulk, 284
    graphite/epoxy, 181
    patterns, 45, 57, 196
    properties, 183, 187
    room temp./high temp., 162, 182, 227, 256
Preparation, masters, 129, 224
Pressure
    bag molding, 259–260
    checking, 394, 396
    pads, 83, 255
    vessel, 373
Prewetting, 196
Pricing materials, 35–36
Primer, silicone bag, 299
Printer, 45
Process
    brake form, 234
    inspection, 389
    roll, 234
Production flow, 370
Productivity design, 43
Program proofing, 44
Proof, mold and tool, 394
    program, 44, 142
Proportional limit, 9
Prototype, mold and tool, 29, 62, 64, 66–67
Pseudoplaner facets, 53
Punch tapes, 49, 59
Purchasing
    CAD/CAM systems, 46
    wood, 90

Quality
    assurance, 383, 392

426 INDEX

Quality *(continued)*
  control, 3, 31, 103, 381–382, 392
  mold and tool, 2
  quick disconnect, 334
  quoting cost, 31
  reusable vacuum bag, 292

Rabbit ear, vacuum bag, 264–265
Radiation, 21, 24
  absorption, 24
Radiography, 384
Rail, tool, 325
Radius
  bag bridge, 262
  form laminate, 280
  intensifier, 324
  lamination, 198
  mold laminate, 279
Reaction injection molding (RIM), 121
Reactive diluent, 103
Recessed mold edges, 38–39
Reduce
  shrinkage, 104
  scale image, 46
Refurbish, 40, 70
Release, 143, 145, 150
  fabric, 57, 199, 255, 302
  film, 64, 66, 147, 150, 285, 321
  high temperature, 149
  internal, 147
  methods, 151
  permanent, 149
  wedges, 150
Reinforced
  mass-cast mold, 174
  plastic lay-up, 92
  reusable vacuum bag, 401
Reinforcement
  fibrous, 104
  gypsum cement, 80
  resin ratio, 107
Reliable
  molds and tools, 22
  removable mandrel, 200
Removal part, 335
Replacing, 40, 70–71
  inserts, 351
  laminate face, 350
  pins, 351
Repair, 71, 40
  backup structure, 348

blistered tool, 352–353
cracks, 350
faces, 350
inserts, 351
material choice, 349
molds and tools, 346, 352–353
pins, 351
porous mold, 352
small tool areas, 355–356
void, 354
Requirements facility, 357
Reservoir, resin, 334
Resin
  bleeder, 254, 260
  epoxy, 92, 98
  heat resistant, 97
  phenolic, 96
  processing, 96
  system, room temp./hi temp., 64
  transfer mold, 406
  trap, 334–335
Resistant, reversion, 290
Resurfacing
  metal-faced tool, 354
  tool, 348, 351
Responsibilities, of designer, 41
Reusable
  seams, 295
  vacuum bag, 38, 83, 253, 290, 401
Reversion
  resistant elastomer, 3, 290, 402
  vacuum bag, 291, 374
RF curing, 407
  heating, 202
Rigid
  foam, 121
  L.T. (Locating Template), 336–337
Robotic optical inspection, 389
RIM, 121, 223
Rolled
  formed
    metal fasteners, 155
    mold and tool, 233
  conveyors, 365
  metal, N/C machined, 238
  mold and tool edge, 39
Room temperature
  prepreg, 182, 256
  resin system, 64
Rough N/C machining, 57

INDEX 427

Routing
    jig, 342
    tool, 335, 339
        vacuum-assisted, 341
    water jet, 340
Roving, 105
    woven, 105
Rubber
    vacuum bag, 290
        silicone, 290
R-RTM, 229
    ceramic mold, 406
RTV (Room Temperature Vulcanizing),
        silicone, 295

Safety, 357, 373, 376–377, 379–380
    chemical, 100, 142, 148
    equipment, 378
    factors, 6, 12, 17
        equation, 9
Salt mandrel, 199
Sand, 104
Sandwich panel backup structure, 206
Satin fabric, 108
Scallops, machining, 177
Scanning device, 47
Scoring cut, 39
Screeding plaster, 79
Scribe lines, 27, 144
Seal, 143–144, 168
    double vacuum bag, 270, 272, 285
    plaster, gypsum cement, 78
    reusable vacuum bag, 306–308, 310–312
Sealant
    tape, application, 268–269
    vacuum, 263, 291, 325
Seam
    butt, 198
    reusable vacuum bag, 293–297
Selection, CAD/CAM system, 46
Selvage edge, 198
Semiautomatic vacuum bag, 290
Separator
    coating, 77–78
    gypsum plaster cement, 78
Shear, 13–15
Sheet metal mold, 237
    wax, 64, 144–145
Shock, thermal, 369
Shop equipment, plaster, 75
Shrinkage, 104

monitoring, 95
Silicone, 84, 179
    uncured rubber bags, 302
Size, bag ear calculation, 266
Skin, vacuum bag, 258, 262
Slab stock, 137
Slip sheet, 39
SMC, 115
Software, 44
Softwood, grades, styles, 86
Solid
    eggcrate laminate, 204
    fluid boundary, 24
    frame shape, 45
    graphite block, 111
Solvent, 148
    swell, 95
Sonics, 390
Spacing back-up boards, 207
Spanish cedar, 88
Specific heat, 24–26
Specimen tensile test, 9
Spheres, 45
Spherical filler, 115
Splash, plaster, 133
Spline, 45
    plaster, 132
Spotting bushings, 145
Spray-on
    metal face tool, 223, 228
    vacuum bag, 242, 303, 305
Spreader bars, mold, 361
Springback, 217–218
SRJ (Standard Routing Jig), 339, 342, 349
S-RIM, mold, 221, 223, 229–230, 400, 410
S-RIM, ceramic mold, 223, 229–231, 406, 410
Stack, vacuum bag sequence, 261
Standard
    equipment, 358
    routing jig, 339, 342
Statics, 5, 6, 8, 22
Steel
    mold, tool, 1, 67, 69, 231
    tool, repair, 335
Stefan-Boltzmann constant, 24
Stiffener, 125, 180, 202
    integral, 128, 219, 221, 314
    laminate face, 39
    mold and tool member, 8, 109
    picture frame, 194

Stiffener *(continued)*
  rolled edge, 221
Storage, 38
  area, 369, 372
  cost, 359
  indoor, 370-371, 373
  mold and tool, 40, 359, 368
  of gypsum cement, 76
  of patterns, 80
  of wood, 91
  outside, 368
Strain equation, 9
Strength
  limits, 13
  of materials, 5-8, 22
    mold design, 5
Stress
  concentration equation, 16
  curve, 10, 11
  equation, 8
  limit, 12
  normal, 13
  region, 14
  safe line, 13
  strain diagram, 5, 8
Stretch, draw form mold, 236
Strings, 45
Stripper, epoxy, 143
Structural
  adhesive, 123
  film, 124
  foam, 66
  member, 21
Structure, backup, 202, 231, 286
Submersibles, ocean, 119
Substructure
  attachment, 286
  eggcrate, 109
  modification, 348
  tie-in, 208
Sugar pine, 86
Superposition, 17
Support
  frame, 137, 215
  reusable vacuum bag, 306
  structure repair, 346
Surface
  area, 45
  casting, 170, 172, 177
    elimination, 193

coat
  film, 124
  liquid, 189, 191
  cracking, 198
  fillers, 40
  finishing, 139
    metal-faced, 228
  geometry analysis, 50, 53
  N/C machining, 54
  preparation adhesive, 125
  repair, 353
  sealing, 144
  vacuum bag, 258, 262
Survey, thermal, 394-395
Syntactic
  ceramic, 409
  foam, 114-115, 119, 153
  mandrel, 246
  properties, 116
  uses, 117
Synthetic wood, 115
System
  handling, 364
  storage, 364

Tallying lumber, 91
Tape, 268
  laying machine, 17, 45
  N/C, 47
  sealant, 254, 291, 325
Tar, 104
TDI, 121, 378
Tearout, 17
Template, 42
  backup structure, 204-205
  drill, 346
  locating, 335-339
  lofting, 130
  ply locating, 336-337, 339
  routing, 339, 346
Tensile
  failure, 15
  strength, 20
  stress usable, 13
Tension, 13
Temperature gradients, 22
Terminal, 45
  tertiary amines, 101
Tests
  cure monitor, 103

final, 383
porosity, 252
shape, 252
Theodolite, 388–389
Theory of elasticity, 17
Thermal
   adjustments, 27
   analysis, 95
   coefficient of expansion (CTE), 26, 68
   conductivity table, 23
   effects, 5, 21–22
   expansion, 41
      foam molding, 255, 327
      molding, 84, 255, 326
      table, 26
   properties, 2
   rubber, 180
   shock, 369
      factor, 27
   shrink point, 27
   shrunk, 27
   survey, 394–395
   tools, 27, 68
Thermocouple, 306, 374, 395
   installation, 305
   through a vacuum bag, 270, 307
Thermoformed, molds and tools, 58, 119
Thermographics, 397
Thermoplastic
   composite, 273, 407
   release, 147
Thermosetting
   adhesives, 123
   plastics, 98
Three-dimensional
   form, 45
   reinforcement, 408
Through the bag resin trap, 335
Through the bag vacuum port, 331, 333
Tie-in, eggcrate, 207–208
Time study, 32
Tolerance, 41, 163
   of loads, 6
Tool
   accessories, 330
   aid, 315
   bag side, 196, 314–316
   cart, 362
   cleaning, 348
   design,

      automated, 3
      system, 47
   edge rolled, 39, 221–222
   fabrication, 3
   face repair, 348
   failure, 13
   fittings, 330
   graphite fabric, 109–110
   handling, 358
   heatup rate, 24
   inspection, 388
   integrally stiffened, 404
   lay-up, 198
   life, 38
   lifting, 360–361
   making, future, 399
   mass, 25
      cast, 179
   masters, 45
   materials, 29, 41
   metal-faced, 222
   moving, 358
   nickel, 250
      vacuum port, 331
   pallet, 362
   picture frame, 194, 290
   post-cure, 192
   porous, 352
   prepreg, 182
   proofing, 394
   rail, 325
   resin trap, 334
   routing, 335, 346
   sealant, 291, 325
   storage, 363, 368, 371
   support, 330
   surface cap, 333
   void repair, 354
   wet lay-up, 182–195
Toroids, 45
Toruses, 53
Total, life, 38
Toxicity, 185
Tracing, 42
Transit jig, 387
Transporter, dolly, 358, 362–363
Trap, resin, 334
Trapped, rubber molding, 326
Treatments, 143
Triaxial fabrics, 108

430 INDEX

Triethylene tetramine, 99
Trim, 335
    diamond wire, 340
    laser, 340
    lines, 27, 144
    parts, 39
    tool, 335
    waterjet, 340
Trimer, foam, 121
Trowel, 123
Trolley for molds and tools, 360
Tube
    attachment, 216
    backup structure, 202
    support frame, 215
Turn key, operation, 44
Two-dimensional form, 45

Uncured silicone rubber bags, 302
Ultimate stress, 12, 13
Undercut
    forming, 70
    master pattern, 134
    part, 335
Unidirectional fabric, 108
Universal vacuum plate, 363
Urethane, 179
    foam, 121

Vacuum
    blaster, 347
    boss, 239, 331
    breather, 254
    check, 252
    forming,
        molds, 119
    leak test, 285
    port, 254, 306, 313
        cap, 332
        nickel tool, 331
        through the bag, 292, 333
        welded, 332
    routing tool, 341
    valve, 306
Vacuum bag
    accessories, 306–307
    bleed, 281
    boat mold, 401, 403
    bridge, 262–263
    cable hoist, 306, 313
    cast, 303, 305
    check, 396
    closure, 306
    disposable, 253
    double seal, 257, 270, 285
    ear, 264–266
    envelope, 272–274
    face, 289
    foil, 256
    frame, 303, 307–308
    high temperature, 255
    hoist brackets, 291–292, 313
    internal heaters, 291
    inside of mold, 275
    laminate mold copy, 262
    master pattern, 262
    material, 264
    new mold surface, 351
    no-bleed, 281, 286
    pad, 304
    paint-on, 292, 303–304
    pattern, 298, 300
    peripheral, 71, 282–283, 286, 288
    pleat, 264
    post-cure, 290
    precautions, 303
    precured, 293, 299
    prepreg, 277–278
    primer, 299
    rabbit ear, 264
    repair, 353
    resin trap, 335
    reusable, 253, 292, 299
    reversion resistant, 291
    rolled up, 308
    seal, 306, 309, 311–312
    sealant tape, 269
    seams, 293–295, 297
        reinforced, 319
    semiautomatic, 255
    size calculations, 267
    spray-on, 292, 303–304
    stack sequence, 261
    stiffeners, 291
    storage, 292
    support, 306–307, 309
    system, 374
    thermocouple, 306
    uncured sheet rubber, 300, 302
    vacuum ports, 313

wet lay-up, 278
Valve, 306
    check, 332
    vacuum, 285
VCHD, 377
Vehicle, wire guided, 365
Veil, mat, 191, 193
Vented bags
    reusable vacuum, 318
    vacuum, 258
Vertical, shear, 18
Vessel pressure, 373–374
Viewing, screen, 45
Viscoelasticity, 17
Viscosity, 104
Void
    mold, 352
    repair, 353–354
    tool, 352
Volatile, escape, 254

Waste disposal, 380
Water
    blown foam, 121
    jet router, 340
Wax
    machinable, 142
    sheet, 144
Wet
    brush, 197

molds, 183, 195
process, 198
vacuum bag,
Wheels, mold, tool, 360
White, master, 396
Wire
    diamond trim, 340
    frame shape, 45
    geometry, 50–52
    guided, storage area, 371–373
        vehicle, 362, 365–366, 371
        work station, 367
    thermocouple installation, 271
Wood, 85
    inpregnated, 85
    kiln drying, 90
    pattern, 88
    purchasing, 90
    seasoning, 90
    select grades, 86
    storage, 91
Work
    flow, 41
    safety, 357
    station, 44, 367
Woven fabric, 107

Yield point, 12–13
Young's modulus, equation, 11